T0406081

Ponnadurai Ramasami (Ed.)

Basic Sciences for Sustainable Development

Also of interest

Basic Sciences for Sustainable Development
Volume 2: Water and the Environment
Ponnadurai Ramasami (Ed.), 2023
ISBN 978-3-11-107089-6, e-ISBN 978-3-11-107120-6

Chemical Sciences in the Focus
Ponnadurai Ramasami (Ed.), 2021
Set ISBN 978-3-11-074105-6
Volume 1: Pharmaceutical Applications
ISBN 978-3-11-071072-4, e-ISBN 978-3-11-071082-3
Volume 2: Green and Sustainable Processing
ISBN 978-3-11-072659-6, e-ISBN 978-3-11-072614-5
Volume 3: Theoretical and Computational Chemistry Aspects
ISBN 978-3-11-073974-9, e-ISBN 978-3-11-073976-3

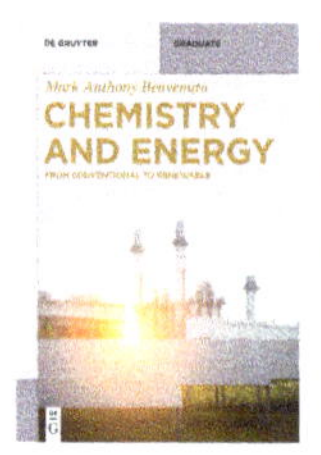

Chemistry and Energy
From Conventional to Renewable
Mark Anthony Benvenuto, 2022
ISBN 978-3-11-066226-9, e-ISBN 978-3-11-066227-6

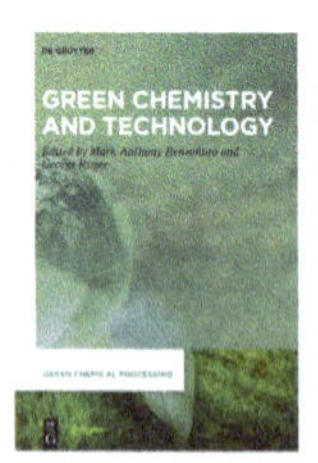

Green Chemistry and Technology
Mark Anthony Benvenuto and George Ruger (Eds.), 2021
ISBN 978-3-11-066991-6, e-ISBN 978-3-11-066998-5

Data Science in Chemistry
Artificial Intelligence, Big Data, Chemometrics and Quantum Computing with Jupyter
Thorsten Gressling, 2021
ISBN 978-3-11-062939-2, e-ISBN 978-3-11-062945-3

Basic Sciences for Sustainable Development

Volume 1: Energy, Artificial intelligence, Chemistry, and Materials Science

Edited by
Ponnadurai Ramasami

DE GRUYTER

Editor
Prof. Dr. Ponnadurai Ramasami
University of Mauritius
Faculty of Science
Department of Chemistry
Computational Chemistry Group
80837
RÉDUIT
MAURITIUS
ramchemi@intnet.mu

ISBN 978-3-11-099097-3
e-ISBN (PDF) 978-3-11-091336-1
e-ISBN (EPUB) 978-3-11-092628-6

Library of Congress Control Number: 2022950756

Bibliographic information published by the Deutsche Nationalbibliothek
The Deutsche Nationalbibliothek lists this publication in the Deutsche Nationalbibliografie; detailed bibliographic data are available on the internet at http://dnb.dnb.de.

Cover image: intararit / iStock / Getty Images Plus
Typesetting: TNQ Technologies Pvt. Ltd.
Printing and binding: CPI books GmbH, Leck

www.degruyter.com

Preface of the book (volume 1) entitled Basic Sciences for Sustainable Development: Energy, Artificial intelligence, Chemistry, and Materials Science

The year 2022 has been declared by the United Nations as the "International Year of Basic Sciences for Sustainable Development". Sustainable development is focused on the UN's 17 Sustainable Development Goals. These require the involvement of basic sciences.

This edited book (volume 1) is a collection of twelve invited and peer-reviewed contributions from chemistry, materials science, energy applications, and artificial intelligence.

Mondal et al. focused on the electronic properties, via computing the band structure, the topological properties through the topological invariants, and the prospects of 2D Dirac and semi-Dirac materials for spintronic applications, by studying the spin polarized transport. Pilcher reported on the development and implementation of two interventions at the first-year undergraduate level: one was designed to integrate systems thinking in first-year organic chemistry using the topic of surfactants and the other in a first-semester service course to engineering students using the stoichiometry of the synthesis of aspirin. Mumuni and co-workers reviewed the strengths, weaknesses, opportunities, and threats analysis for the deployment of artificial intelligence-based diagnostic imaging protocols in low- and middle- income countries. Maji and Sahu carried out a theoretical study to simulate the axial type flow in cylindrical baffled stirred vessel to analyze the flow characteristics for nanofluids. They calculated the mixing time inside the tank using the Partially-Averaged Navier-Stokes turbulence model. Mayanglambam et al. discussed on the organic synthesis of *N*-containing heterocycles in water with a green chemistry perspective. Sharma reviewed on the application of inorganic nanoparticles as nanocatalyst for the synthesis of *N*-containing heterocyclic compounds. Jayasankar elaborated on surfactant-surface active agents behind sustainable living. Bhattacharya and De reviewed on simple β-Carboline alkaloids that are abundantly available in plants, animals and foodstuff with a focus on the sustainable use of naturally occurring compounds in safeguarding human health and protecting our environment. Docrat and co-workers discussed the current knowledge and future prospects of the phytotherapeutic potential of commercial South African medicinal plants. Öztürkkan and Necefoğlu investigated and reported on the crystal structures of transition metal(II) 2-fluorobenzoates with N-donor ligand complexes such as 2,2'-bipyridine, 1,10'-phenanthroline, nicotinamide, isonicotinamide, 4-pyridylmethanol and 2-aminopyridine. Chola et al. reviewed the advances in geopolymeric materials, giving special emphasis

https://doi.org/10.1515/9783110913361-201

to kaolin, metakaolin, zeolite, dolomite, red mud, clay and fly ash based materials. Ganesan and co-workers discussed on biofuel as an alternative energy source for environmental sustainability.

I hope that the chapters of this volume 1 will add to literature and they will be useful references for researchers.

Prof. Ponnadurai Ramasami, CSci, CChem. FRSC, FAAS, FICCE, MMast
UNESCO Chair in Computational Chemistry,
Computational Chemistry Group, Department of Chemistry,
Faculty of Science, University of Mauritius, Réduit 80837, Mauritius
Visiting Professor, Centre of Natural Product Research,
Department of Chemical Sciences, University of Johannesburg,
Doornfontein Campus, Johannesburg 2028, South Africa
Professor Extraordinarius, Department of Chemistry,
University of South Africa, Private Bag X6, Florida 1710, South Africa
E-mail address: p.ramasami@uom.ac.mu

Contents

List of contributing authors

Naeem Sheik Abdul
Applied Microbial and Health Biotechnology Institute
Cape Peninsula University of Technology
Bellville
South Africa

Bamidele Omotayo Awojoyogbe
Department of Physics
Federal University of Technology
Minna
Nigeria

Saurabh Basu
Department of Physics
Indian Institute of Technology Guwahati
Guwahati 781039
Assam
India
E-mail: saurabh@iitg.ac.in

Piyali Bhattacharya
Department of Chemistry
University of Kalyani
Kalyani 741235
India

Basheer Meethale Chelaveettil
Department of Chemistry
Pocker Sahib Memorial Orphanage College
Tirurangadi
Malappuram
Kerala
India

Resha Kasim Vellattu Chola
Department of Chemistry
Pocker Sahib Memorial Orphanage College
Tirurangadi
Malappuram
Kerala
India
E-mail: chempsmotgi@gmail.com

Michael Oluwaseun Dada
Department of Physics
Federal University of Technology
Minna
Nigeria

Swati De
Department of Chemistry
University of Kalyani
Kalyani 741235
India
E-mail: deswati1@gmail.com

Sanduni Anupama De Zoys
Department of Life Science
Jain University
Bangalore
Karnataka
India

Taskeen F. Docrat
Applied Microbial and Health Biotechnology Institute
Cape Peninsula University of Technology
Bellville
South Africa

Füreya Elif Öztürkkan
Department of Chemical Engineering
Kafkas University
Kars 36100
Turkey
E-mail: fozturkkan36@gmail.com

Subbulakshmi Ganesan
Department of Chemistry
Jain University
Bangalore
Karnataka
India
E-mail: g.subbulakshmi@jainuniversity.ac.in

Sudin Ganguly
Department of Physics
School of Applied Sciences
University of Science and Technology
Techno City
Kiling Road
Baridua 9th Mile
Ri-Bhoi
Meghalaya 793101
India

https://doi.org/10.1515/9783110913361-202

Francis Hasford
Department of Medical Physics
University of Ghana
Ghana Atomic Energy Commission
Accra
Ghana

Parvathi Jayasankar
JAIN Deemed to be University
Bangalore
India
E-mail: parvathi.jaysankar@jainuniversity.ac.in

Rajasree KarthyayaniAmma
A M JAIN College
Meenambakkam
Chennai
India

Prajitha Kumari
Department of Chemistry
Pocker Sahib Memorial Orphanage College
Tirurangadi
Malappuram
Kerala
India

Srimanta Maji
Department of Mathematics
Institute of Chemical Technology
Marathwada
Jalna
India
E-mail: s.maji@marj.ictmumbai.edu.in

Mayanglambam Maneeta Devi
Chemistry Department
Manipur University
Canchipur-795003
Manipur
Imphal
India

Jeanine L. Marnewick
Applied Microbial and Health Biotechnology Institute
Cape Peninsula University of Technology
Bellville
South Africa
E-mail: marnewickj@cput.ac.za

Sayan Mondal
Department of Physics
Indian Institute of Technology Guwahati
Guwahati 781039,
Assam
India

Abdul Nashirudeen Mumuni
Department of Medical Imaging
University for Development Studies
Tamale
Ghana
E-mail: mnashiru@uds.edu.gh

Hacali Necefoğlu
Department of Chemistry
Kafkas University
Kars 36100
Turkey
and
International Scientific Research Centre
Baku State University
Baku 1148
Azerbaijan
E-mail: alinecef@hotmail.com

Izegaegbe Daniel Omoikhoje
Department of Life Science
Jain University
Bangalore
Karnataka
India

Gopalakrisnan Padmapriya
Department of Chemistry
Jain University
Bangalore
Karnataka
India

Thasleena Panakkal
Department of Chemistry
Pocker Sahib Memorial Orphanage College
Tirurangadi
Malappuram
Kerala
India

Farsana Ozhukka Parambil
Department of Chemistry
Pocker Sahib Memorial Orphanage College
Tirurangadi
Malappuram
Kerala
India

Sajna Valiya Peedikakkal
Department of Chemistry
Pocker Sahib Memorial Orphanage College
Tirurangadi
Malappuram
Kerala
India

Lynne A. Pilcher
Department of Chemistry
University of Pretoria
Room 1-35
Chemistry Building
Pretoria
South Africa
E-mail: lynne.pilcher@up.ac.za

Akshaya K. Sahu
Department of Mathematics
Institute of Chemical Technology
Mumbai
India
E-mail: ak.sahu@ictmumbai.edu.in

Kongbrailatpam Gayatri Sharma
Chemistry Department
Oriental College (Autonomous)
Imphal
795001 Manipur
India
E-mail: gayatrish83@gmail.com

Okram Mukherjee Singh
Chemistry Department
Manipur University
Canchipur-795003
Manipur
Imphal
India
E-mail: ok_mukherjee@yahoo.co.in

Thokchom Prasanta Singh
Chemistry Department
Manipur University
Canchipur-795003
Manipur
Imphal
India
E-mail: prasantath@gmail.com

Nicholas Iniobong Udeme
Department of Physics
Federal University of Technology
Minna
Nigeria

Sayan Mondal, Sudin Ganguly and Saurabh Basu*

1 Topology and applications of 2D Dirac and semi-Dirac materials

Abstract: Two dimensional (2D) Dirac materials, such as graphene, hold promise of being useful in energy storage, and thus have merged as candidates that are worth exploring through the last couple of decades. In this chapter, we mainly focus on three aspects of these materials, namely, the electronic properties, via computing the band structure, the topological properties through the topological invariants, and the prospects of these 2D materials for spintronic applications, via studying the spin polarized transport. All of these properties are correlated, and hence warrant a thorough discussion. Further, in order to ascertain whether a band deformation induces noticeable effects on the electronic, topological and spintronic properties, we have considered a 2D semi-Dirac system, that does not have Dirac cones, however the conduction and the valence bands touch at an intermediate to the Dirac points in the Brillouin zone. From our studies, we infer that the behaviour of these semi-Dirac systems is quite distinct from their Dirac counterpart. Finally, in order to have noticeable spin polarized transport, we use heavy adatoms (such as, Au) on the graphene matrix which enhances the spin–orbit coupling, and thereby propose a mechanism that will ramify on the spintronic applications.

Keywords: graphene; quantum spin Hall insulator; semi-Dirac system; topological insulator.

1.1 Introduction

Since the discovery of graphene in 2004 [1] by exfoliation technique, the world has realized a perfect two dimensional (2D) material consisting of C atoms placed on the vertices of a honeycomb lattice. Later on C has been replaced by P (phosphorene) [2, 3], Ge (germanium) [4], Si (silicon) [5, 6], transition metal dichalcogenide (TMDC), boron nitride (BN) etc., and similar 2D structures have been realized in experiments. During the last couple of decades, scientists and engineers have discovered a few thousands of

***Corresponding author: Saurabh Basu,** Department of Physics, Indian Institute of Technology Guwahati, Guwahati 781039, Assam, India, E-mail: saurabh@iitg.ac.in
Sayan Mondal, Department of Physics, Indian Institute of Technology Guwahati, Guwahati 781039, Assam, India
Sudin Ganguly, Department of Physics, School of Applied Sciences, University of Science and Technology, Techno City, Kiling Road, Baridua 9th Mile, Ri-Bhoi, Meghalaya 793101, India

As per De Gruyter's policy this article has previously been published in the journal Physical Sciences Reviews. Please cite as: S. Mondal, S. Ganguly and S. Basu "Topology and applications of 2D Dirac and semi-Dirac materials" *Physical Sciences Reviews* [Online] 2022. DOI: 10.1515/psr-2022-0118 | https://doi.org/10.1515/9783110913361-001

atomically thin materials. Some of these actually have buckled structures and have added to the richness of their properties, along with enhanced interest into the research of 2D materials. Several other 2D materials, for example, transition metal oxides, Carbides/Nitrides (which have a special name Mxene), dichalcogenides etc. have been discovered and studied with renewed intensity. The astonishing features of all these materials are extremely high mobility, electrical conductivity, elastic properties, high packing density, surface properties etc. The latter is particularly significant because of the electrocatalytic and photocatalytic utilities, especially in the infrared region. However, the property that stands out especially from the societal perspective is their utility in electrochemical energy storage applications [7]. Also the flexibility of developing various kinds of heterostructure with 2D materials makes them feasible candidates for the design of integrated circuit based devices.

However, these 2D materials suffer from several shortcomings, because of which their applicabilities are limited. For example, in experiments, the limitations of using graphene as electrodes arise owing to its ion-accessible surface area due to stacking one layer over the other. Thus a large quantity of electrolytes is often required to compensate for this. To take remedial measures, 2D materials with porous structures have emerged as a game changer with increased exposure of their surface area to electrolytes and mesoscale sized pores for enhanced charge transfer. The versatility of such technical developments has led to improved stabilities of energy storage devices, including supercapacitors, lithium-ion batteries etc. Currently, more effort is being given to acquire precise control on the number of porous layers and their thickness to suit the intended applications. Moreover, Janus[1] materials in the form of nanosheets, with distinct properties of each face, have led to formation of completely new composites [8].

A very large volume of literature has been devoted to energy storage and conversion, such as design and fabrication of Li/Na-ion batteries, supercapacitors for their large power density and long life cycle, electrocatalytic applications and so on. Simultaneously, the condensed matter physicists across the globe have contributed to the understanding of the electronic and the topological properties of graphene, and its derivatives[2]. Surprisingly, and interestingly, the low energy dispersion of graphene is relativistic (massless Dirac like) in nature. However, the relativistic feature is restricted to the energy varying linearly with the wave vector; the velocity of the electrons is still limited by the Fermi energy. A further dichotomy arises with the effective mass of the electrons, which according to standard definitions of solid state physics can be written as, $m^* = \frac{1}{\hbar^2}\left(\frac{\partial^2 \epsilon}{\partial k^2}\right)^{-1}$ becomes infinity, while the corresponding Hamiltonian is that of a massless particle. However, that is reconciled with a modified definition, namely,

1 Janus is a Greek god with mirror symmetric focus.

2 By derivatives we do not mean material here, but refer to different versions of the low energy model.

$m^{*} = \hbar^2 k\left(\frac{\partial \epsilon}{\partial k}\right)^{-1}$, which yields sensible results. Further, the low energy dispersion shows band touching[3] at six locations of the Brillouin zone (BZ) of which two are distinct (four others can be obtained by adding and subtracting proper reciprocal lattice vectors), that are called as the Dirac points. In standard language, these two Dirac points are called as valleys, and scattering of electrons from one valley to another is strictly prohibited.

With quantum Hall systems unveiling the first realization of a topological insulators, 2D Dirac materials, and in particular graphene was conceptualized as a prospective candidate for showing topological properties. A Berry phase of $\pm\pi$ at the Dirac points added fuel to the speculations. However, the minimum requirement for obtaining a topologically trivial state is that a spectral gap needs to open up at the Dirac points. The question that follows immediately afterwards is whether such a gap is accompanied by conducting edge states. If the answer is yes, then we have a topological insulator.

In a completely different context, Haldane [9] envisioned realizing quantum Hall effect in absence of an external magnetic field, and hence without the requirement of formation of Landau levels. Breaking the time reversal symmetry (TRS) was thought to be supremely important for realizing the quantum Hall effect, which was done by Haldane by introducing a complex direction dependent second neighbour hopping. The scenario is equivalent to a staggered field produced at the vertices and at the centre of the honeycomb structure. This, of course, gives graphene the states of a topological insulator, which is also known as the Chern insulator because of its non-zero Chern number, which is the topological invariant in this case.

A further development took place with the intervention of Kane and Mele [10, 11], where they have repaired the loss of time reversal invariance by explicitly considering the spin degrees of freedom of the electrons, such that spin-$\uparrow$ electrons experience a flux $+\pi$, while that by the spin-$\downarrow$ is $-\pi$. The situation does not support a quantum Hall state, since the TRS is intact, however, gives rise to an important, and the second type of topological insulators, called as the quantum spin Hall insulator. The realization of such insulators, along with spin–orbit coupling (SOC) raised the possibility of using graphene as spintronic devices. Quite frustratingly, the Rashba SOC, which is the main ingredient for such an application is too low and hence impedes graphene's application as a spintronic device. While HgTe-CdTe quantum well structures [12] have conveniently demonstrated realization of the quantum spin Hall phase, we still work on the prospects of enhancing the Rashba SOC in graphene via inclusion of heavy adatoms.

Before we embark on the Hall effect in graphene, let us review the electronic properties. We give a pedagogical description of the tight binding bandstructure and

3 The conduction and the valence bands touch each other.

compute the location of the Dirac points in the BZ, where the low energy dispersion looks relativistic. We further give definitions of all the quantities, such as, Berry phase, Berry curvature, Chern number and the $\mathbb{Z}_2$ invariants (the last two being the topological invariants that characterize the topological properties).

1.2 Basic electronic properties of graphene

The tight binding Hamiltonian of graphene can be written as,

$$\mathcal{H} = -t \sum_{\langle ij\rangle,\sigma} (a^{\dagger}_{i\sigma} \boldsymbol{b}_{j\sigma} + \text{h.c}) \tag{1.1}$$

where the labels i and j denote the lattice sites and $a^{\dagger}$ (b) denote creation (annihilation) operators for electrons at the A(B) sublattice sites. σ denotes the spin of the electrons, however that will be suppressed in the next step onwards owing to no active role played by the spin either in the band structure. We shall make the spin degrees of freedom apparent only when it is needed. Here $t \simeq 2.7$ eV which is considerably large and allows us to ignore electron-electron interaction. The nearest neighbour vectors, $\boldsymbol{\delta}_i$, lattice vectors, a_i and the reciprocal lattice vectors, b_i (see Figure 1.1) are written as,

$$\begin{aligned}
\boldsymbol{\delta}_1 &= \frac{a}{2}(\sqrt{3}\,\hat{x} - \hat{y})\,;\, \boldsymbol{\delta}_2 = \frac{a}{2}(-\sqrt{3}\,\hat{x} - \hat{y})\,;\, \boldsymbol{\delta}_3 = a\hat{y} \\
\boldsymbol{a}_1 &= \frac{a}{2}(\sqrt{3}\,\hat{x} + 3\hat{y})\,;\, \boldsymbol{a}_2 = \frac{a}{2}(-\sqrt{3}\,\hat{x} + 3\hat{y}) \\
\boldsymbol{b}_1 &= \frac{2\pi}{3a}\left(\sqrt{3}\,\hat{k}_x + 3\hat{k}_y\right)\,;\, \boldsymbol{b}_2 = \frac{2\pi}{3a}\left(-\sqrt{3}\,\hat{k}_x + 3\hat{k}_y\right)
\end{aligned} \tag{1.2}$$

with the lattice constant, a = 1.42 Å. Using the nearest neighbour vectors, $\boldsymbol{\delta}_i$, we explicitly write the tight binding Hamiltonian as,

$$\mathcal{H} = -t \sum_{R,\boldsymbol{\delta}} [b^{\dagger}(\boldsymbol{R} + \boldsymbol{\delta}_i) a(\boldsymbol{R}) + a^{\dagger}(\boldsymbol{R}) b(\boldsymbol{R} + \boldsymbol{\delta}_i)] \tag{1.3}$$

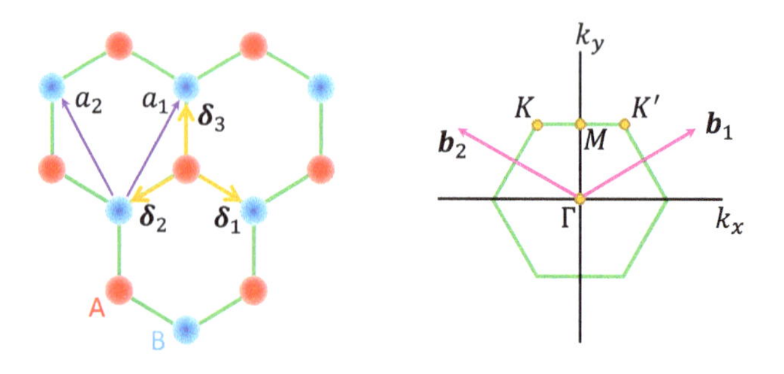

Figure 1.1: Plot showing δ_i, lattice vectors, a_i and the reciprocal b_i lattice vectors for graphene. The figure on right hand side depicts the BZ of graphene.

The lattice vector $\boldsymbol{R}$ at an arbitrary site is given by,

$$\boldsymbol{R} = na_1 + ma_2\, n, m \in N \tag{1.4}$$

The fourier transform for these operators are done using,

$$a_k = \frac{1}{\sqrt{N}} \sum_R e^{-ik \cdot R} a(\boldsymbol{R}) \tag{1.5}$$

This yields the Hamiltonian in the momentum space as follows,

$$\mathcal{H} = -\frac{t}{N} \sum_{k,q} \sum_R \sum_{i=1}^{3} e^{i(k-q) \cdot R} \left[e^{i\boldsymbol{k} \cdot \boldsymbol{\delta}_i} b_q^\dagger a_k + e^{i\boldsymbol{k} \cdot \boldsymbol{\delta}_i} a_q^\dagger \boldsymbol{b}_k \right] \tag{1.6}$$

Using the following definition of the Kronecker delta,

$$\boldsymbol{\delta}_{k.q} = \frac{1}{N} \sum_R e^{i(k-q) \cdot R} \tag{1.7}$$

one gets,

$$\begin{aligned} \mathcal{H} &= -t \sum_k \sum_{i=1}^{3} \left[e^{-i\boldsymbol{k} \cdot \boldsymbol{\delta}_i} b_k^\dagger a_k + e^{i\boldsymbol{k} \cdot \boldsymbol{\delta}_i} a_k^\dagger \boldsymbol{b}_k \right] \\ &= -t \sum_k \sum_{i=1}^{3} (a_k^\dagger b_k^\dagger) \begin{pmatrix} 0 & e^{-i\boldsymbol{k} \cdot \boldsymbol{\delta}_i} \\ e^{i\boldsymbol{k} \cdot \boldsymbol{\delta}_i} & 0 \end{pmatrix} \begin{pmatrix} a_k \\ \boldsymbol{b}_k \end{pmatrix} \\ &= -t \sum_k \sum_{i=1}^{3} (a_{\bar{k}}^{\dagger} b_k^\dagger) h(k) \begin{pmatrix} a_k \\ \boldsymbol{b}_k \end{pmatrix} \end{aligned} \tag{1.8}$$

where $h(\mathbf{k})$ is the Hamiltonian matrix defined by,

$$h(k) = -t \begin{pmatrix} 0 & -\left(e^{i\boldsymbol{k} \cdot \boldsymbol{\delta}_1} + e^{i\boldsymbol{k} \cdot \boldsymbol{\delta}_2} + e^{i\boldsymbol{k} \cdot \boldsymbol{\delta}_2}\right) \\ -\left(e^{-i\boldsymbol{k} \cdot \boldsymbol{\delta}_1} + e^{-i\boldsymbol{k} \cdot \boldsymbol{\delta}_2} + e^{-i\boldsymbol{k} \cdot \boldsymbol{\delta}_2}\right) & 0 \end{pmatrix} \tag{1.9}$$

Since the difference between two nearest neighbour lattice vectors, $\boldsymbol{\delta}_i$ and $\boldsymbol{\delta}_j$ must yield a lattice vector, $\boldsymbol{R}$, we can do a transformation,

$$a_k \to e^{i\boldsymbol{k} \cdot \boldsymbol{\delta}_3} a_k, \quad \text{and} \quad a_k^\dagger \to e^{-i\boldsymbol{k} \cdot \boldsymbol{\delta}_3} a_k^\dagger$$

The above transformation yields a new Hamiltonian matrix,

$$\tilde{h}(k) = -t \begin{pmatrix} 0 & -\left(e^{i\boldsymbol{k} \cdot (\boldsymbol{\delta}_1 - \boldsymbol{\delta}_3)} + e^{i\boldsymbol{k} \cdot (\boldsymbol{\delta}_2 - \boldsymbol{\delta}_3)} + 1\right) \\ -\left(e^{-i\boldsymbol{k} \cdot (\boldsymbol{\delta}_1 - \boldsymbol{\delta}_3)} + e^{-i\boldsymbol{k} \cdot (\boldsymbol{\delta}_2 - \boldsymbol{\delta}_3)} + 1\right) & 0 \end{pmatrix} \tag{1.10}$$

Using the definitions of $\boldsymbol{\delta}_i$, one gets,

$$\tilde{h}(k) = -t \begin{pmatrix} 0 & -\left(e^{i\boldsymbol{k} \cdot a_1} + e^{i\boldsymbol{k} \cdot a_2} + 1\right) \\ -\left(e^{-i\boldsymbol{k} \cdot a_1} + e^{-i\boldsymbol{k} \cdot a_2} + 1\right) & 0 \end{pmatrix} \tag{1.11}$$

One can check that $\tilde{h}(k)$ obeys $\tilde{h}(\boldsymbol{k}) = \tilde{h}(\boldsymbol{k} + \mathbf{G})$, where G is the reciprocal lattice vector, defined as, $\mathbf{G} = p\mathbf{b}_1 + q\mathbf{b}_2$, with p and q being integers. Thus,

$$\tilde{h}(\boldsymbol{k}) = -t\begin{pmatrix} 0 & f(k) \\ f^*(k) & 0 \end{pmatrix} \tag{1.12}$$

where

$$f(k) = -t\left(e^{-ik_y a} + 2e^{ik_y a/2}\cos\left(\frac{k_x\sqrt{3}a}{2}\right)\right)$$

By diagonalizing $\tilde{h}(k)$, one can obtain the tight binding energy as follows,

$$\epsilon_k = \pm t\sqrt{3 + 2\cos(\sqrt{3}\,ak_x) + 4\cos(\sqrt{3}\,ak_x/2)\cos(3ak_y/2)} \tag{1.13}$$

The '+' and the '-' signs in the above dispersion refer to the two bands which touch each other at six points in the BZ (see Figure 1.2). Since graphene has one accessible electron per C atom, one can assume a half filled system where the lower band is completely filled. Further, we wish to discuss the low lying excitations just above the ground state of the system.

This necessitates exploring the low energy theory of graphene. In order to find it, we need to identify the band touching points which can be obtained from the condition, $f(\boldsymbol{k}) = 0$. Separately putting the real and the imaginary parts equal to zero yield,

$$\begin{aligned} \cos(k_y a) + 2\cos(k_y a/2)\cos(\sqrt{3}\,k_x a/2) &= 0 \\ -\sin(k_y a) + 2\sin(k_y a/2)\cos(\sqrt{3}\,k_x a/2) &= 0 \end{aligned} \tag{1.14}$$

Eq. (1.15) can be manipulated as follows,

$$\sin(k_y a/2)\left[-\cos(k_y a/2) + \cos(k_x a\sqrt{3}/2)\right] = 0 \tag{1.15}$$

Thus one is left with two options, namely,

$$\begin{aligned} &\text{either} \quad (i) \quad \sin(k_y a/2) = 0; \text{which means} \cos(k_y a/2) = \pm 1; \\ &\text{or,} \quad (ii) \quad \cos(k_y a/2) = \cos(\sqrt{3}\,k_x a/2) \end{aligned} \tag{1.16}$$

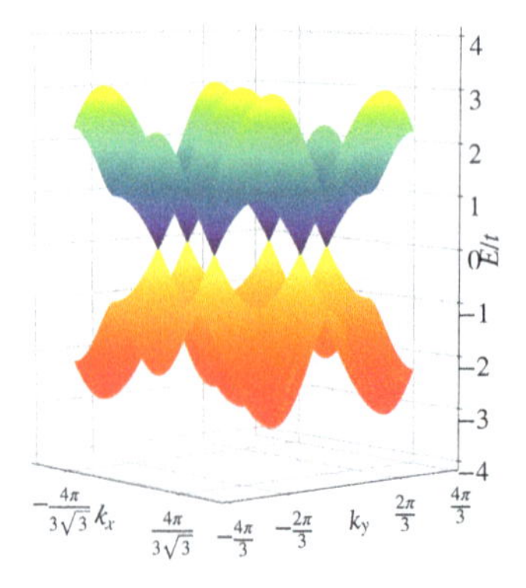

Figure 1.2: Plot showing the two tight binding bands of graphene in the first BZ. The conduction and the valence bands touch each other at six points in the BZ. In the vicinity of those points, the bands are linearly dispersing bearing signatures of (pseudo-)relativistic physics.

Option (i) gives us,

$$1 + 2\cos\left(k_x \sqrt{3}\, a/2\right) = 0$$

which yields the points $\left(\pm\frac{4\pi}{3\sqrt{3}\,a}, 0\right)$ (plus or minus the reciprocal lattice vector, G). Whereas, option (ii) can be written as,

$$\cos\left(k_x a\sqrt{3}\right) + 2\cos^2\left(k_x a\sqrt{3}\,/2\right) = 0.$$

Thus we get four more points, which are, $\pm\frac{2\pi}{3a}\left(\frac{1}{\sqrt{3}}, 1\right)$, and $\pm\frac{2\pi}{3a}\left(1, -\frac{1}{\sqrt{3}}\right)$ (again plus or minus the reciprocal lattice vector, G).

A closer inspection reveals that all the six points are not independent. For example, the set of vectors, namely, $\left(\frac{4\pi}{3\sqrt{3}\,a}, 0\right)$, $\frac{2\pi}{3a}\left(-\frac{1}{\sqrt{3}}, 1\right)$ and $\frac{2\pi}{3a}\left(-\frac{1}{\sqrt{3}}, -1\right)$ can be connected to each other via the combination of the reciprocal lattice vectors, $\boldsymbol{b}_1$ and $\boldsymbol{b}_2$. For example,

$$\begin{aligned}\left(\frac{4\pi}{3\sqrt{3}a}, 0\right) + \boldsymbol{b}_2 &= \frac{2\pi}{3a}\left(-\frac{1}{\sqrt{3}}, 1\right)\\ \left(\frac{4\pi}{3\sqrt{3}a}, 0\right) - \boldsymbol{b}_1 &= \frac{2\pi}{3a}\left(-\frac{1}{\sqrt{3}}, -1\right)\end{aligned} \tag{1.17}$$

The same is true for the other vectors,

$$\left(-\frac{4\pi}{3\sqrt{3}a}, 0\right), \frac{2\pi}{3a}\left(-\frac{1}{\sqrt{3}}, 1\right), \frac{2\pi}{3a}\left(\frac{1}{\sqrt{3}}, 1\right)$$

Thus, only two of them are found to be independent. Traditionally they are called as $\boldsymbol{K}$ and $\boldsymbol{K}'$ and can be written as,

$$\boldsymbol{K} = \frac{2\pi}{3a}\left(\frac{1}{\sqrt{3}}, 1\right), \quad \text{and} \quad \boldsymbol{K}' = \frac{2\pi}{3a}\left(-\frac{1}{\sqrt{3}}, 1\right)$$

Any other independent pair is also a valid choice for $\boldsymbol{K}$ and $\boldsymbol{K}'$.

It should be noted that since the two bands touch at these points, the gap between the conduction and the valence band closes. Thus there are two branches of low energy excitations, namely one of them with momentum close to $\boldsymbol{K}$ and the other close to $\boldsymbol{K}'$. Since $f(\boldsymbol{k})$ becomes zero at $\boldsymbol{k} = \boldsymbol{K}$. Defining $\boldsymbol{q} = \boldsymbol{k} - \boldsymbol{K}$, one can expand $f(\boldsymbol{k})$ near $\boldsymbol{K}$ in Taylor series about $\boldsymbol{q} = 0$,

$$f'(q) = \frac{\partial f(k)}{\partial k_x}\bigg|_{(k_x - K_x)} (k_x - K_x) + \frac{\partial f(k)}{\partial k_y}\bigg|_{(k_y - K_y)} (k_y - K_y) = \frac{3at}{2}\left(q_x + iq_y\right) \tag{1.18}$$

Thus the energy spectrum assumes the form,

$$\epsilon_K(q) = \hbar v_F\left(q_x + iq_y\right) \tag{1.19}$$

where v_F is the Fermi velocity defined by, $v_F = \frac{3at}{2\hbar} \simeq 10^6 ms^{-1}$. Similarly, if we expand around $\boldsymbol{K}'$, one gets,

$$\epsilon_{K'}(q) = \hbar v_F \left(q_x - iq_y\right) \tag{1.20}$$

Thus in a general notation, we can write,

$$\epsilon_{K,K'} = \hbar v_F q \cdot \boldsymbol{\sigma} \tag{1.21}$$

where q is a planar vector (q_x, q_y and σ is the Pauli matrix vector (σ_x, σ_y). The $\boldsymbol{K}$ and the $\boldsymbol{K}'$ points are often called as the Dirac points. The electrons close to those Dirac points are called massless Dirac fermions, as they obey the Dirac equation without the 'mass' term.[4] It may be noted that,

$$\epsilon_{K'}(q) = \epsilon_K^*(q) \tag{1.22}$$

This implies that (as will be seen later) the 'helicity' of the electrons is opposite at $\boldsymbol{K}'$ with respect to $\boldsymbol{K}$.

To sum up our preliminary discussion on graphene, we note that the low energy properties are governed by the electronic dispersion,

$$\epsilon(q) \pm v_F|q| \tag{1.23}$$

which implies that the eigenvalues are only functions of the magnitude of the wave vector q, and does not depend upon its direction in the 2D plane. Also the Hamiltonian on a formal note denotes that of a massless $s = 1/2$ particle, such as a neutrino, however the velocity of the particles are 300 times lower as compared to the speed of light. Further, the handedness (or the helicity) feature of neutrinos are inbuilt, where the electrons behave similar to the 'left handed' neutrino at the Dirac point $\boldsymbol{K}$ and as a 'right handed' neutrino at $\boldsymbol{K}'$ or vice versa.

1.3 Berry phase

According to Berry, if a Hamiltonian of a system is associated with a parameter, and if the parameter changes adiabatically, then after a cyclic evolution, the state of the Hamiltonian will return to its original state along with an additional geometrical phase. This additional phase is called the Berry phase.

A non-zero Berry phase may indicate something interesting going on. For example, consider a'*flag*' transported along a geodesic on the surface of a sphere. A complete

4 The Dirac equation is written in conventional notations as, $\mathcal{H} = c\boldsymbol{\alpha} \cdot \mathbf{p} + \beta mc^2$ where α and β are Hermitian operators which do not operate on the space and time variables. In case of graphene, the second term is absent.

rotation will fail to bring back the '*flag*' in its original configuration. This is what is referred to as the parallel transport of a vector on a curved Riemann surface.

1.3.1 Berry phase of Graphene

We wish to compute the Berry phase of fermions in graphene. For this, we consider the low energy Hamiltonian of graphene given by,

$$\mathcal{H} = \hbar v_F\left(\tau_z\sigma_x q_x + \sigma_y q_y\right) \tag{1.24}$$

where τ_z denotes the valley degree of freedom, that is, $\tau_z = 1$ for $\boldsymbol{K}$-point, while it is -1 for the $\boldsymbol{K}'$-point. As usual, σ denotes the sublattice degree of freedom. The Berry connection is obtained as,

$$\mathcal{A} = \langle\psi_-|\nabla|\psi_-\rangle \tag{1.25}$$

We write down the Dirac Hamiltonian as,

$$h(\boldsymbol{q}) = q\cdot\boldsymbol{\sigma} \quad \text{(the velocity term is dropped).} \tag{1.26}$$

In the polar coordinate $\boldsymbol{q}$ and $h(\boldsymbol{q})$ can be represented as,

$$\boldsymbol{q} = |\boldsymbol{q}|\begin{pmatrix}\cos\phi\\ \sin\phi\end{pmatrix} = q\begin{pmatrix}\cos\phi\\ \sin\phi\end{pmatrix} \tag{1.27}$$

and

$$h(\boldsymbol{q}) = \boldsymbol{q}\begin{pmatrix}0 & \cos\phi - i\sin\phi\\ \cos\phi + i\sin\phi & 0\end{pmatrix} = q\begin{pmatrix}0 & e^{-i\phi}\\ e^{i\phi} & 0\end{pmatrix} \tag{1.28}$$

The normalized eigenvectors are

$$|\psi_-\rangle = \frac{1}{\sqrt{2}}\begin{pmatrix}-e^{-i\phi}\\ 1\end{pmatrix} \text{ and } \quad |\psi_+\rangle = \frac{1}{\sqrt{2}}\begin{pmatrix}e^{-i\phi}\\ 1\end{pmatrix} \tag{1.29}$$

Now we calculate the Berry connection $\boldsymbol{\mathcal{A}}$ for the filled bands using Eq. (1.25). The Gradient operator, ∇_q in Eq. (1.25) in polar coordinate is given by,

$$\nabla_q = \left(\frac{\partial}{\partial q}\hat{q} + \frac{1}{q}\frac{\partial}{\partial\phi}\hat{\phi}\right) \tag{1.30}$$

by the loop C. Note that $|\psi_-\rangle$ does not depend upon q,

If we now introduce a band index, n, the Chern number corresponding to nth band can be written as C_n. The total Chern number is obtained from the contribution from all the bands, namely,

$$C = \sum_n C_n$$
$$C_n = \frac{1}{2\pi} \int_S \boldsymbol{\Omega}_n \cdot dS \tag{1.31}$$

Where S is the surface that encloses the loop and $\boldsymbol{\Omega}_n$ is the Berry curvature of the nth band which is related to $\boldsymbol{\mathcal{A}}$ via, $\boldsymbol{\Omega} = \nabla \times \boldsymbol{\mathcal{A}}$. With $\boldsymbol{\mathcal{A}} = \frac{1}{2q}$, $\nabla \times \boldsymbol{\mathcal{A}} = 0$. So $\mathcal{T} = 0$ and hence $C = 0$ which is not a surprise since the TRS is preserved.

The Berry phase around the Dirac points ($\boldsymbol{K}$ and $\boldsymbol{K}'$) is nothing but the winding number multiplied by π which is then either +1 or −1. This introduces a measure of the topological charge for the Dirac points in the k-space which tells us how the wavefunctions wind around these singular points in k-space differently with respect to each other. The $\boldsymbol{K}$ point carries the topological charge +1 (a vertex) and the $\boldsymbol{K}'$ points carries a topological charge −1.

So the Dirac fermion sitting at $\boldsymbol{K}$ carries a Berry phase π, while the Dirac fermion at $\boldsymbol{K}'$ has a Berry phase $-\pi$, so the overall phase is zero, $\gamma = 0$.

1.4 Graphene as a topological insulator

The low energy Hamiltonian of graphene around the Dirac points (the $\boldsymbol{K}$ and the $\boldsymbol{K}'$) is given by,

$$H_0(k) = \hbar v_F (k_x \sigma_x \tau_z + k_y \sigma_y) \tag{1.32}$$

where σ_x, σ_y pseudospins denote sublattice degrees of freedom and τ_z is the z component of the Pauli matrix which distinguishes the valleys at $\boldsymbol{K}$ and $\boldsymbol{K}'$ points. The important fact here is that the Hamiltonian is independent of the real spin, which will continue to be a valid description till SOC is included. Thus the time reversal operator, $\mathcal{T}$, is taken to be a complex conjugation operator. However, the TRS in graphene implies a transformation from one valley to another, that is $\boldsymbol{K}$ changing over to $\boldsymbol{K}'$. This makes us settle for,

$$\mathcal{T} = \tau_x \mathcal{K} \tag{1.33}$$

where τ_x is the x-component of the Pauli matrix. Note that here $\mathcal{T}^2 = 1$ as we are dealing with spin-less fermions.

Next consider the inversion (or sublattice) symmetry which switches the two sublattices, and also changes the momentum k to $-\boldsymbol{k}$ (remember $\boldsymbol{p} = m\frac{dr}{dt}$) which again implies that the valleys are switched. Thus, $\mathcal{P} = \sigma_x \tau_x$ and

$$\mathcal{T} H_0(k) \mathcal{T}^{-1} = \hbar v_F \tau_x \left(k_x \sigma_x \tau_z + k_y \sigma_y^*\right) \tau_x = H_0(-k)$$

Similarly, under the inversion operator,

$$\mathcal{P} H(k) \mathcal{P}^{-1} = \hbar v_F \sigma_x \tau_x (k_x \sigma_x \tau_z + k_y \sigma_y) \sigma_x \tau_x = H_0(-k)$$

where simple algebra of the Pauli matrices have been used. Thus a combination of the two symmetries leaves the Hamiltonian invariant.

Thus the goal is to transform graphene into a topological insulator so that it exhibits quantum Hall effect with conducting edges and an insulating bulk. Further the edge modes must have a chiral character, which means the current is carried in opposite directions at the two edges. There could be two ways of doing this; either break the inversion symmetry, keeping the TRS intact, or break the TRS, retaining the inversion symmetry. The first case, that is, the breaking of inversion symmetry can be achieved by introducing an on-site potential of the form,

$$\mathcal{H}' = m_I \sum_i \xi_i c_i^\dagger c_i \tag{1.34}$$

where $\xi_i = +1$ and -1 corresponding to the sublattices A and B, respectively. The on-site term, m_I is often referred to as the Semenoff mass. Including this term to the low energy Hamiltonian in Eq. (1.32) becomes

$$\mathcal{H}(q) = \mathcal{H}_0(q) + m_I \sigma_z \tag{1.35}$$

The electronic spectrum is given by,

$$E(q) = \pm\sqrt{\hbar^2 v_F^2 q^2 + m_I^2} \tag{1.36}$$

In a compact notation, one may write it as,

$$E_\mu(q) = \mu\sqrt{\hbar^2 v_F^2 q^2 + m_I^2} \tag{1.37}$$

where $\mu = \pm 1$ represent the conduction and valence bands, respectively.

The spectrum is plotted in Figure 1.3a. As can be seen, gaps of magnitude $2m_I$ open at each of the Dirac points. This gap earns the name Semenoff insulator. However, the nature of the gap is a trivial one in the following sense. The bandstructure plotted for a

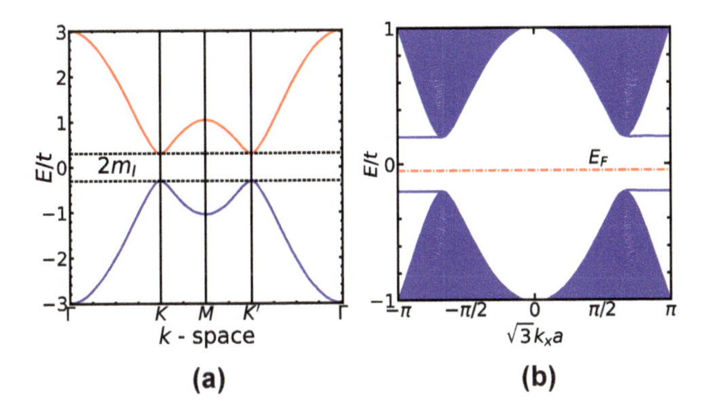

Figure 1.3: The bandstructure of a Semenoff insulator is shown in (a). A gap of magnitude $2m_I$ opens up at the Dirac points. The bandstructure of the nanoribbon is presented in (b), where the red dashed line represent the Fermi energy E_F.

graphene nano-ribbon of size $L_x L_y$[5] in Figure 1.3b shows no crossing of the edge modes across the Fermi energy. Besides the Berry phase and the Chern number also vanish, certifying the trivial nature of the energy gap in the spectrum. This eliminates the possibility of any topological properties of the model induced by m_I. In the following section we have discussed the effect of breaking the TRS in graphene.

1.5 Semi-Dirac system

Earlier we have seen that the dispersion spectrum of graphene is isotropic about the band touching points, that is, the spectrum is linear along all directions. The anisotropy in the band structure of graphene is an interesting aspect. In the tight-binding model for graphene, if we tune the NN hopping along a particular direction (say, $\boldsymbol{\delta}_3$), while keeping the other two same, the two Dirac points move towards each other and finally merge into one at a particular value of t_1, such that, $t_1 = 2t$ where t denotes the hopping energy along the $\boldsymbol{\delta}_1$ and $\boldsymbol{\delta}_2$ directions. The point at which the two Dirac points merge is called the semi-Dirac point (the $\mathbf{M}\left(0, \frac{2\pi}{3a}\right)$ point in the BZ). The band dispersion of the semi-Dirac system is quadratic along one direction and linear along the other direction. Hence the electrons behave as a massive fermions in one direction and massless Dirac particles in other direction simultaneously. The materials that exhibit such anisotropic dispersion are phosphorene under doping and pressure [2, 3], electric fields [13, 14], oxidized silicene layer [15], BEDT-TTF_2I_3 organic salts under pressure [16, 17], in multilayered structures of TiO_2/VO_2 [18, 19] etc. In the following we have discussed the properties of semi-Dirac system.

The tight-binding Hamiltonian that describes the semi-Dirac system is given in the following equation.

$$\mathcal{H} = -\sum_{\langle ij\rangle,\sigma} t_{i,j}\,(a_i^\dagger \boldsymbol{b}_j + \text{h.c.}) \tag{1.38}$$

where $t_{i,j} = t$ when i connects the site j in $\boldsymbol{\delta}_1$ or $\boldsymbol{\delta}_2$ direction, while $t_{i,j} = 2t$ when i connects the site j in the $\boldsymbol{\delta}_3$ direction. The above Hamiltonian in momentum space is as follows.

$$\mathcal{H} = \left[t_1 \cos(\boldsymbol{k}\cdot\boldsymbol{\delta}_3) + \sum_{i=1}^{2} t\cos(\boldsymbol{k}\cdot\boldsymbol{\delta}_i)\right]\sigma_x \left[t_1 \sin(\boldsymbol{k}\cdot\boldsymbol{\delta}_3) + \sum_{i=1}^{2} t\sin(\boldsymbol{k}\cdot\boldsymbol{\delta}_i)\right]\sigma_y \tag{1.39}$$

The band dispersion of the semi-Dirac system is given by,

$$\epsilon_k = \pm\sqrt{2t^2 + t_1^2 + 2t^2\cos\sqrt{3}k_x a + 4tt_1\cos(3k_y a/2)\cos(\sqrt{3}\,k_x a/2)} \tag{1.40}$$

5 This is called as semi-infinite ribbon. It is finite in y-direction and very large (taken to be infinitely large) along the x-direction ($L_x >> L_y$).

where the ± sign refers to the conduction band and valence band, respectively. Note that by putting $t_1 = t$ in Eq. (1.40), we get back the dispersion spectrum of graphene in Eq. (1.13). The electronic dispersion of the semi-Dirac system (put $t_1 = 2t$ in Eq. (1.40)) is plotted along the high symmetry points in the BZ in Figure 1.4a and b. As can be seen that the conduction and the valence bands touch each other at the M point which is intermediate to the $\boldsymbol{K}$ and $\boldsymbol{K}'$ points. Further, the band structure is quadratic along the x-direction ($\boldsymbol{K}\rightarrow M\rightarrow \boldsymbol{K}'$), while it is linear along the y-direction ($\Gamma\rightarrow M\rightarrow\Gamma'$). By fixing $t_1 = 2t$, the low energy Hamiltonian of the semi-Dirac system around the $M\left(0, \frac{2\pi}{3a}\right)$ point can be written as,

$$\mathcal{H}_0^{\mathrm{SD}}(q) = \frac{\hbar^2 q_x^2 \sigma_x}{2m^*} + v_y q_y \sigma_y \tag{1.41}$$

where the Pauli matrices, σ_x and σ_y represent the sublattice degrees of freedom. The effective mass m^* corresponding to the parabolic dispersion along the x-direction and the velocity v_y along the y-direction are expressed as $2\hbar/3ta^2$ and $3ta/\hbar$, respectively. The dispersion relation corresponding the low energy Hamiltonian in Eq. (1.40) is given by,

$$\epsilon_q^{\mathrm{SD}} = \pm\sqrt{\left(\frac{\hbar^2 q_x^2}{2m^*}\right)^2 + \left(\hbar v_y q_y\right)^2} \tag{1.42}$$

The above dispersion verifies the argument that the dispersion is quadratic along the x-direction and linear along the y-direction.

1.6 Haldane (Chern) insulator

In order to pursue the prospects of graphene as a topological insulator, one may break the TRS by including an imaginary second neighbour hopping. Such a term can be written as,

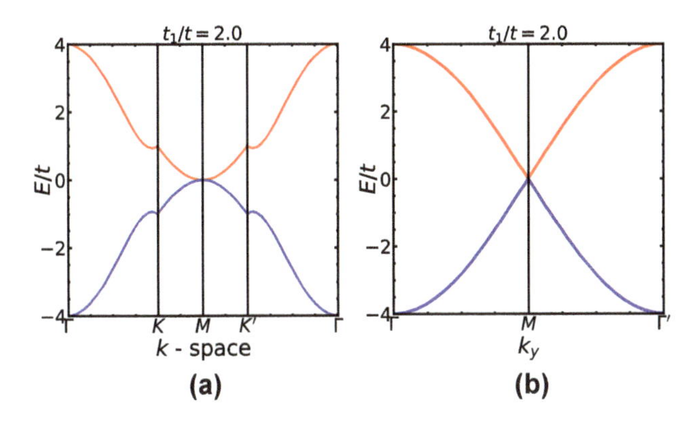

Figure 1.4: The bandstructure of the semi-Dirac system is shown in (a) and (b) along $\Gamma\rightarrow K\rightarrow M\rightarrow K'\rightarrow\Gamma$ and $\Gamma\rightarrow M\rightarrow\Gamma'$ directions, respectively.

$$\mathcal{H}'' = t_2 \sum_{\langle\langle ij \rangle\rangle} e^{i\nu_{ij}\phi} c_i^\dagger c_j \tag{1.43}$$

where the sum runs over the next nearest neighbour (NNN) sites (double angular bracket $\langle\langle ij \rangle\rangle$ imply NNN sites). Here, the term $\nu_{ij} \equiv \text{sgn}(\boldsymbol{d}_{ik} \times \boldsymbol{d}_{kj})_z = \pm 1$, where $\boldsymbol{d}_{ik}$ is the vector pointing from site i to its nearest-neighbor site k. This means that the phase associated with the NNN hopping is $+\phi$ if the electron makes a left turn while travelling to the next-nearest-neighbor and $-\phi$ in the case of a right turn (see Figure 1.5). The phase $e^{i\phi}$ or $e^{-i\phi}$ depending upon the direction of the hopping breaks the time reversal invariance, since the time reversal flips the direction of hopping. Only imaginary part of the phase, ϕ is interesting. Thus to set the real part to zero, we may choose $\phi = \pi/2$. This is known as Haldane model [9], which Haldane had proposed for achieving a (anomalous) quantum Hall state without an external magnetic field, or equivalently the Landau levels. As we shall see shortly, broken TRS implies a finite Chern number, a reason why these insulators are known as Chern insulator. The complex phases are equivalent to a staggered magnetic fields pointing at opposite directions at the center of the honeycomb lattice relative to those at the vertices.

The three NNN vectors are given by $b_1 = \boldsymbol{\delta}_2 - \boldsymbol{\delta}_3$, $b_2 = \boldsymbol{\delta}_3 - \boldsymbol{\delta}_1$ and $b_3 = \boldsymbol{\delta}_2 - \boldsymbol{\delta}_1$, where $\boldsymbol{\delta}_i$ denote the vectors connecting NN sites, the Hamiltonian can be written as,

$$\mathcal{H}'' = t_2 \sum_{i=1}^{3} \left[e^{i\phi} \sum_{r_A} c_A^\dagger(r_A) c_A(r_A + \boldsymbol{b}_i) + e^{-i\phi} \sum_{r_B} c_B^\dagger(r_B) c_B(r_B + \boldsymbol{b}_i) \right] \tag{1.44}$$

In the momentum space, the above Hamiltonian reads

$$\mathcal{H}''(k) = 2t_2 \left[\cos\phi \sum_{i=1}^{3} \cos(\boldsymbol{k} \cdot \boldsymbol{b}_i) 1 + \sin\phi \sum_{i=1}^{3} \sin(\boldsymbol{k} \cdot \boldsymbol{b}_i) \sigma_z \right] \tag{1.45}$$

Till this point, the NNN Hamiltonian is dispersive, that is dependent upon the $\boldsymbol{k}$-vector. However, the low energy Hamiltonian, that is near the $\boldsymbol{K}$ and $\boldsymbol{K}'$ points, it is independent of k at the leading order, where the Hamiltonian assumes the form,

$$\mathcal{H}''_{\pm K} = m_H \tau_z \sigma_z \tag{1.46}$$

where we have combined the forms at the two Dirac points, and

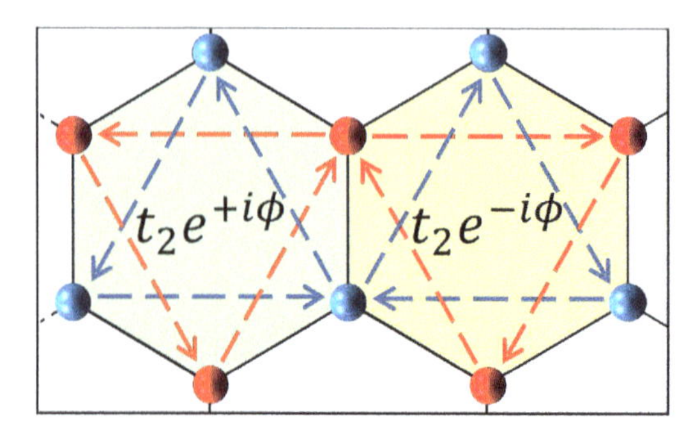

Figure 1.5: The complex next nearest neighbour hopping in the Haldane model.

$$m_H = -3\sqrt{3}\,t_2 \sin\phi \tag{1.47}$$

The above form can be easily obtained by noting that,

$$\sum_{i=1}^{3} \cos(\boldsymbol{k}\cdot\boldsymbol{b}_i) = -\frac{3}{2} \quad \text{and} \quad \sum_{i=1}^{3} \sin(\boldsymbol{k}\cdot\boldsymbol{b}_i) = \mp\frac{3\sqrt{3}}{2}$$

where $k \cdot b_i = \boldsymbol{K}$.

Therefore in the leading order $\mathcal{H}''$ is independent of the momentum k. The term in Eq. (1.46) breaks the TRS (σ_z does not change sign, but τ_z being the valley degree of freedom does). In presence of the complex second neighbour hopping t_2, the energy spectrum of graphene shows gap at the Dirac points similar to case of the Semenoff mass, m_I. In fact, if the Semenoff mass along with the complex NNN hopping is added to the Hamiltonian of graphene, we may open or close the energy gap by tuning the parameter m_I. For example, when $m_I = +3\sqrt{3}t_2$, the energy closes at one of the two Dirac points (either $\boldsymbol{K}$ or $\boldsymbol{K}'$) and opens up at the other Dirac point, while the reverse happens for $m_I = -/3\sqrt{3}t_2$ where the gap closes at the former Dirac point, while opening at the other. We show this in Figure 1.6b and c.

In order to elucidate the topological properties, we calculate the Berry phase and Chern number of the system. The eigenfunctions for the Haldane model can be written as,

$$\Psi^{\mu}(q) = \frac{1}{\sqrt{2}} \begin{pmatrix} \sqrt{1+\beta/E^{\mu}} \\ \lambda\sqrt{1-\beta/E^{\mu}}e^{i\theta_q} \end{pmatrix} \tag{1.48}$$

and the energy spectrum is given by,

$$E^{\mu} = \mu\sqrt{v_F^2 q_x^2 + v_F^2 q_y^2 + \beta^2} \tag{1.49}$$

where $\mu = \pm 1$ and $\beta = 3\sqrt{3}t_2$. This yields a Berry curvature of the form,

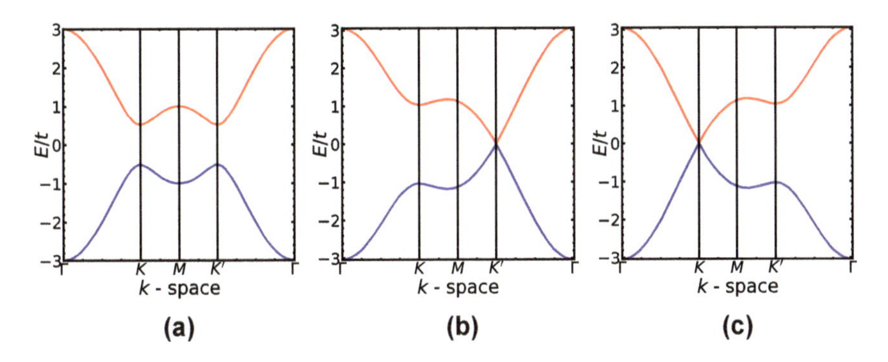

Figure 1.6: The energy band dispersion of graphene with a complex NNN hopping ($t_2 \neq 0$) is shown in (a). The bandstructure is shown in presence of both t_2 and m_I in (b) and (c) for $m_I = 3\sqrt{3}t_2$ and $m_I = -/3\sqrt{3}t_2$, respectively.

$$\Omega^{\mu} = \frac{v_F^2 \beta}{2\tau_z \left[v_F^2 q_x^2 + v_F^2 q_y^2 + \beta^2\right]^{\frac{3}{2}}} \tag{1.50}$$

The corresponding Berry connection is given by,

$$\mathcal{A}^{\mu} = \frac{\mu}{2}\left(1 + \mu \frac{\beta}{|E^{\mu}|}\right) \frac{\hat{\theta}_q}{q} \tag{1.51}$$

It is instructive for the readers to obtain the expressions for the Berry phase, Φ_B using $\Phi_B = \int \mathcal{A}^{\mu}.dq$ ($\mathcal{A}$: Berry connection) and the Chern number using, $C = \oint \Omega^{\mu}(q) d^2q$ ($\mathbf{\Omega}$: Berry curvature). Owing to a non-zero Chern number (C), the model has earned the name 'Chern insulator'. The Chern number is the topological invariant which distinguishes the Semenoff insulator from a Chern insulator. For the Semenoff insulator, $C = 0$. We show the phase diagram in Figure 1.9.

How do we know that this gap is topological in nature instead of a trivial one as seen for a Semenoff insulator? In the following we check for the chiral edge modes in a semi-infinite graphene nanoribbon. In Figure 1.7 we show the bandstructure of a semi-infinite ribbon. It can be noticed that the edge modes from the lower band goes into the upper band and vice versa. If the Fermi energy (shown via the red dashed line) lies in the bulk gap, the currents flow in the opposite directions along either edge of the ribbon. The directions of the current corresponding to the points 'u' and 'v' are shown by the arrows in the yellow panel at the bottom of Figure 1.7. The appearance of such edge modes implies the existence of quantum Hall like state without an external magnetic field.

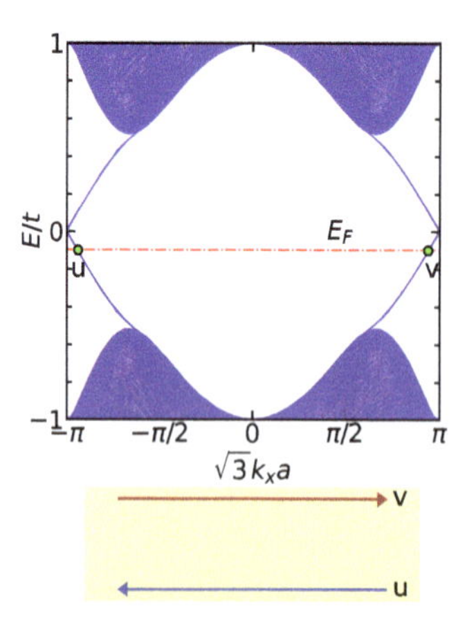

Figure 1.7: The bandstructure of the nanoribbon is presented, where the Fermi energy, E_F is denoted by the red dashed line. The yellow panel represents a part of the semi-infinite ribbon where the directions of edge currents corresponding to the points 'u' and 'v' are shown by the arrows.

1.6.1 Berry Curvature

So far we have looked at the Berry curvature obtained from the low energy Hamiltonian. However, one may also compute it numerically from the full tight binding Hamiltonian using the following relation.

$$\Omega = -\mathrm{Im}\sum_{j\neq i} \frac{\langle\psi_i|\nabla H|\psi_j\rangle \times \langle\psi_i||\nabla H|\psi_j\rangle}{(E_i - E_j)^2} \tag{1.52}$$

where the summation is taken over the occupied bands. It should be noted that if there is any degeneracy at any point in the BZ, the Berry curvature becomes undefined in such a scenario. Hence, in case of graphene, $\boldsymbol{\Omega}$ can not be calculated, because of the degeneracies at the Dirac points. However, in presence of m_I or t_2, there is spectral gap at the Dirac points, and hence the degeneracies are lifted. The plots of the Berry curvature for various values of m_I and t_2 are presented in Figure 1.8.

As can be seen from Figure 1.8a, $\boldsymbol{\Omega}$ is highly concentrated around the Dirac points. The values of $\boldsymbol{\Omega}$ at the $\boldsymbol{K}$ and the $\boldsymbol{K}'$ points are of the same magnitude, but of opposite signs. The reason for having opposite signs is that the Semenoff mass, m_I breaks the inversion symmetry. In Figure 1.8b, the Berry curvature is depicted for a non-zero NNN hopping t_2 and a zero Semenoff mass. In this case $\boldsymbol{\Omega}$ is also concentrated around the Dirac points. Since, the inversion symmetry is present in the system, $\boldsymbol{\Omega}$ has the same magnitude and same sign at the Dirac points. One can create an imbalance in the values of $\boldsymbol{\Omega}$ at the $\boldsymbol{K}$ and the $\boldsymbol{K}'$ points by adding the Semenoff mass. This is shown in Figure 1.8c and d. It is noticed that with the addition of m_I, $\boldsymbol{\Omega}$ at the $\boldsymbol{K}$ and the $\boldsymbol{K}'$ points becomes unequal. For $m_I = 0.1t$, $\boldsymbol{\Omega}$ at $\boldsymbol{K}'$ point is greater in magnitude than that at the $\boldsymbol{K}$ points. Because, the band gap at the $\boldsymbol{K}'$ points is lesser than that at the $\boldsymbol{K}$ points. This phenomenon can be inferred from Eq. (1.52), where the denominator is small for small band gap and consequently the value of $\boldsymbol{\Omega}$ increases. For $m_I = -0.1t$, the Berry curvature is identical to the case of $m_I = 0.1t$, except, the values at $\boldsymbol{K}$ and $\boldsymbol{K}'$ points are interchanged.

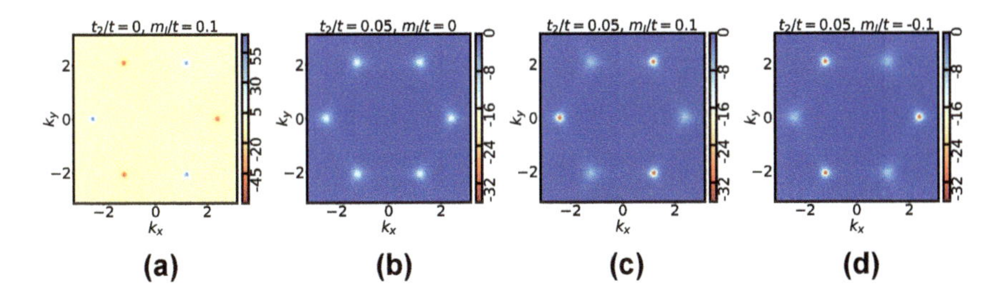

Figure 1.8: The Berry curvature of graphene is depicted in (a) for a non-zero Semenoff mass keeping the NNN hopping zero, while the Berry curvature in presence of t_2 is shown in (b), (c) and (d) for $m_I = 0$, $m_I = 0.1t$ and $m_I = -0.1t$, respectively. The values are encoded in th respective legends.

1.6.2 Chern number phase diagram

The Chern number of a system can be calculated using the relation as follows,

$$C = \frac{1}{2\pi} \iint_{\mathrm{BZ}} \Omega(k_x, k_y) \mathrm{d}k_x \mathrm{d}k_y, \tag{1.53}$$

where the integration is carried out in the first BZ and $\mathbf{\Omega}$ is the z-component of the Berry curvatur,e which can be calculated using Eq. (1.52). Here the Hamiltonian of the system for which the Chern number has been calculated numerically is given by,

$$\begin{aligned}\mathcal{H}(k) = t \sum_{i=1}^{3} [\cos(\boldsymbol{k} \cdot \boldsymbol{b}_i)\sigma_x + \sin(\boldsymbol{k} \cdot \boldsymbol{b}_i)\sigma_y] + 2t_2 \cos\phi \sum_{i=1}^{3} \cos(\boldsymbol{k} \cdot \boldsymbol{b}_i)1 \\ + \left[m_I - 2t_2 \sin\phi \sum_{i=1}^{3} \sin(\boldsymbol{k} \cdot \boldsymbol{b}_i)\sigma_z \right]\end{aligned} \tag{1.54}$$

In absence of t_2, the Chern number is always zero, since the TRS remains intact in the system. So, upon the addition of t_2, by breaking the TRS, one can find a non-zero Chern number which however depends on the values of m_I and ϕ. The variations of Chern number as a function of m_I and ϕ have been demonstrated in Figure 1.9. As can be seen, there are two coloured regions which represent non-trivial topological phases. One with the red colour has $C = +1$, while the blue one has $C = -1$. The white region implies trivial insulating regime with vanishing Chern number. The topological regions are the enclosed regions between two curves which are given by,

$$m_I / t_2 = \pm 3\sqrt{3} \sin\phi \tag{1.55}$$

Here, the constant term $\pm 3\sqrt{3}$ is the value of m_I/t_2 required to close the gap in the band dispersion either at the $\boldsymbol{K}$ point or at the $\boldsymbol{K}'$ point for $\phi = \pi/2$. Thus we see a non-zero Chern number for $m_I < |3\sqrt{3}|$ as long as there is a gap in the band dispersion.

At $m_I = |3\sqrt{3}t_2|$, the gap vanishes either at the $\boldsymbol{K}$ or at the $\boldsymbol{K}'$ point and the system no longer remains an insulator and the Chern number vanishes. For $m_I > |3\sqrt{3}|t_2$ the gap at both $\boldsymbol{K}$ and $\boldsymbol{K}'$ points open up and the system becomes an insulator again but the Chern

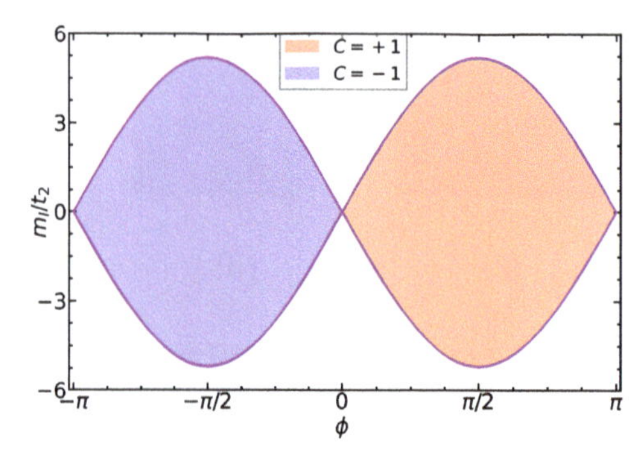

Figure 1.9: The phase diagram of the Haldane model is shown. The coloured regions denote Chern insulating phases with non-zero Chern numbers, while the white region signifies the trivial phase of the system.

number remains zero. Hence, the curve in Eq. (1.55) represents the semi-metallic state of the system and inside the region enclosed by the curves shows Chern insulating phase, while the outside one depicts trivial insulating phase.

1.6.3 Anomalous Hall conductivity

The anomalous Hall conductivity of a system can be obtained using the following relation.

$$\sigma_{xy} = \frac{\sigma_0}{2\pi} \sum_{\lambda} \int \frac{dk_x dk_y}{(2\pi)^2} f\left(E^{\lambda}_{k_x,k_y}\right) \Omega(k_x, k_y) \tag{1.56}$$

where $f(E) = [1 + e^{(E-E_F)/K_B T}]^{-1}$ is the Fermi–Dirac distribution function with the Fermi energy E_F and the absolute temperature T. Here $E^{\lambda}(k_x,k_y)$ signifies the energy dispersion with $\lambda = \pm 1$ being the conduction and valence bands, respectively. Using Eq. (1.56) and (1.52), σ_{xy} have been calculated numerically as a function of E_F as shown in Figure 1.10.

When the Fermi energy lies in the band gap, the Hall conductivity has values quantized at $\sigma_0 = e^2/h$ (see the red curve). Therefore, the e^2/h plateau has a width, which is equal to the band gap in the dispersion spectrum. As the Fermi energy goes away from the bulk gap, the value of σ_{xy} gradually decreases. In Figure 1.10, the blue curve represents the Hall conductivity for $m_I = 0.1t$. Since the band gap at $\boldsymbol{K}'$ point decreases, the plateau width in σ_{xy} also decreases. Therefore, the plateau width vanishes at $m_I = 3\sqrt{3}t_2 \simeq 0.2598t$ (for $t_2 = 0.05t$), since the gap vanishes at $\boldsymbol{K}'$ point. Now, with the further increase of m_I, say, for example $m_I = 0.3t$, gaps open up in the band dispersion, however, the plateau in the Hall conductivity vanishes. Since, the Chern number is zero for $m_I > 3\sqrt{3}t_2$, that is the system is a trivial insulator. Thus, a plateau in the Hall conductivity of a system can be observed for $m_I < |3\sqrt{3}t_2|$, that is, as long as the system remains a Chern insulator.

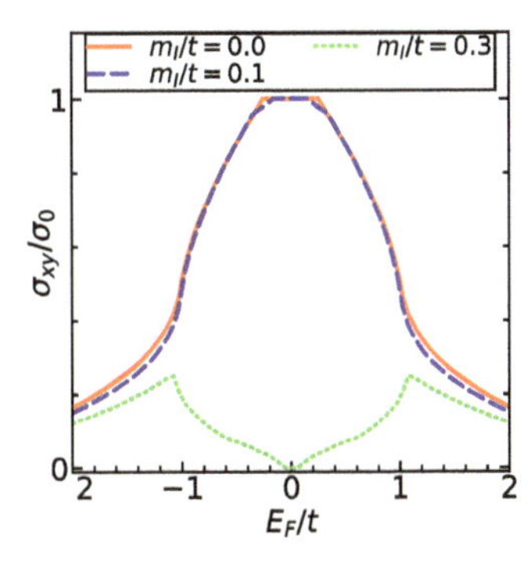

Figure 1.10: Anomalous Hall conductivity of Haldane model.

1.7 Semi-Dirac Haldane model

Now, we shall present results for the semi-Dirac model with broken TRS, which we call as the semi-Dirac Haldane (SDH) model. Earlier we have seen that the addition of the complex NNN hopping in the Hamiltonian of graphene creates energy gap in the bandstructure, that is, the system becomes an insulator. Now, if we add the complex NNN hopping the to the Hamiltonian of semi-Dirac system, no gap opens. Hence the system remains a semi-metal even in presence of the Haldane flux. The reason can be explained from the expression of Haldane term in Eq. (1.45). One can verify that $\mathcal{H}''(k)$ has values $|3\sqrt{3}t_2|$ and zero at $\boldsymbol{K}$ (or $\boldsymbol{K}'$) and M points, respectively. That is why the gap opens at the $\boldsymbol{K}$ (or $\boldsymbol{K}'$) point, while no gap is created at the M point. The bandstructure of the semi-Dirac Haldane model is shown in Figure 1.11a and b. As can be seen, the dispersion is linear about the M point along both the k_x and k_y directions. The low energy form of the bandstructure of the SDH model is written as,

$$\epsilon_q^{\text{SDH}} = \pm\sqrt{\left(\frac{\hbar^2 q_x^2}{2m^*}\right)^2 + \left(\hbar v_y q_y\right)^2 + \left(v_x \hbar q_x\right)^2} \tag{1.57}$$

where $m^* = 2\hbar/3ta^2$, $v_y = 3ta/\hbar$ and the term $v_x = -4\sqrt{3}at_2/\hbar$ is the velocity along the x-direction. At very low energies the q_x^2 term dominates the q_x^4 term and hence the first term in the square root of Eq. (1.57) becomes negligible. Therefore, at very low energies, we observe anistropic linear dispersion, since the velocities are different in the x- and y-directions. The ratio of velocities along x and y directions is $|v_x/v_y| = 4t_2/\sqrt{3}t$.

If we add a Semenoff mass m_I to the SDH model, a gap opens up in the bandstructure. One may think that since the TRS is broken in the SDH model, and there is a gap in the system due to a non-zero m_I, the system must possess a non-zero Chern number. However in this system the Chern number is always zero no matter whatever value of m_I we take. This can be verified from the bandstructure of the ribbon as

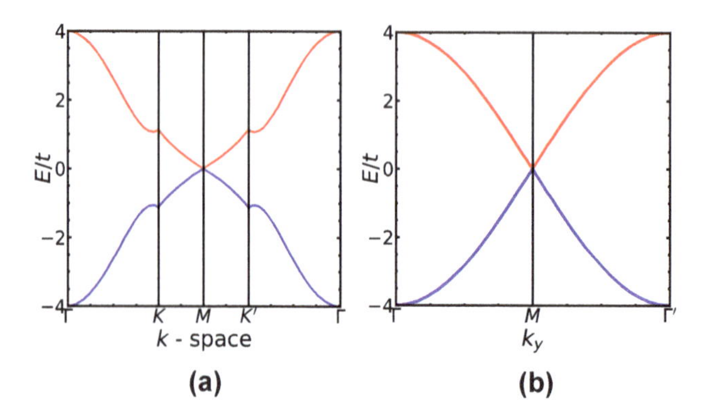

Figure 1.11: The bandstructure of the SDH model is shown in (a) and (b) along $\Gamma \to K \to M \to K' \to \Gamma$ and $\Gamma \to M \to \Gamma'$ directions, respectively.

presented in Figure 1.12. As can be seen for $m_I/t = 0.3$, there are four points where the Fermi level (E_F) intersects the edge modes (see Figure 1.12a). Corresponding to those points, there is a pair of counter-propagating edge modes along each edge of the ribbon (see Figure 1.12b). As a result, the net edge current along each edge vanishes.

Now let us look at the Berry curvature of the SDH model, the expression of which is given by

$$\Omega^{\mu}_{\text{SDH}}(q) = \frac{v_x v_y \alpha q_x^2}{2\mu\left[v_x^2 q_x^2 + \alpha^2 q_x^4 + v_y^2 q_y^2\right]^{\frac{3}{2}}} \tag{1.58}$$

and the corresponding Berry connection in given by,

$$\mathcal{A}^{\mu}_{\text{SDH}} = -\frac{1}{2}\frac{v_y^2}{|\epsilon|(|\epsilon| - \mu v_x q_x)}\left(q_x^2\hat{q}_y - 2q_x q_y \hat{q}_x\right) \tag{1.59}$$

where $\alpha = (2m^\star)^{-1}$. Since the bands touch at M point, the Berry curvature of the SDH model is undefined. It is instructive to the readers to compute the Berry phase of this system using $\Phi_B = \int \mathcal{A}^{\mu}.dq$, which we leave it to interested readers.

1.8 Kane–Mele model

Earlier it has been shown that the Dirac points are protected by the time reversal and the inversion symmetries. Breaking any one of them opens a gap at the Dirac points by splitting the degeneracy. Throughout our discussion thus far, the spin of the electron never played a role.[6] Kane and Mele included the spin [10, 11], thereby writing two

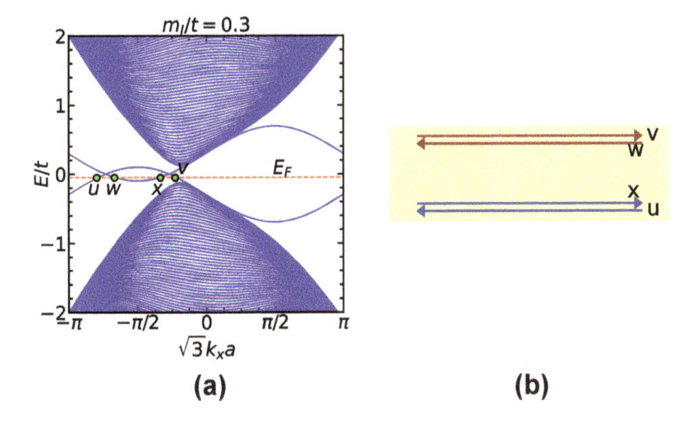

Figure 1.12: The edge states of the SDH model is shown for (a) $m_I/t = 0$ and the edge currents are shown in the yellow panel in figure (b), which represent a part of the semi-infinite ribbon.

6 The Pauli matrices denote sublattice and valley degrees of freedom.

Haldane Hamiltonians one for spin-↑ fermions and the other for spin-↓ fermions. This results in,

$$\mathcal{H} = -t \sum_{\langle ij\rangle,\alpha} c_{i\alpha}^{\dagger} c_{j\alpha} + it_2 \sum_{\langle\langle ij\rangle\rangle,\alpha\beta} v_{ij} c_{i\alpha}^{\dagger} (s_z)_{\alpha\beta} c_{j\beta} \tag{1.60}$$

Where s_z is the real spin of the electron and (α, β) denote the spin indices. Although s_z denotes the z-component of the Pauli matrices $\boldsymbol{\sigma}$, it is written with s_z to distinguish it from the sublattice and the valley indices. The NNN hopping term (the second term) describes a SOC that couples the chirality of the electrons, described by v_{ij} with $s_z = \pm 1$ being the z-component of spin. It is as if the orbital angular momentum vector L is associated with the chirality in a familiar $\boldsymbol{L}\cdot\boldsymbol{S}$ term. The spin s_z in NNN hopping causes a sign difference between the spin-↑ and spin-↓ electrons.

The second term, even though it resembles the Haldane term, respects all the symmetries of graphene. Time reversal flips the direction of hopping, that is reversing the motion, but simultaneously it also flips the spin, thereby yielding another negative sign. Since this term respects all the symmetries of graphene, it can be present there. It should be noted that in this model we get four bands (the conduction band and the valence bands and each of them for each spin). The conduction and the valence bands are degenerate. Even though the bands for each spin are identical to that of the Haldane model, in this case they behave distinctly. Because the phase of the complex NNN hopping for ↑-spins and ↓-spins are set to be $\phi = +\pi/2$ and $-\pi/2$, respectively, and hence have opposite masses at the $\boldsymbol{K}$ point (and the $\boldsymbol{K}'$ point). Therefore, the Chern number for spin-↑ and spin-↓ electrons have opposite signs, that is, $C_{\uparrow} = +1$ and $C_{\downarrow} = -1$. Please note that we have brought in a spin index to the Chern number. The total Chern number, $\sum \sigma C\sigma = 0$ which is a consequence of TRS being preserved.

Although the total Chern number of the system is zero, the difference of the Chern numbers does not vanish. This implies that we can assign the system with a non-zero topological invariant, which is termed as $\mathbb{Z}_2$ invariant. This type of insulators fall into the class of $\mathbb{Z}_2$ topological insulator which has non-zero $\mathbb{Z}_2$ index. The difference between the Chern numbers $C_{\uparrow}$ and $C_{\downarrow}$ is given by $C^s = C_{\uparrow} - C_{\downarrow} = 2$. The $\mathbb{Z}_2$ invariant [20, 21] can be calculated using $C^s/2$ mod 2, which is 1. Now we add the Semenoff mass m_I, and Rashba [23] SOC (λ_R) to the Hamiltonian of the Kane–Mele model and vary them to observe the variations of the $\mathbb{Z}_2$ invariant. Such variation is depicted in Figure 1.13, where the region enclosed by the curve (the blue region) has $\mathbb{Z}_2$ invariant 1, while it vanishes outside. Therefore, the system has non-trivial topological phases ($\mathbb{Z}_2 = 1$) inside the curve, while in the outside region, it is a trivial insulator ($\mathbb{Z}_2 = 0$).

Now let us look at the dispersion spectrum of a ribbon. In Figure 1.14, the spin filtered edge states of the Kane–Mele insulator are shown for two different values of m_I. $m_I/t = 0.1$ lies inside the curve in Figure 1.13, that is, the $\mathbb{Z}_2$ invariant is 1 and we observe a pair of edge modes from conduction bands goes into the valence bands and vice versa (see Figure 1.14a). As a result, if the Fermi level lies in the bulk gap, the spin currents will flow along both the edges of the ribbon (shown in Figure 1.14c) and hence we will observe

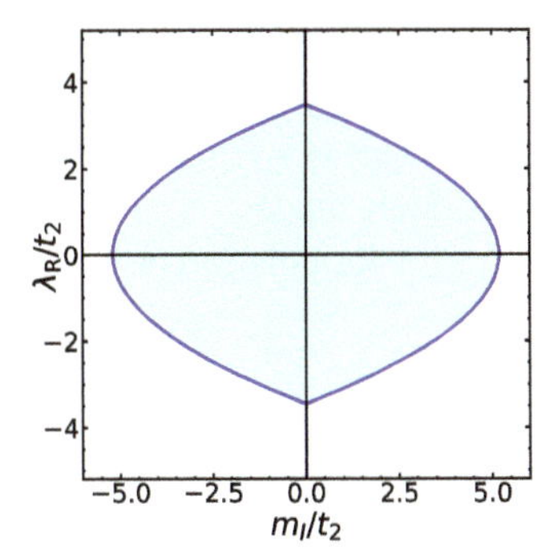

Figure 1.13: The phase diagram of the Kane–Mele model is shown. The blue region represents non-trivial topological phase, while the trivial insulating phase is denoted by the white region.

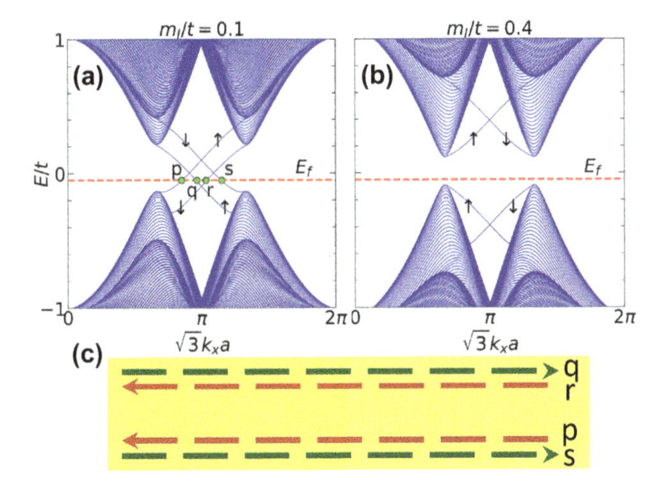

Figure 1.14: The edge modes for the Kane–Mele model on a nanoribbon showing (a) the topological and (b) the trivial phases. The red dashed lines in each figure represent the Fermi energy E_F. The yellow panel in figure (c) represents a part of the semi-infinite ribbon, where the arrows indicate the edge currents corresponding to the green dots in figure (a).

spin Hall conductivity. However, for $m_I/t = 0.4$, that is, when $\mathbb{Z}_2 = 0$, there is no crossing of the edge modes (see Figure 1.14b) and hence we will not observe a finite spin Hall conductivity.

1.8.1 Spin Hall conductivity

Earlier we have seen that although the Chern number of bands corresponding to each spin is non-zero, the total Chern number, C still vanishes because TRS of the system remains intact. As a result, the charge Hall conductivity, σ_{xy} is also zero. However, the spin Hall conductivity does not vanish in this case. The spin Hall effect is the generation of spin current at the edges of a sample perpendicular to the applied charge current. This results in the accumulation of opposite type spins at the edges of a sample. In the following we discuss the spin Hall conductivity of the Kane–Mele model.

In order to calculate the spin Hall conductivity the following formula can be used,

$$\sigma_{xy}^{\text{spin}} = e\hbar \sum_{\lambda} \int \frac{d\boldsymbol{k}}{(2\pi)^2} f(E_n) \Omega_{xy}^{z}(\boldsymbol{k}) \tag{1.61}$$

where $f(E_n) = [e^{\beta(E_n - E_F)} + 1]^{-1}$ is the Fermi–Dirac distribution function with the electronic energy E_n and $\beta(= (K_B T)^{-1})$ being the inverse temperature. The Berry curvature, Ω_{xy}^{z} in Eq. (1.61) can be calculated using the following equation.

$$\Omega_{xy}^{z}(\boldsymbol{q}) = i \sum_{m \neq n} \frac{\langle n|v_{sx}^{z}|m\rangle\langle m|v_y|n\rangle - (x, \leftrightarrow, y)}{(E_n(\boldsymbol{q}) - E_m(\boldsymbol{q}))^2} \tag{1.62}$$

where $x \leftrightarrow y$ interchanges the x and y variables in the first term of the numerator. Here, $v_i = \frac{1}{\hbar} \partial H / \partial k_i$ is the velocity operator and $v_{si}^{z} = \frac{1}{2}\left\{v_i, \frac{\hbar}{2} s_z \sigma_0\right\}$ is the velocity operator corresponding to the spin current. Using Eq. (1.61), one can compute the spin Hall conductivity as a function of the Fermi energy, E_F as shown in Figure 1.15.

As can be seen, a plateau quantized at $\sigma_0 = e/2\pi$ exists in the conductivity spectrum for a certain range of values of E_F, such that E_F lies in the gapped region of the dispersion spectrum. However, σ_{xy} decreases as E_F goes away from the gapped region. This conductivity spectrum is shown for $\tilde{m} < 0$. For $\tilde{m} > 0$, the spin Hall conductivity vanishes since the system is a trivial insulator for such values of $\tilde{m}$.

1.9 Semi-Dirac Kane–Mele model

The semi-Dirac Kane–Mele (SDKM) is obtained by including the spin dependent Haldane flux to the Hamiltonian of semi-Dirac system, such as,

$$\mathcal{H} = -\sum_{\langle ij \rangle, \alpha} t_{i,j} c_{i\alpha}^{\dagger} c_{j\alpha} + it_2 \sum_{\langle\langle ij \rangle\rangle, \alpha\beta} v_{ij} c_{i\alpha}^{\dagger} (s_z)_{\alpha\beta} c_{j\beta} \tag{1.63}$$

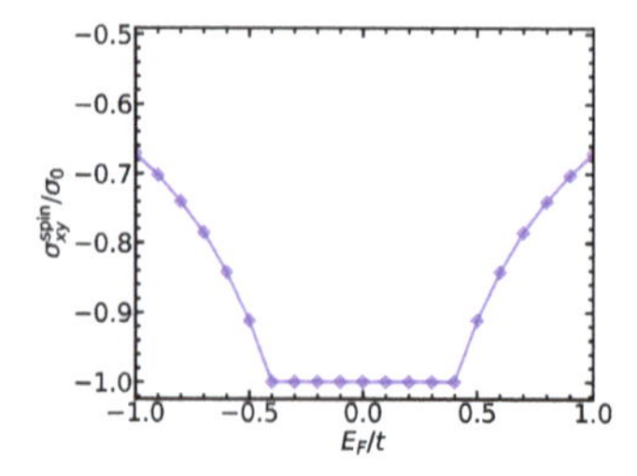

Figure 1.15: The spin Hall conductivity of the Kane–Mele model is shown. There is a non-zero value and a plateau region in the vicinity of zero bias ($E_F/t \simeq 0$).

where $t_{i,j} = t$ when i connects the site j in $\boldsymbol{\delta}_1$ or $\boldsymbol{\delta}_2$ direction, while $t_{i,j} = 2t$ when i connects the site j in the $\boldsymbol{\delta}_3$ direction (see Figure 1.1). The bandstructure of the SDKM model is identical to that of the semi-Dirac Haldane model, except the conduction and valence bands are degenerate, that is, the contuction and valence bands for spin-↑ and spin-↓ electrons coincide with each other. However in presence of the Semenoff mass m_I or the Rashba SOC (λ_R), the degeneracy is lifted, which implies the bands corresponding to the different spins are separated as shown in Figure 1.16a, and also a gap opens up between the conduction and valence band. The Chern numbers corresponding to each spin vanishes which is evident from the SDH model. Since, the bands of the SDH model have zero Chern number, in the SDKM model the Chern numbers of the bands corresponding to each spins also vanish. This implies that the $\mathbb{Z}_2$ invariant of the SDKM model also vanishes. In Figure 1.16b, we have shown the vanishing of $\mathbb{Z}_2$ invariant as one increases the hopping t_1. As can be seen, for the Dirac system, the value of the $\mathbb{Z}_2$ invariant is 1 in the region enclosed by the red curve, that means, inside the enclosed region, the system shows non-trivial topology, and the system is a trivial insulator outside the enclosed region ($\mathbb{Z}_2 = 0$). With the increase of t_1, the region enclosed by the curve, gradually decreases. At $t_1 = 2t$, that is for the SDKM model, the $\mathbb{Z}_2$ invariant vanishes for all values of λ_R and m_I.

The vanishing of the $\mathbb{Z}_2$ invariant for the SDKM model can also be verified from the edge state of the model which is shown in Figure 1.17. When the Fermi energy (the red dashed line) lies in the bulk gap, it intersects the edge modes at eight points. As a result, there are a total of two four edge modes along each edge of the ribbon, where the edge currents for both the spins flow in the opposite direction. Therefore, the net edge currents along both the edges vanishes. Hence the system is a trivial insulator.

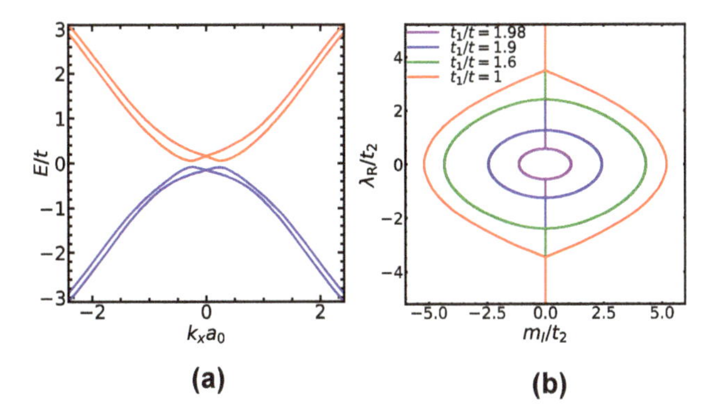

Figure 1.16: The bandstructure of the SDKM model is depicted in (a) for $\lambda_R = 0.06t$ and $m_I = 0.1t$. The phase diagram is shown in (b) for various values of t_1. The $\mathbb{Z}_2$ invariant is one inside the region enclosed by the curve, while it vanishes in the outside region.

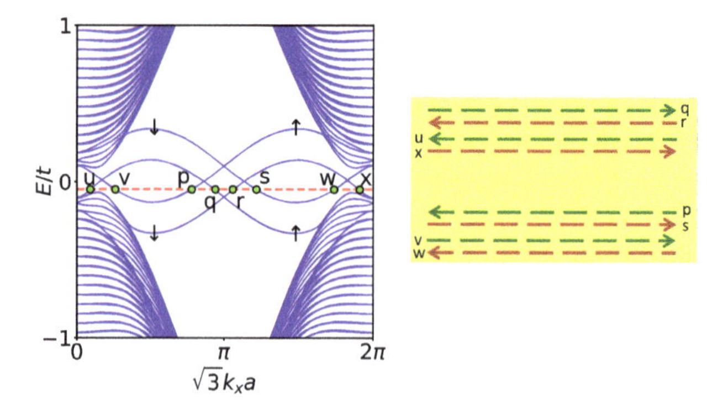

Figure 1.17: The edge state of the SDKM model is shown. The red dashed line denotes the Fermi energy. The yellow panel represents a part of the semi-infinite ribbon, where the edge currents corresponding to the intersecting points between the edge modes and E_F are shown by the arrows.

1.10 Spin polarized conductance

Theoretically, it was predicted by Kane and Mele [10, 11] that the quantum spin Hall (QSH) phase can be observed in graphene. However, the QSH effect cannot be observed in clean graphene since the intrinsic SOC strength is very weak [22, 24]. So, in order to increase the SOC, one may introduce heavy adatoms, such as, indium (In ^{49}), thallium (Th ^{81}), gold (Au ^{79}) etc. in the lattices of graphene. The heavy adatoms create a local imbalance of the charge density, which acts like a local electric field. The mobile charges couple to this field, and hence enhance the SOC. In this section, we shall discuss the transport properties of Au adatom decorated graphene. The Au adatoms induce a significant amount of Rashba SOC, which may provide additional clues in transport properties.

A schematic diagram of the ribbon is shown in Figure 1.18, where the Au atoms are placed randomly at the centre of the hexagons. The Au atoms are surrounded by six nearby carbon atoms with whom it interacts with, and hence the intrinsic SOC or the Rashba SOC increases. In presence of such adatoms, the Hamiltonian of the system can be written as,

$$\begin{aligned} \mathcal{H} = &-\sum_{\langle ij\rangle,\alpha} t_{i,j} c_{i\alpha}^{\dagger} c_{j\alpha} + i\lambda_{\mathrm{SO}} \sum_{\langle\langle ij\rangle\rangle\in\mathcal{R},\alpha\beta} \nu_{ij} c_{i\alpha}^{\dagger} (s_z)_{\alpha\beta} c_{j\beta} \\ &+ \lambda_{\mathrm{R}} \sum_{\langle i,j\rangle\in\mathcal{R},\alpha\beta} c_{i\alpha}^{\dagger} \left(s \times \hat{d}_{i,j} \right)_z c_{j\beta} - \mu \sum_{i\in\mathcal{R}} c_{i\alpha}^{\dagger} c_{i\alpha} \end{aligned} \tag{1.64}$$

where the first term is the nearest neighbour hopping with the hopping strength $t_{i,j}$ (already described earlier). The second term represents the SOC with the strength λ_{SO}. The SOC is enhanced by the Au atoms placed at the hexagons at site $\mathcal{R}$. The third term denotes the Rashba SOC with the strength $\lambda_{\boldsymbol{R}}$. In the fourth term, μ is the onsite energy of carbon atoms, which are located at the hexagons at $\mathcal{R}$ hosting the adatoms.

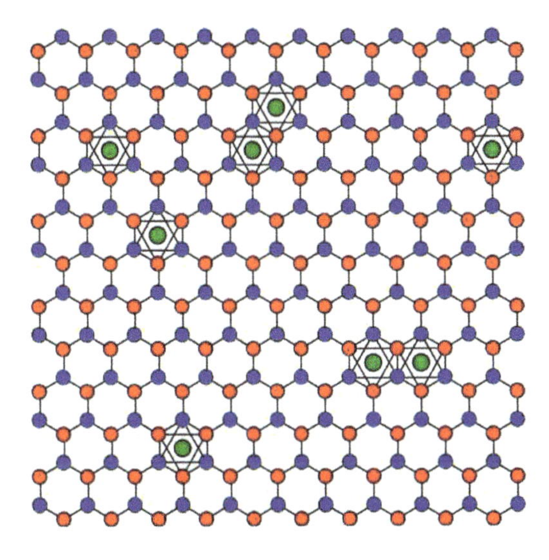

Figure 1.18: A schematic diagram of Au decorated graphene is shown. The green circles at the centre of hexagons represent the Au atoms, while the red and blue circles represent the C atoms of graphene.

Now let us calculate the spin polarized conductances. First we attach two leads, one in the left side and the other one in the right side of the ribbon. We put potentials V_L and V_R in the left and right leads, respectively. Then, the spin polarized conductance is calculated via,

$$G_\alpha^s = \frac{\hbar}{2e} \frac{I_\alpha^s}{V_L - V_R} \tag{1.65}$$

where I_α^s is the spin current polarized in a particular direction, α ($\alpha = x, y, z$) and flowing from left lead to right lead. Using the Landaur–Büttiker formula, the spin currents can be computed using the following expression.

$$I_p^\alpha = \frac{e^2}{h} \sum_q \mathrm{Tr}\left[\hat{\sigma}_\alpha \Gamma_q \mathcal{G}_r \Gamma_p \mathcal{G}_a\right](V_q - V_p) \tag{1.66}$$

where $\hat{\sigma}_\alpha = (\sigma_x, \sigma_y, \sigma_z)$ are the Pauli matrices.

Before we go into the result we first mention the parameters used in the calculation. The ribbon we have taken has 45 and 24 unit cells along the x- and the y-directions, respectively. The lattice constant is taken to be unity. Two different values of the adatom concentration, n_{ad} is taken, namely, $n_{\text{ad}} = 0.1$ and 0.3. From the first principles calculations [25, 26], the values of λ_{R}, λ_{SO} and μ are taken as $0.007t$, $0.0165t$ and $0.1t$, respectively.

Without any external magnetic field, one can have a non-zero spin polarized conductance, provided Rashba SOC is present in the system. It has been shown that the y-component of the spin polarized conductances can be obtained in a graphene nanoribbon [27]. Zhang [28, 29] et al. have shown that the x- and z-components of the spin polarized conductances (G_x^s and G_z^s) vanish in a graphene ribbon, since the longitudinal mirror symmetry (LMS) is present in the system. However, presence of the Au adatoms, breaks the LMS in the ribbon, and hence non-zero G_x^s and G_z^s are expected. Further, since the LMS is broken in the y-direction, we must have a non-zero G_y^s. The spin polarized conductances are shown in Figure 1.19 both for graphene ($t_1 = t$) and the semi-Dirac systems ($t_1 = 2t$). It is evident from Figure 1.19a and b, the value of G_y^s increases with the increase of n_{ad}. As we have introduced the random distribution of the adatoms, the LMS gets broken in the system. Therefore we see finite values in G_x^s and G_z^s. If we notice the spectrum for the semi-Dirac systems (Figure 1.19c and d), it becomes clear that the magnitude of the conductivities decreases both for $n_{\text{ad}} = 0.1$ and $n_{\text{ad}} = 0.3$. This is so, since the semi-Dirac system is not a QSH insulator, and hence we are getting non-zero values of the conductivity because the ribbon has a finite width in the y-direction.

Thus, the spin polarized conductivity can be increased by increasing the Au concentration. Theoretically, the same effect can be realized by increasing λ_{R} and λ_{SO}, where one can observe a significant increase in the spin polarized conductivity. So, let us consider an hypothetical adatom which enhances the Rashba SOC (λ_{R}) and the intrinsic SOC (λ_{SO}). Specifically, we take $\lambda_{\text{R}} = 0.1t$ and $\lambda_{\text{SO}} = 0.08t$ which are higher in magnitude by one order than that created by Au adatoms. In Figure 1.20 we have

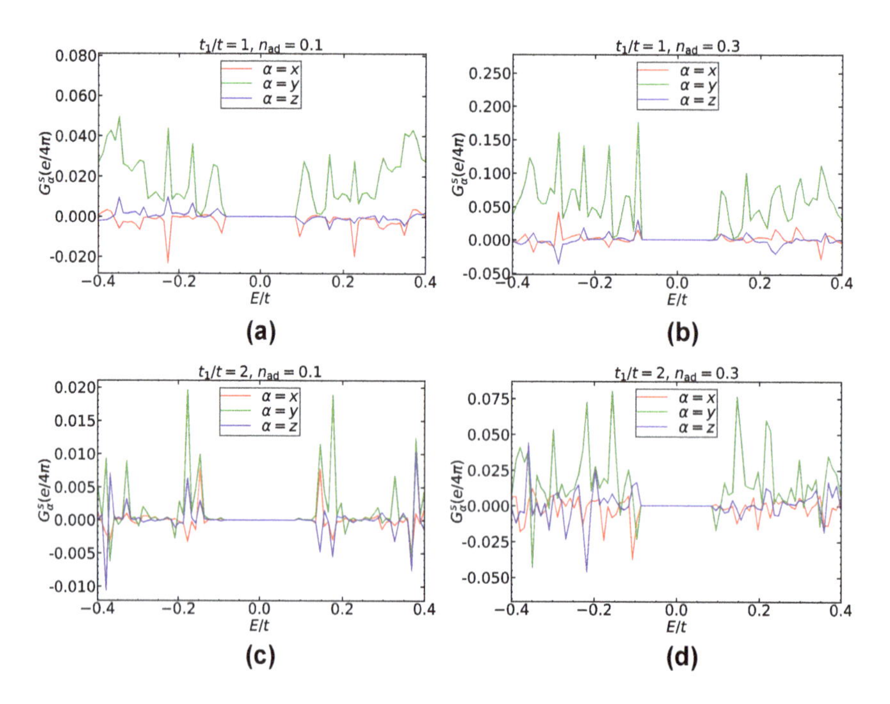

Figure 1.19: The spin polarized conductivity is shown for the Dirac systems ($t_1 = t$) in (a) and (b) for $n_{\text{ad}} = 0.1$ and $n_{\text{ad}} = 0.3$, respectively, while the same for the semi-Dirac systems ($t_1 = t$) are presented in (c) and (d) for $n_{\text{ad}} = 0.1$ and $n_{\text{ad}} = 0.3$, respectively. The values of λ_{R} and λ_{SO} are $0.0165t$ and $0.007t$, respectively.

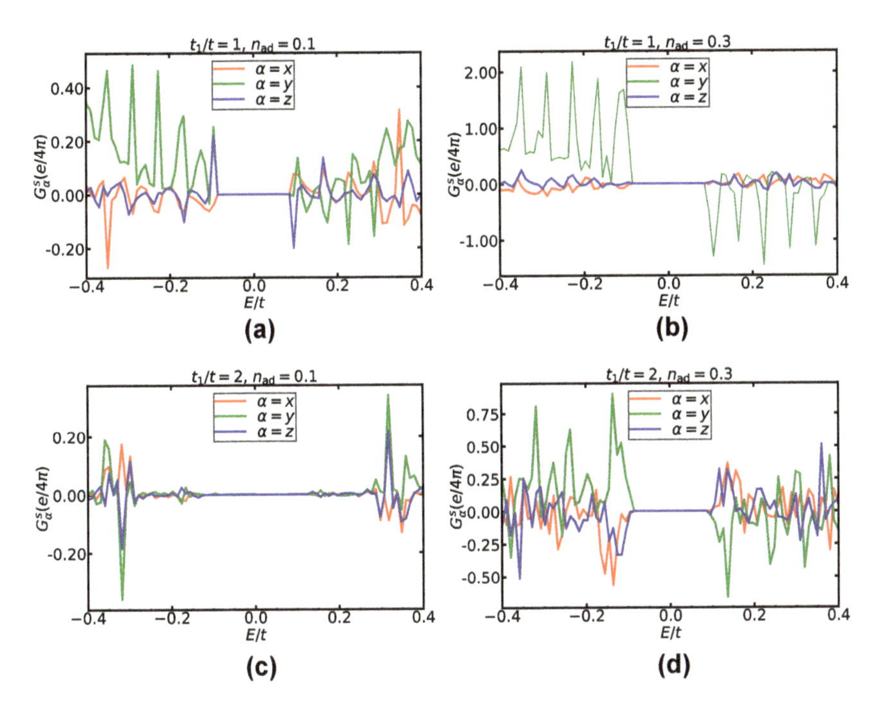

Figure 1.20: The spin polarized conductivity is shown for the Dirac systems ($t_1 = t$) in (a) and (b) for $n_{ad} = 0.1$ and $n_{ad} = 0.3$, respectively, while the same for the semi-Dirac systems ($t_1 = t$) are presented in (c) and (d) for $n_{ad} = 0.1$ and $n_{ad} = 0.3$, respectively. The values of λ_R and λ_{SO} are $0.1t$ and $0.08t$, respectively.

presented the spin polarized conductances both the for graphene and semi-Dirac system for two different values of n_{ad}. As can be seen, the magnitudes of the spin polarized conductivities are one order higher than the previous case. Similar to the previous case, the values of the spin polarized conductivity of semi-Dirac system is less than that of the graphene. Further, for higher concentrations of adatoms, G^s_α ($\alpha = x, y, z$) increases.

1.11 Conclusions

In this chapter, a pedagogical study of the electronic dispersion of graphene has been done. The bandstructure of a prototype 2D material, namely, graphene reveals that it is a semi-metal because the conduction and the valence bands touch each other at the two non-equivalent Dirac points, namely, $\boldsymbol{K}$ and $\boldsymbol{K}'$ points in the BZ. One can make it an insulator, that is, open a gap either by introducing a Semenoff mass, or by a direction dependent complex NNN hopping. Graphene in presence of the NNN hopping is very interesting since the system becomes a topological insulator. The calculation of the

Chern number shows the system has a non-zero Chern number. Further, the calculation of the Berry curvature shows that they are highly concentrated at the Dirac points. The edge state spectrum reveals that the ribbon possess helical edge current. So, the anomalous Hall conductivity is calculated where a plateau at e^2/h can be observed. The semi-Dirac physics is introduced by changing one of the hopping energies. In such a situation, all the topological properties, which were present in the case of graphene, vanishes completely. Thus, a machinery for studying topological properties, such as, Berry connection, Berry phase, Berry curvature and Chern number are developed and applied to graphene and the semi-Dirac systems. Moreover, electron spin is also included in the calculations related to graphene. It is found that the Chern number in such scenario vanishes but another topological invariant, the $\mathbb{Z}_2$ invariant does not vanish. The Rashba SOC is also added to get a $\mathbb{Z}_2$ invariant phase diagram. The presence of the topological phase is confirmed via calculating the spin Hall conductivity which shows a plateau at $e/2\pi$. The $\mathbb{Z}_2$ invariant is found to be zero for the semi-Dirac Kane–Mele model, that is, the system remains a normal insulator. To make these 2D materials useful for applications, the spin polarized transport of Dirac and semi-Dirac materials is calculated, which can be useful from the spintronics perspective. Owing to an extremely weak SOC, our calculations reveal that one can introduce adatoms in the graphene matrix and by increasing the adatom concentration, one can enhance the spin polarized conductivities. The spin polarized conductivities should have applications in the spin dependent transport properties. This raises rich prospects of manipulating a particular kind of spin for processing information and communication. Spin transport may become possible in the near future, which will facilitate dissipationless transfer of information. Though we agree that 'spintronics' replacing 'electronics' is a far fetched idea, however with galloping technological development, hopefully sooner, rather than later, the society will derive benefits from such fundamental studies as shown by us in this work.

References

1. Novoselov KS, Geim AK, Morozov SV, Jiang D, Zhang Y, Dubonos SV, et al. Electric field effect in atomically thin carbon films. Science 2004;306:666.
2. Rodin AS, Carvalho A, Castro Neto AH. Strain-induced gap modification in black phosphorus. Phys Rev Lett 2014;112:176801.
3. Guan J, Zhu Z, Tománek D. Phase coexistence and metal-insulator transition in few-layer phosphorene: a computational study. Phys Rev Lett 2014;113:046804.
4. Acun A, Zhang L, Bampoulis P, Farmanbar M, van Houselt A, Rudenko AN, et al. Germanene: the germanium analogue of graphene. J Phys Condens Matter 2015;27:443002.
5. Guzmán-Verri GG, Voon LCLY. Electronic structure of silicon-based nanostructures. Phys Rev B 2007;76:075131.
6. Cahangirov S, Topsakal M, Aktürk E, Şahin H, Ciraci S. Two- and one-dimensional honeycomb structures of silicon and germanium. Phys Rev Lett 2009;102:236804.

7. Wang J, Malgras V, Sugahara Y, Yamauchi Y. Electrochemical energy storage performance of 2D nanoarchitectured hybrid materials. Nat Commun 2021;12:3563.
8. Xu B, Qi S, Jin M, Cai X, Laib L, Sun Z, et al. 2020 roadmap on two-dimensional materials for energy storage and conversion. Chin Chem Lett 2019;30:2053.
9. Haldane FDM. Model for a quantum hall effect without landau levels: condensed-matter realization of the "parity anomaly". Phys Rev Lett 1988;61:2015.
10. Kane CL, Mele EJ. Quantum spin hall effect in graphene. Phys Rev Lett 2005;95:146802.
11. Kane CL, Mele EJ. Quantum spin hall effect in graphene. Phys Rev Lett 2005;95:226801.
12. Bernevig BA, Hughes TL, Zhang SC. Quantum spin hall effect and topological phase transition in HgTe quantum wells. Science 2006;314:1757–61.
13. Rudenko AN, Yuan S, Katsnelson MI. Erratum: toward a realistic description of multilayer black phosphorus: fromGWapproximation to large-scale tight-binding simulations [Phys. Rev. B92, 085419 (2015)]. Phys Rev B 2015;92:085419.
14. Dutreix C, Stepanov EA, Katsnelson MI. Laser-induced topological transitions in phosphorene with inversion symmetry. Phys Rev B 2016;92:241404.
15. Zhong C, Chen Y, Xie Y, Sun Y-Y, Zhang S. Semi-Dirac semimetal in silicene oxide. Phys Chem Chem Phys 2017;19:3820.
16. Suzumura Y, Morinari T, Piéchon F. Mechanism of Dirac point in α type organic conductor under pressure. J Phys Soc Jpn 2013;82:023708.
17. Hasegawa Y, Konno R, Nakano H, Kohmoto M. Zero modes of tight-binding electrons on the honeycomb lattice. Phys Rev B 2006;74:033413.
18. Pardo V, Pickett WE. Phys Rev Lett 2009;102:166803.
19. Pardo V, Pickett WE. Compensated magnetism by design in double perovskite oxides. Phys Rev B 2010;81:035111.
20. Kane CL. Topological band theory and the invariant. Amsterdam: Elsevier; 2013, 6:3–34 pp.
21. Fu L, Kane CL. Time reversal polarization and aZ2adiabatic spin pump. Phys Rev B 2006;74:195312.
22. Min H, Hill JE, Sinitsyn NA, Sahu BR, Kleinman L, MacDonald AH. Intrinsic and Rashba spin-orbit interactions in graphene sheets. Phys Rev B 2006;74:165310.
23. Rashba EI. Spin-orbit coupling in condensed matter physics. Sov Phys Solid State 1960;2:1109.
24. Yao Y, Ye F, Qi X-L, Zhang S-C, Fang Z. Spin-orbit gap of graphene: first-principles calculations. Phys Rev B 2007;75:041401.
25. Van Tuan D, Marmolejo-Tejada JM, Waintal X, Nikoli´c BK, Valenzuela SO, Roche S. Spin hall effect and origins of nonlocal resistance in adatom-decorated graphene. Phys Rev Lett 2016;117:176602.
26. Weeks C, Hu J, Alicea J, Franz M, Wu R. Publisher's note: engineering a robust quantum spin hall state in graphene via adatom deposition. Phys Rev X 2011;1:021001.
27. Chico L, Latge A, Brey L. Symmetries of quantum transport with Rashba spin-orbit: graphene spintronics. Phys Chem Chem Phys 2015;17:16469.
28. Zhang Q, Chan KS, Li J. Spin-polarized transport in graphene nanoribbons with Rashba spin-orbit interaction: the effects of spatial symmetry. Phys Chem Chem Phys 2017;19:6871.
29. Zhang Q, Lin Z, Chan KS. Spin polarization switching in monolayer graphene through a Rashba multi-barrier structure. Appl Phys Lett 2013;102:142407.

Lynne A. Pilcher*

2 Embedding systems thinking in tertiary chemistry for sustainability

Abstract: In response to the IUPAC call to introduce systems thinking in tertiary chemistry education, we have developed and implemented two interventions at the first-year undergraduate level: one was designed to integrate systems thinking in first-year organic chemistry using the topic of surfactants and the other in a first-semester service course to engineering students using the stoichiometry of the synthesis of aspirin. We demonstrate how the systems thinking approach in both interventions did not lose the focus of the chemistry content that needed to be covered, exposed students to the concept of systems thinking, started to develop some systems thinking skills, and made a case for the contribution that chemistry can and should make to meet the UN sustainable development goals. Through both the design and the implementation process, it has become clear that introducing systems thinking is complex and it remains a challenge to keep the complexity manageable to avoid cognitive overload. Both interventions leveraged the power of group work to help students deal with the complexity of the topics while also developing participatory competence required for sustainability. The development of systems thinking skills and a capacity to cope with complexity requires multiple opportunities. Infusing syllabus themes that relate to real chemical systems with a systems thinking perspective can provide such an opportunity without compromising chemistry teaching. We believe that skills development should continue throughout the undergraduate chemistry degree to deliver chemistry graduates who can make a difference to global sustainability.

Keywords: chemistry education; collaborative learning; curriculum development; graduate attributes; visualization.

2.1 Introduction

2.1.1 Chemistry and systems thinking for sustainability

The UN General Assembly proclaimed 2022 as the International Year of Basic Sciences for Sustainable Development. Their purpose is to raise awareness of the importance of the basic sciences in addressing the Sustainable Development Goals (SDGs). Meeting

***Corresponding author: Lynne A. Pilcher**, Department of Chemistry, University of Pretoria, Room 1-35, Chemistry Building, Pretoria, South Africa, E-mail: lynne.pilcher@up.ac.za. https://orcid.org/0000-0003-3382-8536

As per De Gruyter's policy this article has previously been published in the journal Physical Sciences Reviews. Please cite as: L. A. Pilcher "Embedding systems thinking in tertiary chemistry for sustainability" *Physical Sciences Reviews* [Online] 2022.
DOI: 10.1515/psr-2022-0119 | https://doi.org/10.1515/9783110913361-002

the SDGs such as zero hunger, good health, clean water, and affordable and clean energy requires an understanding of chemical substances, their transformations and their interactions within the earth system. In addition, the planetary boundaries framework has identified nine processes that regulate the stability and resilience of the earth system [1]. These processes have direct links to chemistry either through the measurement of substances or management of chemical transformations. Clearly, chemistry underpins considerations of how present and future generations can live within the limits of our natural world. This is described as chemistry providing "the molecular basis of sustainability" [2]. The practice of chemistry without consideration of the implications for the environment has caused significant harm. To address the negative impact of the chemical industry on the environment, green chemistry tools and metrics have been developed by chemists and engineers to promote the sustainable manufacture of products. These encourage practitioners to think holistically about manufacturing across the life cycle of products, from sourcing raw materials to reducing harmful waste. However, despite these advances, the majority of chemicals are synthesized without full consideration of their toxicity to humans and the environment, and their ability to be reused or recycled or sourced renewably [2]. For chemists to ensure that the basic science of chemistry contributes to sustainability, a systems thinking perspective is required [3]. This perspective needs to permeate all chemists' thinking and practice.

A system refers to a set of interdependent or interrelated components that function as a whole. Systems thinking goes beyond fragmented knowledge to a holistic engagement with complex systems. In the context of sustainability, the systems of interest include different domains such as the environment, the economy and society; and different scales from the molecular and microscopic to the macroscopic and global. Systems thinking includes the analysis of the components and considers their interactions and interdependence. Systems thinking has been identified as a key competence needed to build transition strategies towards sustainability [4]. Identifying intervention points, anticipating future trajectories, and planning transition processes all require a holistic view of complex chemical, ecological and social systems, and an understanding of their dynamics and molecular basis.

2.1.2 Systems thinking in chemistry education

In 2016 a group of "Chemists for Sustainability" in the International Organization for Chemical Sciences in Development (IOCD) proposed that chemistry teaching needs to be reoriented as a science for the benefit of society, tackling global challenges, and contributing to sustainable development [5]. To do so, they argued that chemistry needs to incorporate systems thinking into the curriculum. This means that in addition to learning the principles and practice of chemistry, students need to develop the capacity for thinking and working across disciplinary boundaries.

Mahaffy and Matlin brought together ca. 20 global leaders in chemistry education to develop learning objectives and strategies to infuse systems thinking into the teaching of introductory chemistry (or general chemistry) at the post-secondary level. This project was conducted under the auspices of the International Union of Pure and Applied Chemistry (IUPAC) and was funded by the IUPAC (Project No.: 2017-010-1-050).[1] They chose to focus on post-secondary general chemistry courses because these courses serve both future chemists and many other future STEM and health professionals.

Chemistry is not just a fundamental science; it is also very complex. In making chemistry accessible to students, chemistry education has taken a reductionist approach, considering topics of chemistry in isolation. This reductionist approach to science education and scientific research has resulted in a significant increase in our knowledge of the natural world and great technological advances, but it does not prepare students adequately to address global world challenges [6]. Furthermore, students often experience learning chemistry as isolated and fragmented disciplinary knowledge. The majority of students in gateway general chemistry courses see little relevance of chemistry to their range of future careers in science, technology, engineering, and health care [2]. Reorienting chemistry education with systems thinking would provide a framework for connecting chemistry knowledge at the molecular level with the needs of society and the sustainability of the earth and would make the relevance of chemistry obvious thereby advancing meaningful learning. It would also provide an opportunity to develop domain-specific critical thinking and problem solving as topics of discussion are expanded beyond the facts of chemistry to issues which have normative, moral, ethical, or public policy dimensions.

While introducing systems thinking in chemistry education presents many benefits, dealing with complexity on multiple scales is no small task. Educators and curriculum developers will have to develop strategies to introduce complexity with appropriate scaffolding and consideration of learning progressions. They have to be mindful of cognitive overload in setting reasonable goals for learning outcomes, tasks and assessments [7]. They will need to use tools to frame and manage the complexity that will help students to develop the capacity to navigate complexity [8]. Research from STEM education suggests that systems thinking is not a "natural" way for humans to think and opportunities to develop systems thinking skills will have to be intentionally built into the curriculum [9].

Not specifically designed with the sustainability agenda in mind, the systems thinking hierarchical (STH) model (Table 2.1) outlines the components of systems thinking. It presents a hierarchy of difficulties for the development of systems thinking

---

1 A key output of the project was the special issue of the prominent Journal of Chemistry Education on Reimagining Chemistry Education: Systems Thinking, and Green and Sustainable Chemistry published in 2019.

Table 2.1: The systems thinking hierarchical (STH) model [10].

	The ability to …
Level I: Analysis	1. identify the components of a system and processes within the system,
Level II: Synthesis	1. identify simple relationships between or among the system's components, 2. identify dynamic relationships within the system, 3. organize the systems' components, processes and their interactions within a framework of relationships, 4. identify cycles of matter and energy within the system,
Level III: Implementation	1. recognize hidden dimensions of the system – the patterns and relationships not seen on the surface that give rise to natural phenomena, 2. make generalizations, 3. think temporally: retrospection and prediction.

skills which were empirically derived in an earth science context [10]. It can be used as a guide for designing teaching interventions with appropriate scaffolding and for managing expectations in chemistry.

In the context of chemistry, the advanced level systems thinking skill "to recognize hidden dimensions of the system" readily translates to the domain-specific skill of molecular level reasoning, drawing in the "molecular basis for sustainability". This analysis of the properties, interactions, and spatiotemporal effects of chemical compounds leading to emergent system properties is termed mechanistic reasoning by Talanquer [11]. He advocates that chemical systems thinking should at least contain three core components: mechanistic reasoning based on chemical principles, a context-based focus and a sustainable action perspective.

Another valuable contribution to articulating the essential characteristics of systems thinking in chemistry education is the "Characteristics Essential for designing or Modifying Instruction for a Systems Thinking approach" (ChEMIST) table by York and Orgill [9]. The table (Table 2.2), proposed as a guide for designing, analysing, and optimizing teaching activities, distilled five essential characteristics of a systems thinking approach: (I) recognizing the system as a whole, (II) Identifying relationships between parts, (III) identifying causal variables, (IV) examining behavior over time, and (V) identifying interactions of the system with the environment. For each characteristic, activities could be classified on a continuum from a more analytical approach to a more holistic approach and they propose that instruction should include activities that develop both analytical and holistic thinking skills.

In contrast to STEM disciplines such as engineering, environmental science and biology, there is little experience incorporating systems thinking in chemistry education, and consequently, resources for teaching and assessment are limited. Chemistry instructors have themselves been educated with the reductionist approach and with the pressures they face within the education system are unlikely to adopt novel

Table 2.2: The characteristics essential for designing or modifying instruction for a systems thinking approach (ChEMIST) table [9].

A systems thinker in chemistry education should.	Less holistic more analytical/elaborative	⇄	More holistic less analytical/elaborative
I. Recognize a system as a whole, not just as a collection of parts	Identify the individual components and processes within a system	Examine the organization of components within the system	Examine a system as a unified whole
II. Examine the relationships between the parts of a system and how those inter-connections lead to cyclic system behaviors	Identify the ways in which components of a system are connected	Examine positive and negative feedback loops within a system	Identify and explain the causes of cyclic behaviors within a system
III. Identify variables that cause system behaviors, including unique system-level emergent behaviors	Identify the multiple variables that influence a given system level behavior; consider the potential effects of stochastic and "hidden" processes on the system-level behavior	Examine the relative, potentially nonlinear, effects that multiple identified variables have on a given system-level behavior	Identify, examine, and explain (to the extent possible) emergent system-level behaviors
IV. Examine how system behaviors change over time	Identify system-level behaviors that change over time	Describe how a given system-level behavior changes over time	Use system-level behavior-over-time trends under one set of conditions to make predictions about system-level behavior-over-time trends under another set of conditions
V. Identify interactions between a system and its environment, including the human components of the environment	Identify and describe system boundaries	Consider possible effects of a system's environment on the system's behaviors; consider how the system under study might be a component of and contribute to the behaviors of a larger system	Consider the role of human action on current and future system-level behaviors

teaching approaches without being furnished with detailed teaching materials. We describe two projects in chemistry education in which we implemented systems thinking teaching interventions for large first-year chemistry courses; one for science students and one for engineering students.

2.2 Two systems-thinking teaching interventions for chemistry

2.2.1 Project 1: systems thinking for first-year organic chemistry – surfactants

There have been several contributions to developing systems thinking teaching resources for general chemistry, thus our first project was directed at introductory organic chemistry, which in our setting, is taught in the last quarter of the first year of university. Project 1, the MSc Science Education project of Micke Reynders, was initiated during the Covid-19 pandemic and hence was conceptualized for implementation online or face-to-face.

The topic of the intervention was the complex system centered on the surfactant linear alkylbenzene sulfonate (LAS, Figure 2.1). This surfactant forms the active ingredient of common commercial laundry detergents and hence satisfied the criterion of being a chemical that students encounter in their everyday lives. Several connections to the syllabus were identified building on the general chemistry topics of intermolecular forces, solution chemistry and acid–base chemistry. It also fitted in with the early topics in the organic chemistry syllabus: molecular structure and bonding, use of skeletal structures, functional groups, the alkanes sourced from crude oil and reaction types. These represent the hidden dimensions at the molecular level scale of the system. Furthermore, surfactants, water and air, yield the emergent properties of micelle formation critical to the practical functions of detergency and of foaming at the microscopic scale.

The ability of surfactants to disrupt lipid bilayers of viruses fitted the context of hygiene during the pandemic through connection to the washing of cloth face masks [12]. Detrimental associations with human health could be made due to the cytotoxicity of LAS to the skin and its ability to stimulate the growth of colon cancer when ingested at a low level [13]. The latter possibility needs to be considered in the light of evidence that when "brown" water containing LAS is used directly for crop irrigation LAS is taken up by carrots in measurable proportions [14]. Of additional local relevance

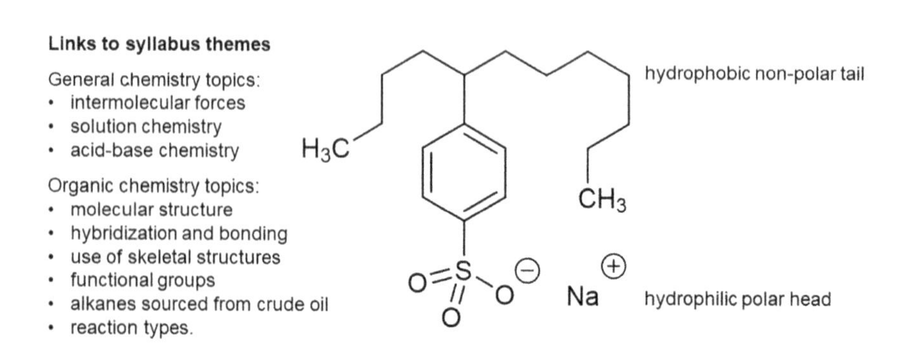

Figure 2.1: Linear alkyl benzene sulfonate (LAS) and the connections to the syllabus.

considering societal and environmental factors is the use of rivers for washing laundry in rural communities without an alternative water supply and the effect on the river system [15]. Similarly, the failure of sewage systems has resulted in the release of large quantities of LAS into river systems with consequent foaming cutting out light that facilitates the biodegradation of LAS [16]. Delayed biodegradation extends the ecotoxicity of LAS. The industrial manufacture of LAS provided touchpoints to environmental and economic issues and drew attention to the roles of catalysts and energy systems. Thus, LAS can illustrate the point that chemicals have both hazards and benefits that must be considered together [2].

Criteria applied for choice of Topic for both projects
- organic compound
- chemical of daily life – personal relevance
- fit with syllabus – molecular level scale
- relevance at scale within the domains of society, environment, economy or to their future professions.

2.2.1.1 Design

A Systems-Oriented Concept Map Extension (SOCME) diagram formed the centre of the intervention design. The SOCME diagram was chosen as the tool to visualize the greater LAS system with its economic, societal and environmental subsystems and their components and relationships. Concept mapping has been used extensively in science education to integrate complex ideas, and SOCME diagrams are said to contain the complexity of systems thinking [17]. Visualizations are particularly useful when topics are too complex for internal cognition alone [18].

Group work was incorporated as another important design feature to help manage the cognitive load inherent to both chemistry and systems thinking [19], to engage social constructivist practices for learning and to build interpersonal competence, another key competence for sustainability [4]. The jigsaw cooperative learning approach was used to structure the group work. This approach consists of two sets of group configurations for the participants, called home groups and subsystem groups in our case (Figure 2.2). It works well where sub-topics of similar complexity can be allocated to individual home group members to investigate. Since each sub-topic has inherent complexity, the subtopic is explored within a group setting by members, from different home groups allocated to that subtopic. After working in subsystem groups, members return to the home group equipped to make an informed contribution to the overall picture. In this intervention, each home group had three members, each allocated to a specific subsystem – economic, societal, or environmental. Initially home

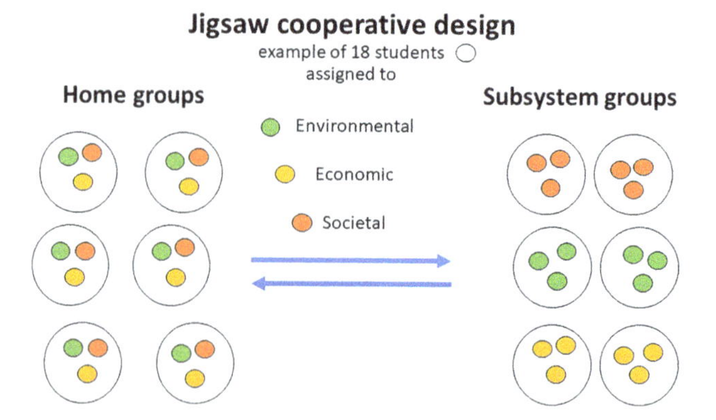

Figure 2.2: The Jigsaw cooperative learning approach: each student is a member of a home group and a subsystem group. Within a home group each student is allocated one of the three subsystems to study. Students focus on their subsystem with others similarly allocated within the subsystem group. Appropriately informed, students return to their home group where subsystems are integrated and the system is studied as a whole.

groups engaged with a chemistry concept map, then the subsystem groups elaborated partial SOCME diagrams for their subsystem and finally, home groups integrated and elaborated partial SOCME diagrams for the whole surfactant system with its various subsystems.

Key design elements – Project 1
- Visualization tool (SOCME diagrams)
- Cooperative learning approach (Jigsaw design with individual roles inspired by POGIL)
- Learning outcomes based on the ChEMIST table
- Scaffolding guided by the STH model levels of difficulty.

To ensure that the intervention covered all bases of chemical systems thinking, the ChEMIST table was used to inspire the development of learning outcomes for the intervention. Corresponding activities were designed for students to engage with the five characteristics of systems thinking using analytical and holistic perspectives. An alignment of the ChEMIST table outcomes with the STH model allowed us to anticipate the difficulty level of the tasks. To avoid losing students along the journey, more difficult aspects of systems thinking were scaffolded by presenting them in the partial SOCME, providing prompts or requiring simpler activities. Yet, there was an opportunity for students to be creative and make their contributions across the range of systems thinking skills. This would allow the level of systems thinking skill development to be monitored.

After getting students to consider the unsustainable aspects of LAS production and use, the intervention concluded with a presentation of chemistry's contribution to the development of sustainable surfactants. In general, surfactants do not perform well in hard water and a number of chelating additives are added to traditional laundry detergents to circumvent the problem. Chemists are designing surfactants with molecular structures that reduce the need for these additives which themselves have a negative impact on the environment. In addition, these surfactants were prepared from renewable resources [20].

2.2.1.2 Implementation

The surfactant systems project was implemented online due to Covid-19 restrictions with a group of about 240 first-year students in the second-semester general chemistry course serving students in the Faculty of Natural and Agricultural Sciences of the University of Pretoria. The intervention was implemented in two 3-h practical time slots a fortnight apart of which 2 h each were spent on group work. Groups of 60 students met virtually in a class with two facilitators via the learning management system. They watched short instructional videos during the plenary sessions but worked on group quizzes and their SOCME diagrams in breakout rooms with their home or subsystem groups. Marks were awarded for the quizzes and the final group SOCME diagram. The latter was assessed using a rubric based on the SOLO taxonomy. The marks contributed only a small proportion (<1%) to their final course grade.

2.2.1.3 Preliminary findings

While detailed findings will be published elsewhere, our preliminary findings revealed that the students found the intervention accessible. Students were authentically confronted with a complex system and despite expressing that it was difficult for them because they were "not used to it", they embraced the teaching intervention. Although the core chemistry foundation of the system was presented as facts and did not require further extension or application at a molecular level, the research participants declared that they found molecular level reasoning difficult indicating that they had returned to the chemistry to make sense of it once they had understood its relevance to the systems of everyday life. Students also found it difficult to organize systems' components into a framework of relationships. This was the highest-level systems thinking skill that was not specifically prompted. Research participants reflected the development of a range of system thinking skills with more students demonstrating proficiency at the more analytical level and fewer at the more holistic level with a few showing that they had taken the message to heart and demonstrating their wish to advocate for a sustainable action perspective in their daily lives and that of their friends or families.

Challenges of group work in the specific online setting detracted from the experience for some students in the bigger group, but for others it presented the first opportunity that year to engage with their peers at university. The students who volunteered to participate in the research study, being fully committed to the intervention, valued the opportunity for collaborative learning.

The assessment of the group SOCME diagrams proved challenging for our teaching assistants (TAs). While we could identify scope to refine the assessment rubric, this finding revealed that the TAs were capable of grading the lower level systems thinking skills consistently, but not the higher-level systems thinking skills. We attribute this discrepancy to their insufficient competence in systems thinking. However, the assessment was low stakes and the impact on final grades was insignificant.

2.2.2 Project 2: systems thinking in chemistry for first-year engineering students – sustainable aspirin manufacture

The second project, the MSc Science Education project of Cathrine Chimude, involved collaboration with chemical engineering researchers. A final year chemical engineering student completed a life cycle analysis (LCA) of aspirin production via various routes, some of which started from renewable feedstocks. This LCA provided the data for our systems thinking teaching intervention in a first-semester general chemistry service course for first-year engineering students. It has been proposed that the introduction of green chemistry and life cycle analysis could provide entry points for considering overlaps between the boundaries of different systems [21]. A comparison of different routes can show that how one chooses to design a synthetic pathway to a final product can have significant consequences. It can open an opportunity to discuss how these impacts, especially as chemistry is currently practiced, can lead to unintended outcomes that are unsustainable [8]. This would highlight the relevance of chemistry to their future profession in engineering and the contribution that they, as engineers, could make to sustainability.

Aspirin was chosen as the signature drug because of its fit with the curriculum – the synthesis of aspirin is a routine laboratory experiment in many introductory chemistry courses. Furthermore, they are likely to have encountered aspirin within their extended families for its use as an antipyretic/analgesic or anticoagulant. The intervention focused on the calculation of green chemistry metrics for three routes to aspirin production and developing a life cycle inventory (LCI) for the two most promising routes. It was placed shortly after the topic of stoichiometry, with students having completed the study themes of bonding, the mole concept and chemical reaction equations. These serve as a necessary foundation for engaging with the synthesis routes for aspirin manufacture. During the intervention, the theory course was moving from the topic of thermochemistry to that of equilibrium and thermodynamics, priming students to think about energy requirements associated with chemistry. In our experience, engineering students are comfortable with calculations and like to use numbers to make decisions. Thus, the calculation of green chemistry metrics and the use of the life cycle engineering tool presented an appropriate starting point for the introduction of systems thinking concerning the consequences of manufacturing choices within the environmental, economic, and societal domains.

2.2.2.1 Design

This project, unlike the first, was limited to a single 3-h practical session. The design elements specific to this project were individual preparation for a cooperative learning approach, reflection on learning before and after the contact session, a graphic to explain system boundaries (LCI gates), a zoom out approach moving from green chemistry metrics to the LCI to extended systems, and role play. Thus, project 2 also made use of group work to keep the cognitive load manageable. To equip students to make a meaningful contribution to the group, as individuals, they had to calculate the green chemistry metrics for the synthetic routes ahead of the dedicated project session based on an information document that presented the aspirin synthesis routes and the calculations of green chemistry metrics. They also had to complete a reflective questionnaire that prompted thinking about criteria for making decisions for sustainable manufacturing processes.

At the start of the 3-h session, which was conducted face-to-face, the principles of LCA and the role and working of LCIs were taught. Students were then divided into groups of three or four and had to complete an LCI worksheet using the data provided. This gave them a taste of this chemical engineering practice [22] appropriate to a first-year undergraduate level. Graphical visualization of the cradle to grave product life cycle (Figure 2.3) was provided to assist students in initially limiting the complexity within the boundaries of the manufacturing process. Throughout the intervention, the system boundaries were expanded incrementally, from the exemplary foreground system to the exemplary background system, for students to consider the effects of different routes on the sustainability of the process. This could be termed a "Zoom out" approach to dealing with complexity [7].

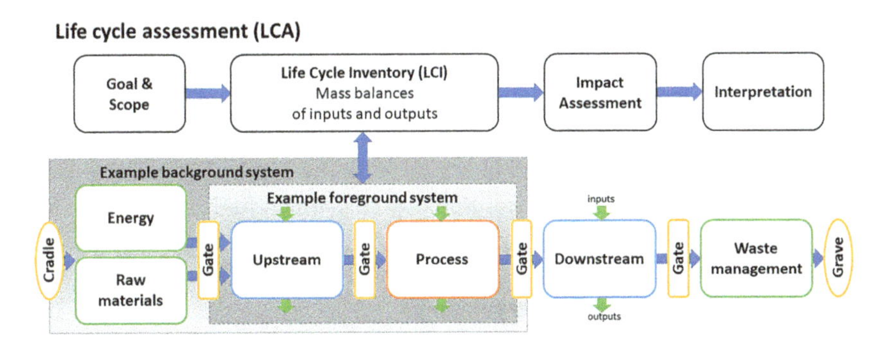

Figure 2.3: Life cycle assessment for chemical manufacturing processes. The assessment considers the full life cycle of the product from raw materials (cradle) to disposal after use (grave). To conduct a detailed life cycle inventory (LCI), which includes the mass balances for the processes under consideration, appropriate system boundaries (gate to gate) are chosen. The LCI will provide data to inform the choice, optimization, and implementation of the process within the LCA [23].

Key design elements – Project 2
- Individual preparation for a cooperative learning approach
- Reflection in learning
- Visualization tool (LCI – gates)
- A zoom out approach moving from green chemistry metrics to the LCI to extended systems
- Role play.

After completing the group LCI worksheet, the concept of systems thinking was explained to students and groups then had to discuss a question each on raw materials and waste management. Students also learned that making judgements of the most sustainable route based on a single part of the process alone could lead to less sustainable manufacturing in the long term e.g. the harvesting of salicylic acid from oil of wintergreen has very poor atom economy and carbon efficiency and the use of agricultural land to produce sufficient feedstock of the particular herbs might not be sustainable. An alternative would be to look at renewable sources of phenol from bio-waste. At the other end of the scale, aspirin is fully biodegradable which is important for sustainability [24]. By contrast, other painkillers such as Brufen have been reported to have post-use negative effects on the environment [25]. It is important to sensitise future scientists and engineers to the need to consider the full life-cycle of the product in order to make sustainable choices.

To conclude the session, the groups were given a scenario in which an engineer had found a more sustainable production route requiring an upgrade of the plant. They had to take on the roles of the Engineer, Financial Manager, and Environmental manager on the company board and debate the way forward. This exercise prompted students to envisage potential long-term financial implications of not choosing the more sustainable routes in order to persuade the financial manager that it was worth investing in sustainability.

Within a week after the exercise, they had to again complete a reflective questionnaire on decision making for sustainable manufacturing processes. Reflection in learning helps students build structural connections in their knowledge, personalize, and contextualize their knowledge and hence increase the depth of knowledge [26]. Reflection is a core component of critical thinking [27], which itself is necessary for systems thinking.

2.2.2.2 Implementation and preliminary findings

There were approximately 575 students enrolled in the course. They were allocated to one of four sessions and one of two venues per session based on venue capacity. The class was composed of students of chemical, civil, electrical, electronic, metallurgical, and mechanical engineering. Each student attended a 3-h laboratory session in the laboratory as a "dry" practical activity. This project has only recently been implemented and data collection and analysis are still in progress.

Groups were formed on the spot by the teaching assistants. It took a little time for the group members to trust one another to conduct the calculations to the satisfaction of all members. However, through working towards a common goal the groups formed and by the time it came to the role play, they were performing well. The students experienced the tasks as achievable with minor prompts from the teaching assistants.

Student feedback was overwhelmingly positive. At the start of the laboratory session, it appeared that female students were more interested in the topic than their male counterparts, but by the end of the session, all students were enthusiastically engaged. The role-play question resulted in animated discussions with students owning their roles. The course lecturer and teaching assistants endorsed the values communicated and the objectives of the project.

2.3 Discussion

We have developed two different systems thinking instructional approaches that center on chemistry and the importance of considering chemistry in addressing global sustainability challenges. Our criteria for the choice of systems thinking topic and our key design principles, for stand-alone systems thinking interventions, have been outlined. During and after implementation, we witnessed significant gains in terms of student engagement, development of positive perceptions of the relevance of chemistry and saw hints of ownership of the challenge of sustainability. We also achieved constructive group work because of clear expectations for the contribution of each student. Students endorsed the experience, suggesting that complexity was sufficiently managed. Instructors were similarly convinced of the value of the interventions. We experienced the challenges associated with authentic assessment of skills development, specifically for large student groups, the handicap that teaching assistants and instructors have in terms of their own level of systems thinking skills, and the limit to what can be achieved with a single intervention. This suggests the importance of embedding systems thinking throughout the undergraduate curriculum.

Several barriers to educational reform have been noted [2]. One barrier - the time required to develop teaching tools for implementing a systems thinking approach within a chemistry context - was overcome by assigning two MSc Science Education projects to the topic. With teaching material purpose-built for the local context, we experienced no resistance to its use by chemistry lecturers. Instead, the lecturers embraced the opportunity to implement our designs because they are advocates for chemistry and wanted to communicate that chemistry has an important contribution to make to the sustainability agenda. After experiencing the interventions with the students, the lecturers were willing to make these interventions permanent within the programme.

For both groups of students, we found that students were motivated to engage with the chemistry content that was presented within the context of systems and sustainability. This finding fits the expectation that when the relevance of the chemistry is demonstrated, students move toward meaningful learning [28, 29]. It also stimulated the professional identity development of first-year science students who normally see themselves as students and not as budding scientists with a contribution to make [30]. Authentic assessment of development of systems thinking skills, specifically for large student groups is challenging [31]. To display their achievement levels in the full range of skills, students need the opportunity to respond to less algorithmic questions and tasks. Such assignments cannot be graded using simple marking schemes or examples typical of textbook questions. Furthermore, because systems thinking has not been a part of chemistry education, instructors and teaching assistants have not had sufficient time or opportunity to develop systems thinking skills themselves. In a short training session, they can learn to assess basic systems thinking skills but a reliable assessment of more complex systems thinking skills, even with the guidance of a rubric, is challenging for teaching assistants delegated the task of grading large groups. This presents a validity challenge to assessment and we therefore recommend assessment to be low stakes and that extra time and care be invested in providing systems thinking training to the TAs beyond training in the use of the rubric.

While we were gratified to see the students embracing the topic and growing systems thinking skills, there is a limit to what can be achieved in the first year through single interventions. First-year tertiary chemistry is appropriate for communicating the importance of chemistry to sustainability and introducing students to the concept of and need for systems thinking, but is insufficient to fully develop systems thinkers. Complex skills such as those that constitute systems thinking take time and repeated opportunities to develop and should be provided throughout the undergraduate programme. In making the case to future scientists, health professionals, and engineers that chemistry's, and hence chemists', contribution to the sustainability agenda is key, it is important that the tertiary system develops chemists as systems thinkers so that they are equipped to make the contribution advocated for. This implies that there is a need to embed repeated opportunities to develop systems thinking skills throughout the curriculum. Systems are inherent to chemistry. Molecules and complex ions are systems of atoms with their own emergent behavior, which is so much more than the sum of the atoms. An emergent property is a novel property of the system, which is not possessed by its constituent parts nor is the sum of them. Chemical reactions, even the small localized reaction in a laboratory flask, are dynamic systems giving rise to new products. Yet, chemistry is not taught as a series of systems and the teaching of the emergent characteristics of chemical entities, their properties and reactions, is neglected [32]. A number of student misconceptions have been attributed to this neglect. Despite chemistry not being taught with a systems lens, chemists grow in their systems understanding of chemistry as they build expert knowledge structures.

The recognition of the importance of systems thinking to sustainability has stimulated a re-look at the molecular level with a systems lens. Systems thinking gives us a vocabulary to better communicate the nature of chemical entities to students. By using appropriate terminology when teaching the chemistry content, two goals can be achieved: (i) the nature of chemistry becomes explicit and (ii) systems thinking concepts represented by the terminology become part of the chemists' resources. For example, we are currently working on a third project to infuse systems focus into the teaching of the competition between substitution and elimination reactions in organic chemistry. This project is yet another example of possibilities for continuous development of systems thinking skills through the undergraduate curriculum by a reorientation and enrichment of the teaching of chemistry content rather than the replacement of content. Where appropriate, connections to sustainability should continue to support meaningful learning and grow the skill of thinking at different scales. For example, in the competition between substitution and elimination reactions, competing reaction pathways lead to unwanted byproducts and increased waste, which can have an impact on the sustainability of manufacturing processes.

2.4 Conclusions

It is a big task for a lecturer to embed systems thinking into chemistry education because it is counter to their chemistry teaching and learning experience. However, the benefits are worth the effort. Notably, the learning of chemistry is not sacrificed, it is enriched. Systems thinking becomes a tool to cope with the complexity of chemistry so that students can better understand the subject. By making the sustainability perspective central to first-year teaching, students leave a service course with a fresh view of the relevance of chemistry to their daily lives and the centrality of chemistry to addressing the big global challenges. This is so much richer than the current predominant view that chemistry is just a course to be passed on the path to acquiring a degree. Furthermore, considering chemistry in the context of biological, ecological, societal, economic, and other systems reveals beneficial and harmful effects. This provides a context for discipline-specific critical thinking instruction because topics are no longer restricted to the factual matters of chemistry, which a first-year student is not well placed to critique, but includes issues that have normative, moral, ethical or public policy dimensions [33]. Systems thinking and critical thinking are graduate attributes to be developed. Because chemistry provides the molecular basis for sustainability, chemistry graduates should be empowered to stand up and be counted in big multidisciplinary teams tackling the grand challenges of the planet. Our two interventions contribute to the growing set of resources for teaching and assessment, being developed by chemists dedicated to quality education. Quality chemistry education will serve as a solid foundation for the contribution that chemistry can and should make to sustainable development.

Acknowledgements: I acknowledge valuable contributions from colleagues and students in the Department of Chemistry: Marietjie Potgieter for her contribution towards conceptualisation, critical reading and co-supervision of one of the MSc projects; Dorine Dikobe for co-supervision of the second MSc project; lecturers Anita Botha and Natasha October for the opportunity to implement our teaching designs in their courses; and MSc students, Micke Reynders and Cathrine Chimude, for their contribution to the intervention design, data collection and analysis. I thank Philip de Vaal (supervisor) and Henri le Roux (final year undergraduate) from the Department of Chemical Engineering for conducting an LCA of aspirin manufacture to generate data for the Aspirin-LCA teaching intervention.

References

1. Steffen W, Richardson K, Rockstrom J, Cornell SE, Fetzer I, Bennett EM, et al. Planetary boundaries: guiding human development on a changing planet. Science 2015;347:1–10.
2. Mahaffy PG, Matlin SA, Holme TA, MacKellar J. Systems thinking for education about the molecular basis of sustainability. Nat Sustain 2019;2:362–70.
3. Matlin SA, Krief A, Hopf H, Mehta G. Re-imagining priorities for chemistry: a central science for "freedom from fear and want". Angew Chem Int Ed 2021;60:25610–23.
4. Wiek A, Withycombe L, Redman CL. Key competencies in sustainability: a reference framework. Sustain Sci 2011;6:203–18.
5. Matlin SA, Mehta G, Hopf H, Krief A. Repositioning chemistry for the 21st century. Atlas Sci 2016: 1–2.
6. Orgill M, York S, MacKellar J. Introduction to systems thinking for the chemistry education community. J Chem Educ 2019;96:2720–9.
7. Pazicni S, Flynn AB. Systems thinking in chemistry education: theoretical challenges. J Chem Educ 2019;96:2752–63.
8. Constable DJC, Jiménez-González C, Matlin SA. Navigating complexity using systems thinking in chemistry, with implications for chemistry education. J Chem Educ 2019;96:2689–99.
9. York S, Orgill M. ChEMIST table: a tool for designing or modifying instruction for a systems thinking approach in chemistry education. J Chem Educ 2020;97:2114–29.
10. Ben-Zvi Assaraf O, Orion N. System thinking skills at the elementary school level. J Res Sci Teach 2010;47:540–63.
11. Talanquer V. Some insights into assessing chemical systems thinking. J Chem Educ 2019;96: 2918–25.
12. Jahromi R, Mogharab V, Jahromi H, Avazpour A. Synergistic effects of anionic surfactants on coronavirus (SARS-CoV-2) virucidal efficiency of sanitizing fluids to fight COVID-19. Food Chem Toxicol 2020;145:111702.
13. Bradai M, Han J, Omri AE, Funamizu N, Sayadi S, Isoda H. Effect of linear alkylbenzene sulfonate (LAS) on human intestinal Caco-2 cells at non cytotoxic concentrations. Cytotechnology 2016;68: 1267–75.
14. Sablayrolles C, Montréjaud-Vignoles M, Silvestre J, Treilhou M. Trace determination of linear alkylbenzene sulfonates: application in artificially polluted soil—carrots system. Int J Anal Chem 2009;2009:404836.

15. Gordon AK, Muller WJ, Gysman N, Marshall SJ, Sparham CJ, O'Connor SM, et al. Effect of laundry activities on in-stream concentrations of linear alkylbenzene sulfonate in a small rural South African river. Sci Total Environ 2009;407:4465–71.
16. Mungray AK, Kumar P. Fate of linear alkylbenzene sulfonates in the environment: a review. Int Biodeterior Biodegrad 2009;63:981–7.
17. Aubrecht KB, Dori YJ, Holme TA, Lavi R, Matlin SA, Orgill M, et al. Graphical tools for conceptualizing systems thinking in chemistry education. J Chem Educ 2019;96:2888–900.
18. Hoffmann MHG. Cognitive conditions of diagrammatic reasoning. Semiotica 2011;186:189–212.
19. Kirschner F, Paas F, Kirschner PA. A cognitive load approach to collaborative learning: united brains for complex tasks. Educ Psychol Rev 2009;21:31–42.
20. Park DS, Joseph KE, Koehle M, Krumm C, Ren L, Damen JN, et al. Tunable oleo-furan surfactants by acylation of renewable furans. ACS Cent Sci 20162016;2:820–4.
21. Mahaffy PG, Krief A, Hopf H, Mehta G, Matlin SA. Reorienting chemistry education through systems thinking. Nat Rev Chem 2018;2:0126.
22. Reichmanis E, Sabahi M. Life cycle inventory assessment as a sustainable chemistry and engineering education tool. ACS Sustainable Chem Eng 2017;5:9603–13.
23. Kleinekorte J, Fleitmann L, Bachmann M, Kätelhön A, Barbosa-Póvoa A, von der Assen N, et al. Life cycle assessment for the design of chemical processes, products, and supply chains. Annu Rev Chem Biomol Eng 2020;11:203–33.
24. Choudhary J, Chauhan P, Gahlout M, Prajapati H. Biodegradation of salicylic acid from soil isolates. Int J Drug Res Tech 2016;6:107–14.
25. Cuthbert R, Parry-Jones J, Green RE, Pain DJ. NSAIDs and scavenging birds: potential impacts beyond Asia's critically endangered vultures. Biol Lett 2007;3:90–3.
26. Chang B. Reflection in learning. Online Learn 2019;23:95–110.
27. Facione PA. Critical thinking: a statement of expert consensus for purposes of educational assessment and instruction. Millbrae, CA: T. C. A. Press; 1990:28 p.
28. Habig S, Blankenburg J, van Vorst H, Fechner S, Parchmann I, Sumfleth E. Context characteristics and their effects on students' situational interest in chemistry. Int J Sci Educ 2018;40:1154–75.
29. Mayer RE. Rote versus meaningful learning. Theory into practice 2002;41:226–32.
30. Pilcher LA, Riley DL, Mathabathe KC, Potgieter M. An inquiry-based practical curriculum for organic chemistry as preparation for industry and postgraduate research. S Afr J Chem 2015;68:236–44.
31. Fox J, Freeman S, Hughes N, Murphy V. "Keeping it real": a review of the benefits, challenges and steps towards implementing authentic assessment. AISHE-J. 2017;9:2801.
32. Tümay H. Reconsidering learning difficulties and misconceptions in chemistry: emergence in chemistry and its implications for chemical education. Chem Educ Res Pract 2016;17:229–45.
33. Facione PA. Critical thinking: a statement of expert consensus for purposes of educational assessment and instruction. Millbrae, CA: T. C. A. Press; 1990:33 p.

Abdul Nashirudeen Mumuni*, Francis Hasford,
Nicholas Iniobong Udeme, Michael Oluwaseun Dada and
Bamidele Omotayo Awojoyogbe

3 A SWOT analysis of artificial intelligence in diagnostic imaging in the developing world: making a case for a paradigm shift

Abstract: Diagnostic imaging (DI) refers to techniques and methods of creating images of the body's internal parts and organs with or without the use of ionizing radiation, for purposes of diagnosing, monitoring and characterizing diseases. By default, DI equipment are technology based and in recent times, there has been widespread automation of DI operations in high-income countries while low and middle-income countries (LMICs) are yet to gain traction in automated DI. Advanced DI techniques employ artificial intelligence (AI) protocols to enable imaging equipment perceive data more accurately than humans do, and yet automatically or under expert evaluation, make clinical decisions such as diagnosis and characterization of diseases. In this narrative review, SWOT analysis is used to examine the strengths, weaknesses, opportunities and threats associated with the deployment of AI-based DI protocols in LMICs. Drawing from this analysis, a case is then made to justify the need for widespread AI applications in DI in resource-poor settings. Among other strengths discussed, AI-based DI systems could enhance accuracies in diagnosis, monitoring, characterization of diseases and offer efficient image acquisition, processing, segmentation and analysis procedures, but may have weaknesses regarding the need for big data, huge initial and maintenance costs, and inadequate technical expertise of professionals. They present opportunities for synthetic modality transfer, increased access to imaging services, and protocol optimization; and threats of input training data biases, lack of regulatory frameworks and perceived fear of job losses among DI professionals. The analysis showed that successful integration of AI in DI procedures could position LMICs towards achievement of universal health coverage by 2030/2035. LMICs will however have to learn from the experiences of advanced settings, train critical staff in relevant areas of AI and proceed to develop in-house AI systems with all relevant stakeholders onboard.

***Corresponding author: Abdul Nashirudeen Mumuni,** Department of Medical Imaging, University for Development Studies, Tamale, Ghana, E-mail: mnashiru@uds.edu.gh
Francis Hasford, Department of Medical Physics, University of Ghana, Ghana Atomic Energy Commission, Accra, Ghana
Nicholas Iniobong Udeme, Michael Oluwaseun Dada and Bamidele Omotayo Awojoyogbe, Department of Physics, Federal University of Technology, Minna, Nigeria

As per De Gruyter's policy this article has previously been published in the journal Physical Sciences Reviews. Please cite as: A. N. Mumuni, F. Hasford, N. I. Udeme, M. O. Dada and B. O. Awojoyogbe "A SWOT analysis of artificial intelligence in diagnostic imaging in the developing world: making a case for a paradigm shift" *Physical Sciences Reviews* [Online] 2022.
DOI: 10.1515/psr-2022-0121 | https://doi.org/10.1515/9783110913361-003

Keywords: artificial intelligence; computer aided diagnostics; developing countries; diagnostic imaging; machine learning; SWOT analysis.

3.1 Introduction

3.1.1 Diagnostic Imaging

Traditional diagnostic approaches rely on a patient's symptoms and associate them with clinical conditions that may have symptoms similar to those observed [1]. In contrast to clinical pathology which involves diagnosis using biopsy samples from patients, diagnostic imaging techniques provide images of patients that can be reviewed to make a clinical diagnosis. Diagnostic imaging is therefore largely non-invasive while clinical pathology is invasive.

Diagnostic imaging utilizes ionizing (plain X-rays, computed tomography (CT), and nuclear medicine) and non-ionizing (magnetic resonance imaging (MRI) and ultrasound) radiation-based technologies [2] to generate images of the internal parts of the human body and organs for purposes of evaluation to diagnose, monitor, or treat medical conditions. Diagnostic imaging plays a central role in disease management and helps to ensure that countries achieve the objectives of sustainable development goal 3 (SDG 3). However, there is an acute shortage of diagnostic imaging equipment in low and middle-income countries (LMICs). Current estimates hold that only 1 CT scanner serves about 1 million patients in LMICs compared to about 40 scanners serving about 1 million patients in high-income countries. The disparity is even greater for MRI and nuclear medicine equipment, in addition to a huge shortage of personnel in these specialties [3].

In the face of these shortages, the demand for imaging services is on the increase in the developing world in areas such as radiotherapy planning, image-guided interventions, and tumor sampling for pathology work-up [2], in addition to the core roles of accurate and timely diagnosis and disease monitoring. Advances in imaging technologies that meet the healthcare demands of developing countries are therefore urgently needed.

3.1.2 Artificial Intelligence

Artificial intelligence (AI) was first conceptualized by Alan Turing in 1950 when he developed the concepts of machine learning, genetic algorithms, reinforcement learning, and the Turing test [4]. The term "artificial intelligence" was subsequently created by a team of scholars in 1956 at a Dartmouth College conference [5].

Artificial intelligence is a computer science specialty devoted to developing systems with capabilities of simulating human behavior and intelligence [6] based on programmed algorithms. Within specific contexts [7], for a given set of human-defined objectives [8], an AI system can be trained to autonomously simulate human learning, memory, analysis, predictions, and make recommendations, innovation, and decisions (which usually need human intelligence) without further programming [8–11]. Machine learning is a type of AI that involves the creation of computer algorithms that can detect patterns in large data sets and make intelligent predictions without the need for repeated programming. The Institute of Electrical and Electronics Engineers (IEEE) regards AI systems as "autonomous and intelligent systems" [12]. The goal of any AI is to make a clever computer system comparable to humans to solve complex problems, influencing real or virtual environments [8].

According to Mitchell [13], experience is gained through more data exposure, with AI systems acquiring great problem-solving abilities and the ability to acquire knowledge from data that humans cannot. There are three primary forms of machine learning: supervised learning, unsupervised learning, and reinforcement learning [6]; however, many other learning protocols exist. Supervised training teaches the model using a labeled input dataset to produce a proper output. Unsupervised training recognizes patterns within data in an uncertain result, allowing the machine to train itself on the data. Reinforcement training investigates each action in turn and then labels the results with positive or negative weights for reinforcement, thereby learning from the outcomes of interactions [14].

Deep learning (DL) discoveries have resulted in several machine learning applications. DL is a subset of machine learning in which algorithms are organized into several processing levels using an artificial neural network according to how the human brain is structured [7]. As demonstrated by LeCun et al. [15], these structures develop data representations with several layers of patterns and have the capability to connect complex nonlinear relationships and predictions without human involvement.

Even though there has not been an agreement among specialists in AI research about the actual description of "intelligence" [6], some AI experts have suggested that a computer system is said to be acting "intelligently" when: (1) its actions are deemed appropriate for its intended purpose; (2) it is able to adapt to changing conditions and purposes; (3) it automatically learns from experience; and (4) it independently makes decisions within the limits of its computational algorithms [16].

Generally, AI systems fall into two categories [6]. Early AI research focused on "artificial general intelligence"; however, developing this kind of AI is both challenging and complicated. Current focus is therefore on "artificial narrow intelligence", which develops systems with the ability to perform a well-defined single task extremely well [17, 18]. Therefore, almost all modern AI-based healthcare solutions are considered "artificial narrow intelligence". There are however no universally accepted sub-types of AI relevant to health [6].

Many AI systems have been installed in high-income countries compared to developing countries, where the use of AI in healthcare is currently being evaluated [5,8]. There are however indications that efforts are being made by stakeholders to reduce the disparity margin. For instance, the United Nations (UN) convened a global meeting in 2017 to discuss the development and deployment of AI solutions aimed at reducing poverty and providing critical public services in countries [6]. Another meeting was held subsequently to assess the role of AI in achieving the UN's Sustainable Development Goals (SDGs). The UN believes that the safe deployment of AI applications can help the world to achieve its SDGs, including the health-related objectives under SDG 3 (universal health coverage) [8].

The deployment of AI applications in healthcare settings should be consistent with six core principles of ethics recommended by the World Health Organization [8]: (1) protection of autonomy; (2) promotion of well-being, safety and interest of the public; (3) maintain transparency, explainability, and intelligibility; (4) encourage responsibility and accountability; (5) provide inclusiveness and equity; and (6) deploy AI that is responsive and sustainable.

3.1.3 The role of artificial intelligence in diagnostic imaging

AI was first applied in the field of health as a medical diagnostic decision support (MDDS) system for the diagnosis of congenital heart diseases in 1961 [19]. By 1969, an MDDS system for simulations and diagnosis was developed at the University of Leeds, United Kingdom [20, 21]. An expert system known as MYCIN was developed in the early 1970s to identify infectious pathogens and suggest possible antibiotics for treatment [22]. By the 1990s, many more MDDS systems were developed for use in many areas of healthcare, including internal medicine, veterinary medicine, clinical pathology, psychiatry, radiology, and diagnostic imaging [20–24]. The rapid development of AI applications in recent times is being promoted by advances in technology, such as high-speed broadband, computers, smartphones and watches, and cloud computing [4].

Despite the many years of training in diagnostics, unfortunately, errors are still extremely common, and they can have life and death consequences [1]. There are at least 5% cases of misdiagnosis and 10% of deaths associated with misdiagnosis in the United States [1, 25]. These estimates are expected to even be higher in developing countries where there are staff and equipment shortages. This situation justifies the implementation of AI technologies in such settings to close the deficit gaps, where necessary.

AI applications in diagnostic imaging technologies are developed based on the following algorithms [26–30]: computer aided diagnosis (CAD), radiomics, convolutional neural networks (CNN), and artificial neural networks (ANN).

CAD methods are the oldest form of AI widely used in radiology, where outputs generated aid clinicians to make diagnosis. Recent CAD approaches have been modified to detect hidden characteristics in diagnostic images and label these apparently aberrant parts, in order to minimize probability of the clinician missing disorders of interest. They are generally used to examine and classify pathologies in medical images. The commonest CAD software are those used for mammography breast cancer detection and chest CT pulmonary nodule detection.

Sahiner et al. [31] examined nodules overlooked by radiologists and discovered that CAD identified 56–76% of these lesions. According to the study, CAD can detect early-stage lung cancer. Additional studies, such as that of van Ginneken et al. [32] have evaluated the efficacy of CAD in diagnosing other lung diseases such as emphysema, pleural effusion, and pneumonia, with varying results. Morton et al. [33] assessed the utility of CAD in mammography for the identification and delineation of microcalcifications, lesions, and architectural distortion. CAD therefore functions effectively as a secondary assessment to that of the radiologist.

Developed from several decades of research into CAD applications in diagnosis, prognosis and therapeutics, radiomics utilizes characterization techniques to extract quantitative information from a medical image and analyzes this information to assist in decision-making. Clinical symptoms that otherwise cannot be visually detected easily are thus revealed. Radiomics can be used to reveal information about tumor biological behavior such as characterization of tumor aggressiveness, viability, and response to chemotherapy and/or radiotherapy. Other applications of radiomics include estimation of prognosis in identified lung cancers, estimation of risk of distant metastasis, prediction of histology, and profiling of mutations of lung tumors.

CNN algorithms, the commonest deep learning model in imaging, assign biases and learnable weights to various points in an image for purposes of discretely identifying features independently [34]. They enable imaging technologies to learn pattern recognition with fewer processing steps and training. A CNN learns new knowledge through supervised, unsupervised and reinforcement paradigms. Such AI systems are programmed to perceive the world in the same way as humans do, and apply knowledge learned in various functions such as image and video recognition, image analysis and classification, media creation, language processing, and decision-making. The design of CNNs is based on the connectivity of the neurons in the human brain. The neurons individually respond to stimuli in defined fields which are networked to occupy an entire perceptible field. CNN, unlike CAD, does not require training before using a pre-taught algorithm and is being integrated and built on new knowledge without the need for human re-training. All deep learning algorithms like CNN have been successfully applied in specialized medical imaging procedures [35].

ANN algorithms operate through several input nodes (called neurode or perceptron), which vary in signal weighting from the superficial to deep-lying nodes.

Each node is only activated when there is a cumulative input signal from all the other nodes. The development of ANNs was inspired by how the nervous system and the brain interact through electrical activity originating from interconnected neurons. Even though modern ANN systems only have a passing resemblance to this physiological system, the foundation of their operating principle remains the same.

Advances in AI applications in diagnostic imaging have not actually focused on new imaging technologies, but on the integration of these algorithms with increasing complexities within equipment operation and image review processes [36]. The large volumes of diagnostic images produced by imaging units using technology-driven equipment fit well with the data-reliant characteristic of AI [36]. A significant surge in proposed AI applications for diagnostic imaging has therefore been reported in recent years; many AI tools are now available to support image acquisition and reporting processes [37, 38].

Most diagnostic imaging equipment use CNN-based deep learning approaches in applications such as dose optimization, image processing, selection of tube potential and current in CT, image noise reduction, automatic protocol selection, multiparametric reporting, image analysis in cardiac sonography, and patient positioning at the isocenter of CT and MRI scanners. Radiation treatment systems also use CNN-based algorithms to delineate organs at risk from tumors while some linear accelerators are able to automatically move some of their parts to ensure the precision of dose delivered to the treatment volume [39].

Given these exciting and important roles of AI in healthcare delivery, coupled with the rising demand for diagnostic imaging services at the core of healthcare systems in developing countries, it is expected that many AI-based imaging equipment will eventually find their way into these settings in the near future.

3.1.4 SWOT analysis

SWOT analysis is a framework used to assess the competitiveness of an entity in terms of its internal strengths (S) and weaknesses (W), as well as its external opportunities (O) and threats (T). The goal of SWOT analysis is to guide prudent planning by the entity towards achievement of its goals. It is thus a tool designed to facilitate a realistic, fact-based, evidence-based, or empirical look at an entity's internal and external factors, as well as its current and future potentials [40, 41].

Strengths describe the competitive advantages of the entity that make it excel or standout from similar entities (e.g., uniqueness of AI applications); weaknesses are challenges that negatively impact on the efficient performance of the entity and must be overcome to make the entity perform at its optimum level (e.g., an AI application that requires broadband to operate). Opportunities are favorable external factors that can offer the entity some competitive advantage (e.g., availability of technologies to support the deployment of AI applications); threats are external factors that have the potential to harm the entity (e.g., biases in input data to train an AI application could affect its acceptance in other settings) [40, 41].

3.1.5 Objectives

Based on the established role of AI in diagnostic imaging, the SWOT analysis is used in this narrative review to qualitatively examine the critical factors related to the competitive advantages (strengths), challenges (weaknesses), opportunities for adoption, and threats to the successful deployment of AI solutions within diagnostic imaging practice in developing countries. Drawing from this analysis, a case is then made to justify the need for widespread AI applications in diagnostic imaging in resource-poor settings.

3.2 Strengths of AI in diagnostic imaging

AI has been used in medicine to provide accurate real-time medical data and optimize important clinical decisions. The technology has provided cost-saving medical approaches and helped streamline tasks such as medical history tracking [42], translation of clinical details [43], and appointment scheduling [44, 45]. Artificial intelligence technology has accelerated clinical operations to save valuable time in healthcare facilities and has been used to analyze extensive data collection for accurate diagnoses, treatment and prediction of disease outbreaks through research [46]. In a survey of 500 senior healthcare executives in the United States, 96% of them believed that AI has a critical role in efforts to reach health equality goals [45]. While growing concerns are shared by some that AI technology may soon replace radiologists, Curtis Langlotz, a radiologist at Stanford University believes that "AI will not replace radiologists, but radiologists who use AI will replace radiologists who don't" [47].

The use of AI in medical imaging has revolutionized radiation medicine, especially for cancer care. Machine learning tools have been applied in cancer management to detect diseases, assist in decision making, and recommend treatment approaches [48]. Responsible use of AI has enabled medical imaging centers and departments around the world to enhance and expand key functions, reduce administrative burden, and allow physicians to focus on their core patient care mission [45]. In recent years, the number of published articles on the use of AI in medical imaging has grown from a few hundreds to thousands of articles per year. Magnetic resonance imaging (MRI) and computed tomography (CT) together constitute over 50% of recent research articles on AI in radiology [49].

In developing regions of the world such as Africa, where a clear inadequacy of radiologists, medical physicists and other healthcare professionals exists, the workload in interpretation and reporting of medical images could be daunting and lead to delays in clinical service delivery to patients as well as associated unintended accidents [50]. An important relevance of artificial intelligence in medical imaging is the augmentation of the abilities of physicians in developing countries [47]. Areas for which AI is being extensively used in diagnostic imaging modalities like ultrasound, magnetic resonance imaging, plain X-ray, computed tomography, nuclear medicine, etc. are detection of medical conditions, monitoring of diseases, image acquisition, segmentation, and analysis [51]. The strengths of AI in diagnostic imaging that is transforming radiation medicine in the developing world include, but are not limited to, classification of tumors of the brain and other organs, recognition of cancers of the breast, detection of hidden fractures, detection of neurological abnormalities, and offering of secondary opinion [47].

Cancer is one of Africa's leading causes of death due to inadequate diagnostic and therapeutic options [51]. The most prevalent cancers in Northern Africa are breast cancer, bladder cancer, and liver cancer. Esophageal cancer and cervical cancer are dominant in East Africa, while prostate cancer, lung cancer and colon cancer dominate in Southern Africa [52]. The application of AI and deep learning tools to aid the diagnosis of these conditions through medical imaging have been widely studied in high-income countries [53–63] but very few of such studies have been reported in Africa and other low-income countries [2,64–67]. The two ways of using AI for detection and analysis of abnormalities in diagnostic images are through programming of algorithms with predefined criteria to execute clinical tasks and by allowing an algorithm to acquire knowledge from huge amounts of data through either supervised or unsupervised techniques. AI brings several advantages to healthcare, some of which are discussed below.

3.2.1 Identification of medical conditions

Deep learning systems in AI have excelled by learning ranked representations of some categories of diagnostic images from huge numbers of clinical scans and used these to aid medical diagnosis and revolutionized radiation medicine [68–71]. By employing AI automated detection, physicians report diagnostic images based on the reading priority, hence speeding the reporting time and improving patient outcomes [51]. The AI technology pulls similar images from the acquired database in a fraction of seconds to a few minutes for review and reporting. Deep learning algorithms have been used for several medical imaging procedures such as assessment of the risk of malignancy of lung nodules, classification of liver masses, estimation of pediatric skeletal maturity from hand radiographs, and eliminating the need for breast and thyroid biopsies [72–74]. Some other ways by which machine learning has been applied in medical imaging include characterization of carotid plaque through CT, prediction of lesion-specific ischemia through quantitative coronary CT and angiography detection of fatty liver through ultrasound [75].

3.2.2 Monitoring and prediction of diseases

Monitoring of medical conditions has been boosted in recent years with the introduction of deep learning tools. Common example is the application in breast cancer diagnosis, where studies indicate that doctors miss up to about 40% of breast lesions during routine screening through conventional means [47]. In Africa, breast cancer is the commonest cancer and the leading cause of death among women [76, 77]. Through the use of AI simulation tools in radiology, misdiagnosis of breast cancers which often leads to conduct of surgical and invasive procedures on healthy asymptomatic women are avoided. Artificial intelligence tools help in identifying cancers earlier than other means and also produces improved detection accuracy. An AI-based study conducted in Korea indicates that the intelligence tool aids in mammography screening by radiologists and increases the accuracy of detection from 75.3 to 84.8% [78]. Radiology AI uses algorithms to analyze previously acquired mammograms to help predict patients likely to develop breast cancer in the future [79]. Through this means, women identified as high risk are encouraged and recommended for routine screening.

In assessing the accuracy of an AI tool to make diagnosis using 44,755 ultrasound images, the tool was found to increase radiologists' ability to accurately identify breast cancer by 37 percent. In addition, the application assisted clinicians to reduce the number of tissue samples and biopsies needed to confirm tumors by 27 percent. The study demonstrated how AI has the ability to help radiologists to read breast ultrasound images and identify only those that show real signs of breast cancer and to circumvent confirmation by biopsy in cases that turn out to be benign [48]. Radiologists have traditionally used X-ray imaging over many years to identify bone fractures, especially in elderly patients. Such fractures are tiny, difficult to visualize, and they mostly hide under soft tissues. The potential of using Deep Convolutional Neural Networks (DCNN) to help radiologists identify fractures has been studied [47]. The DCNN tool has been found to identify invisible fractures in CT and MRI images [80]. Research has shown that radiologists could identify 83% of fractures while DCNN has a 91% accuracy [81]. The early day AI tools such as OsteoDetect are employed for detecting distal radius fractures in wrist scans [82].

In the sector of neurological studies, AI has been used in radiology to diagnose neurodegenerative disorders such as Parkinson's, amyotrophic lateral sclerosis (ALS), and Alzheimer's by tracking retinal movements. A speech analysis AI tool, developed based on convolutional neural networks, has also been developed to perform analysis of

Alzheimer's condition of patients. The AI tool recognizes early sign of Alzheimer's in elderly patients mainly based on their speech pattern with 95% accuracy [83].

During the peak of the COVID-19 pandemic, some African countries employed AI technology to aid clinical diagnosis of patients in diagnostic radiology. The North African countries, forming part of the Middle-East and North Africa (MENA) region, were first in Africa to implement the use of AI diagnostic tool for COVID-19 using medical images. Radiologists and clinicians used AiRay, an AI radiology tool, to aid in detecting COVID-19 from chest CT examinations [84]. South Africa also implemented CAD4COVID, a software developed to help triage and monitor COVID-19 patients through X-ray imaging [85, 86]. In 2020, Nigeria introduced ChestEye AI diagnostic imaging tool into healthcare institutions for improving the detection of pathologies and tuberculosis [87]. A few other countries including Rwanda, Kenya, Nigeria, and Ghana have on small scales implemented AI tools for aiding the diagnosis of COVID-19 and tuberculosis from medical images [88].

3.2.3 Classification of tumors

The conventional way of classifying tumors has been to perform surgical procedure and remove tissues of the cancerous organ for laboratory tests. Aside from being invasive, this procedure can last several hours to days before results are received and good diagnosis could be made [89]. Introducing AI into diagnostic imaging reduces tumor classification time to about 2–5 min and can be conveniently performed anywhere, either in the consulting or operating room. A study performed in the United Kingdom indicates that machine learning techniques in radiology and MRI are very important non-invasive methods for classification of brain tumors in children. The technique can distinguish between the three major types of brain tumors in the posterior fossa of the brain. Such tumors are the leading cause of cancer-related deaths in children [90–92]. In another study conducted at Tulane University, AI was used by researchers on medical images to detect and diagnose colorectal cancer and its classification more accurately than pathologists. The study group also found that AI diagnoses the disease in a cost-effective way, ultimately reducing workload of pathologists [93].

3.2.4 Image acquisition

Artificial intelligence technology has helped address challenges of high-quality image acquisition. In diagnostic imaging, precision of clinical decision-making is highly dependent on the quality of images produced by imaging systems [94]. This is achieved by defined imaging protocols and modalities. To achieve the best quality images for medical diagnosis, AI is used as optimizing tool for assisting technologists, radiologists and/or nuclear physicians in choosing personalized patients' protocols, in tracking the patients' dose parameters, and providing estimates of radiation risks associated with cumulative doses and the patients' susceptibility [51].

3.2.5 Image processing, segmentation and analysis

Deep learning algorithms in AI are gaining marked popularity in diagnostic imaging and are widely being used for reconstruction of medical images and enhancement of their quality. Deep learning technology has shown great potential of being used to address the inefficiency of manually isolating areas of interest in diagnostic images and thereby reducing the subject of variability in image interpretation [95]. The European Society of Radiology (ESR) has described AI's automated image segmentation as crucial for reducing the burden on radiology workflow. Deep learning algorithms provide high consistency levels at unmatched speeds [51]. They provide robust registration needed for multimodality imaging and accurate registration [96]. Deep learning networks have produced strong models for medical image radiomics, by quantitatively extracting peculiar features. Data gathered from radiomics studies, such as shape, brightness, wavelength and contour are extracted from medical images using machine learning approaches, providing important information for assessing cancer genetics, differentiating benign and malignant tumors and predicting treatment response [28, 97–100].

3.3 Weaknesses of AI in diagnostic imaging

The intrinsic difficulties of machine learning, the imperfections of ethics and regulations, and society's lack of acceptance have all hampered AI advancement in the developing world. Understandably, the excitement around AI has led to concerns that computers would eventually replace radiologists. However, the roles of radiologists will be extended in response to increasing demands for their services and advances in AI-based imaging techniques. The weaknesses of AI-based systems are discussed below.

3.3.1 Data Issues

Deep learning algorithms heavily rely on access to big data sets to be successfully trained. Despite the availability of most clinical data in electronic format, access to sufficient amount of this data may be a challenge in many settings. An AI algorithm requires access to this sensitive data (without the patient-identifying information) in a secure and anonymous manner, for purposes of learning and getting experienced.

There is a hurdle in developing countries in generating a generally defined picture format across the Picture Archiving and Communications Systems (PACS) of hospitals to create a national imaging data bank accessible to AI systems. PACS have been used in most developed settings for over a decade now, making millions of pictures in a standard format available across many facilities. Few other sectors have this volume of imaging data gathered for specific reasons, digitally available in well-structured archives [101]. In advanced settings, PACS are not as widely utilized in other medical specialties, such as ophthalmology and pathology. This is however changing as imaging is becoming more common across disciplines, resulting in massive data sets being made available. This is not however the case in developing countries since the availability of hardware or a cloud-based system is a concern and these countries have inadequate funds to purchase or maintain the storage capacity required to keep the vast size of diagnostic images captured during imaging. AI is dependent not just on enormous data sets for training, but also on precise data classification. To ensure accuracy and reliability, this must be complemented by well-qualified readers. Incorrect data labeling will significantly influence the learning process of the algorithm. Future learning will then be incorrect, lowering the machine's total accuracy and posing potential harm to patients. For example, mis-labeling 'malignancy' as 'normal' will not only influence the learning process but may be lethal to patients.

3.3.2 High installation and maintenance costs

Expensive and complex imaging equipment is not suitable for use in developing countries, particularly in rural and semi-rural areas. These machines need steady energy supply, which is sometimes lacking even in urban areas and must be put in specialized structures that necessitate air conditioning and a unique design for radiation protection. Their regular maintenance is required; failure is not rare and repairs are sometimes delayed because they are performed by specialist organizations that typically cover a vast geographical region. The expenses of acquiring equipment are substantial, and long-term maintenance and operating expenditures constitute a significant portion of the overall costs and must be budgeted for ahead of time when planning imaging center operations. Furthermore, due to prevailing conditions peculiar to the context of the developing world, estimating maintenance and operating expenses is difficult. Ultrasound machines are substantially less expensive to acquire and maintain compared to CT and MRI scanners. Portable ultrasounds are even far cheaper and are therefore well-suited for use in resource-poor settings. Another significant barrier to implementing AI in health care in these countries, particularly in the Sub-Saharan African region, is the lack of reliable and accessible internet services among the population, with the number of internet users believed to be far lower than in the advanced countries.

3.3.3 Technical Training

Due to low numbers of AI systems developers in developing countries, the concept of AI is relatively new in these regions compared to European and Western countries [102]. A professionally qualified crew, comprising radiologists, radiographers, and medical physicists, is required to operate radiology equipment. Ultrasound training in low-income countries is frequently tailored to local needs and is significantly less extensive than in developed countries. Ultrasound is the most often utilized imaging modality in poor countries due to its low cost, ease of maintenance, and high practical output, and a pragmatic approach with reduced education of the scarce personnel is comprehensible and required to spread its usage. Significant emolument disparities between the private and public sectors contribute to the public sector's personnel shortfall. Brain outflows to either outside the public sector or even outside the resource-poor countries are a big concern to clinical AI implementation in these regions [103].

3.3.4 Ethical issues

Data ethics is the cornerstone of AI, according to Mittelstadt et al. [104], and its core topics include informed consent, privacy and data protection, ownership, objectivity, and openness. Patients, as data owners, have the right to know the extent of information stored about them and how this information will be used [105]. AI systems are created based on existing data to learn and draw conclusions. The most prevalent ethical concern is unfairness driven by prejudice in data sources. Any data set will be skewed to some extent depending on gender, sexual orientation, race, or sociologic, environmental, or economic variables [105].

According to research done in the United States, doctors may have overlooked favorable outcomes for African-Americans because they felt the model's positive predictive value for African-Americans was poor. The low positive rate may be caused by the small number of African- Americans who took part in the first trial, and false-positive findings were more common [106]. Rajkomar et al. [107] argue that these injustices may occur throughout model creation, implementation, and assessment. Clinicians always stop treating patients who have specific results when offering medical services. AI will learn about these disparities in human preferences, which will lead to major ethical issues that might be disastrous for people [108]. In terms of resource distribution, the model will forsake inhabitants in impoverished nations who have protracted hospital stays owing to poverty or long distances from healthcare facilities.

The approach may assign case management resources disproportionately to patients from developed countries. Because there is no local expert to reject these outcomes in developing countries, the danger of automation bias may be amplified. The occurrence of automation deviations will also impede the practical use of AI. This problem is compounded by AI judgments based on minor traits that humans are unable to detect. Furthermore, clinicians with minimal expertise in machine learning may now be confronted with new ethical issues since AI-based models will be difficult to explain by most clinicians in the developing world.

3.3.5 Patient data security

Patient data security is the most essential concern in the application of AI to the medical profession in underdeveloped nations and it also demands the most thorough assessment. Most AI devices now require a variety of electronic devices to operate, such as computers, mobile phones, and wristbands. As a result, there are three major concerns about the security of such devices. For starters, even the most physically unclonable operations will be influenced by factors like cost, temperature change, and electromagnetic interference. Secondly, due to the work demands of clinicians, they may find it difficult to employ AI that combines numerous technologies. They need to be trained to access and handle medical system data, which may interrupt medical workflow and result in data leakage. Furthermore, doctors may have poor knowledge of the concepts and application techniques of AI products in real-world settings, resulting in issues such as decreased efficiency and more mistakes. Thirdly, there is the question of AI network security to consider. If crucial nodes in the complicated network transmission process are attacked or fail, a worldwide cascade response may develop. Not to mention that even the most secure software algorithm is vulnerable to attack.

Access to sensitive health data is required for the implementation of AI systems. Accessing this sensitive information poses a risk of cyber-attacks, putting patients' privacy at risk. Despite its strong performance in the first design examination, the AI system's performance in a targeted design confrontation is frequently inadequate. A false-positive attack can be used to create a negative sample, or a false-negative assault can be used to provide a positive sample, leading to system classification confusion. Errors will arise in the system even if there is no external intervention. Due to changes in disease patterns and insufficient training data, the initial algorithm will eventually deviate from expected operation [109]. According to Belard et al. [110], the focus of any AI system is on efficacy rather than security. AI is now accomplished through software codes. Engineers will certainly make blunders while dealing with thousands of codes. Frequent patches and upgrades can help to enhance an AI system. However, with AI algorithms employed in medicine, such mistakes may directly harm patients' health.

3.3.6 Communal Acceptability

Although most people are inclined to believe in an AI-based diagnosis, they prefer to trust doctors more frequently when the AI-based diagnosis varies from that of the doctor. Furthermore, medical personnel in developing countries are concerned about being replaced by AI in the future [111]. The lack of acceptance is linked to a potential fear of technological unemployment; broad adoption of AI is thought to render certain jobs obsolete. Even in AI-advanced nations, a worldwide poll of 3000 AI-proficient CEOs from 10 countries found that the majority of enterprises did not anticipate AI to considerably reduce the size of their workforce [112]. Thus, contrary to popular belief, the purpose of AI would be to supplement rather than replace the human team if properly implemented in developing countries.

3.4 Opportunities for AI in diagnostic imaging

Successful deployment of AI applications in the healthcare system will certainly change a lot of job roles of professionals in the sector [113], including the roles of diagnostic imaging professionals. Skillsets that were hitherto very essential may no longer be relevant. How well professionals prepare to embrace this revolution will determine how relevant they can be in the face of AI [114]. Indeed, there are uncertainties about the extent to which AI can impact on diagnostic imaging and job roles of professionals in that area. This is particularly the case because AI is yet to gain widespread deployment and adaptation in the developed world, not to mention the developing world [115]. However, the diagnostic imaging community in the developing world should put in place solid strategies to prepare themselves for this revolution which is not too distant from now. Already, there are reported anxieties over job losses, loss of critical thinking skills and clinical competencies of diagnostic imaging professionals if AI is fully integrated into diagnostic imaging in developing countries [8, 12, 39, 64, 116, 117]. The following opportunities for the diagnostic imaging community in the developing world should encourage professionals in the field to fully accept, adopt and integrate AI into their practice.

3.4.1 Service extension to underserved communities

In many parts of developing countries, most communities have limited access to healthcare services due to significant gaps in distribution of healthcare professionals and equipment [8]. In such situations, AI tools can be carefully designed to fill this gap; such tools must, in their design, incorporate the diversity of socio-economic and healthcare settings of the communities [8]. For a successful implementation of the AI tools, diagnostic imaging professionals should be trained on how to use them, and the community should be engaged throughout the implementation process through sensitization.

Smart mobile phone penetration in most parts of developing countries presents a great opportunity for the deployment of AI tools [6]. Diagnostic imaging could make use of mobile phone applications that employ the inbuilt camera of mobile phones to capture images and transmit them to targeted professionals in the healthcare workflow. In addition, developments in cloud computing and digitization of health information can promote the deployment of expert systems [6] to enhance the processes of providing diagnostic imaging services to populations in hard-to-reach communities in the developing world through telemedicine. In some cases, these expert systems can perform most of the tasks required to be undertaken by the professional, such as scanning an anatomy to capture and transmit the image. Depending on the task, this can be performed by a non-professional at the patient's end, under the guidance of the professional over phone, and transmitted to the imaging professional over the phone using the internet.

3.4.2 Evaluation and reporting of images

The quest for high efficiency and efficacy in patient care has been the major motivation for AI integration in diagnostic imaging systems [118]. Fortunately, the diagnostic imaging community has always readily received such advances in imaging technologies that are consistent with the ethics of the profession [36]. This has over the years been associated with increasing amounts of diagnostic imaging data at far higher rate than the number of available trained image readers (radiologists) [118]. There is therefore a dramatic increase in the workload of radiologists, and there are reports that an average radiologist is required, in extreme cases, to interpret one image every 3–4 s in an 8-h workday to meet demands of their workload [118]. This task largely involves making clinical diagnosis and decisions through visual inspection in the face of uncertainties; errors are therefore inevitable especially under these constrained conditions.

AI tools used in diagnostic image interpretation use algorithms that are either hand-crafted feature-based or deep neural network-based [26, 118], and outcomes of measure normally relate to detection of pathology, disease or tumor characterization, monitoring, image reconstruction, registration and reporting. The application of AI in image reporting is a role that can be played by diagnostic imaging professionals with basic knowledge in the applied algorithms. Indeed, this is a practice in the United Kingdom for over 20 years now where qualified diagnostic imaging professionals have an established role in image interpretation predominantly in the areas of projection radiography and mammography [36]. There is also evidence to show that in such arrangements, AI-assisted image interpretation done by diagnostic imaging professionals across various imaging modalities yields earlier cancer diagnosis, greater cancer screening turn-around times and addresses reporting backlogs in radiology units [119] at a relatively cheaper cost when compared to the task of a radiologist especially for high volume examinations or modalities like chest radiography, CT lung and mammography screening [39].

In order for diagnostic imaging specialists to tap into this opportunity, they need to train in pathology, radiology, Bayesian logic, statistics and data science [120]. The goal of this training should be towards pattern recognition and should not take as much time as the traditional training period for radiologists. There may be initial resistance to merging specialties in radiology and diagnostic imaging due to the uniqueness of each profession in pedagogy, tradition and salary structure [120]. AI is however predicted to naturally lead the redefinition of the roles of these professions, displacing the professionals to tasks needing human element while fusing human intelligence with artificial intelligence to execute more challenging cognitive tasks. It is believed that an AI-led diagnostic task could empower a trained diagnostic imaging professional to evaluate imaging data for an entire town in Africa [120]. Thus, contrary to the fear of job losses among diagnostic imaging professionals in the era of AI [8, 12, 39, 64, 116, 117], jobs will not be lost but will rather be modified [120] and expanded in scope, if professionals in the developing world are well positioned to embrace this revolution.

3.4.3 Synthetic modality transfer

AI promises to present an opportunity for interconversion of images between modalities; for example, it is possible to create a CT image from an MRI scan or vice versa through AI [121, 122]. Clearly, successful implementation of this algorithm in the developing world will obviate the need for a second imaging procedure completely, thus reducing the cost of diagnostic imaging investigation to the patient, increasing diagnostic efficiency and accuracy, and saving cost to the hospital for having to procure the various types of scanners installed in their imaging unit. This however does not suggest the exclusion of the imaging professional in the investigation process. An oversight responsibility and diligence will be required to ensure that the imaging data is not corrupted [36] and that AI protocols are consistent with expected outputs.

In the current arrangement of the diagnostic imaging profession, there are modality-specific job positions and limited cross-modality expertise [36]; flexibility of switching roles may therefore be challenging in most cases. Meanwhile, requests for imaging investigations from physicians is on the rise, placing more demands for diagnostic imaging support to diagnosis, treatment and disease monitoring. Therefore, synthetic modality transfer implemented through AI will offer diagnostic imaging professionals an opportunity to achieve a range of modality and technology-interfacing competencies with much ease and flexibility.

3.4.4 Imaging protocol optimization

Contrast-based imaging, treatment planning in radiotherapy, dose optimization and imaging protocol optimization to generate quality images of diagnostic value are key tasks performed by diagnostic imaging professionals. AI holds great prospects to offer opportunities to increase the accuracy of these tasks and ensure the full implementation of personalized and individualized healthcare and potentially reducing unintended treatment outcomes. For example, specific to patients' needs, contrast volume and injection rate can be adjusted and synthetic contrast enhancement can be generated without the use of real contrast using an AI-based system [36].

The choice of imaging protocol and application are not often consistent within or across imaging departments or modalities [123–126]. This justifies the need for AI-based protocol selection procedures. Currently, AI-based applications are available for dose reduction in mammography [127], CT and PET/CT [128, 129], as well as MRI scan time reduction [130]. These offer quicker image acquisitions and increased patient throughput. There is a generally held view that image quality in sonography largely depends on the operator's competence without much attention to technology considerations [36]. AI-based ultrasound positioning and measurement tools could provide sonographers an opportunity to produce high-quality sonographic assessment reports with lower error rates [131] in investigations such as fetal measurements and kidney function assessment [132, 133]. Opportunities are also available for image quality improvements [134] and automated image evaluation [135] using AI-guided techniques.

3.4.5 Improved patient care

Despite the infiltration of AI applications in diagnostic imaging, the Ionizing Radiation (Medical Exposure), IR (ME) regulations [136] still emphasize that ultimate responsibilities of medical image acquisition and processing lie in the hands of the imaging professional regardless of the level of automation [39]. It is unlikely that this regulatory provision will change anytime soon because AI tools have still not been developed to a point where human intervention will completely be eliminated even in advanced settings [39]. There are therefore rather new roles expected to be played by the imaging professional in the era of AI applications. They will be expected to lead in the establishment of best practices for utilization of semi-automated [36] and fully automated AI systems within established legal frameworks or entirely new legal frameworks as may be recommended by the profession. Under these best practices, they will be required to play critical roles of educating patients on radiation protection [136], the risks of AI-based systems to them, how AI will influence human interaction and clinical decision-making, as well as roles in the acquisition of patients' consent for AI research protocols [36].

In a nuclear medicine study, Kim et al. [137] reported the application of deep learning algorithms in independently diagnosing Parkinson's disease using dopamine active transporter (DAT) single-photon emission CT (SPECT) images and recorded a sensitivity of 96%, while a 94% sensitivity of predicting Parkinson's disease with AI-based SPECT method was also observed elsewhere [138].

It is reported that AI tools are already presenting opportunities to diagnostic imaging professionals in some developing countries to positively influence the patient management pathway [39]. For example, in Ghana, Tanzania and Zambia [39, 139], the CAD4TB software installed on X-ray machines enables diagnostic imaging professionals trained in its use to provide first-line diagnosis of suspected TB cases using computer-aided diagnosis of pulmonary tuberculosis from chest radiographs. The results are reported to compare with expert diagnosis [139]. This eliminates the need to transfer the images to the radiologist for interpretation, which normally takes between 3 days and 1 week or even more to complete [39], depending on the workload of the radiologist. Training is being provided to the diagnostic imaging staff in some units in Ghana on how to interpret pathological heatmaps of TB and COVID-19 using specialized AI tools, and communicate the results to the referring physician through virtual platforms in real time. Undoubtedly, waiting time for patients is being reduced while clinical care is improved at the imaging departments [39].

3.4.6 Audit of AI-based imaging systems

The European medical device regulations (MDR 2017/745) [140] recommends that all automated systems must have quality control checks to ensure acceptability of their outputs. Diagnostic imaging professionals will therefore be required to perform regular audits and reviews of the outputs and decisions of AI-based diagnostic imaging systems. There is a huge chance that a huge fraction of all AI-assisted imaging outputs will need some sort of audit to establish sensitivity, specificity and accuracy of the systems. The diagnostic imaging community in developing countries must therefore take this opportunity to lead the process and explore the options of establishing standardized systems to review AI-based imaging data [36]. This will however require immediate training in the relevant areas of AI applications in diagnostic imaging [36, 39].

3.4.7 Availability of technological support systems

Even though AI applications are currently in the early stages in most developing countries, there are huge amounts of clinical data being collected using various technologies in health facilities. Most hospitals employ electronic healthcare databases for evaluating and organizing pictures, such as the Radiological Information

System (RIS) for handling medical imaging and associated data, and Electronic Medical Records (EMR) for collating clinical data such as notes, pathology and laboratory data [141].

In addition to this large amount of available data, other enabling factors available for efficient AI implementation in the developing world include developments in cloud computing, considerable investments in digitizing health information, and robust mobile phone penetration [115, 142]. In fact, some vendors of diagnostic imaging equipment (such as the CAD4TB software in Ghana, Tanzania and Zambia [39, 139]) have already installed AI software on their machines; many AI solutions are also open source [2] and can be accessed from any location in the world. There are indications of steps taken by data scientists in AI to pool human resources together to systematically integrate AI into diagnostic imaging [27].

Furthermore, high-profile meetings have been convened by intergovernmental agencies to discuss the development and liberalization of AI solutions to address specific global challenges [143, 144]. Coupled with this, the United Nations has also acknowledged the key role of AI in achieving its Sustainable Development Goals [145]. Subsequently, the National Institutes of Health in the United States, under its Harnessing Data Science for Health Discovery and Innovation in Africa (DS-I Africa) program, is reported to have invested about US$74.5 million over 5 years to support data science, innovation and encourage discoveries in the health sector in Africa [144].

The diagnostic imaging community in developing countries that use AI solutions have opportunities to access funds for research and training in the areas of AI-based diagnostic imaging. This is because funding agencies (government, donors and commercial entities) will more likely invest in systems whose outputs can easily be measured and can be bench-marked against available resources [115]. Thus, given these resources and investments, the impact of AI solutions in healthcare, including diagnostic imaging, is imminent and professionals in this area should not miss this opportunity.

3.5 Threats to AI in diagnostic imaging

External factors that will likely influence the successful integration of AI into routine diagnostic imaging applications need to be identified in the early stages of AI adoption in developing countries. Developing economies that aim to tap into investment opportunities for AI integration into the practice of diagnostic imaging must necessarily put in measures to mitigate the negative impacts of these threats. Some of the core threats to efficient utilization of AI in diagnostic imaging services in the developing world are discussed below.

3.5.1 Input training data biases

Successful implementation of AI in diagnostic imaging will depend on clear definition of clinical contexts in which such applications will be used [146]. Due to the lack of inherent accuracy checks in specialized AI algorithms [6], the AI tool needs to be constantly trained by feeding the appropriate quality and amount of imaging data into the system [6, 8, 27]. Meanwhile, expert systems are often built to target high-income countries [27] and may not therefore adequately address diagnostic imaging issues in low and middle-income countries. Input data used to train these AI systems are often biased in race, ethnicity, age, and gender [8] raising questions about acceptability of their outcome measures in the context of the developing world. Moreso, not many countries in the developed world also contribute to the training data [27]. It is therefore possible to misdiagnose or correctly diagnose conditions from imaging data that may not be treatable in the context of poor settings [6]. Even more serious is the fact that most AI systems are 'black box' technologies offering no opportunity for bias assessment [6].

The situation is further exacerbated by inequalities in distribution and access to diagnostic imaging services and other relevant technologies in developing countries [8]. Such a digital divide does not only limit the application of AI systems but also threaten the quality and representativeness of the input data if at all the AI systems require to be calibrated and trained.

3.5.2 Lack of accountability and regulatory frameworks

The need for large amounts of data as input to AI-based technologies raises questions about data ownership and who can have access to which specific data for any purpose. In 2014, a policy framework known as the *Data Sharing Principles in Developing Countries* [147] was drafted but there is yet to be a widespread adoption of this framework in member countries [6]. The policy was crafted to address the principle that data generated with public funds should be regarded as a public good [6, 147]. However, from both patient and AI developer standpoints, privacy laws and data access and ownership agreements are perceived to be potential threats to successful implementation of AI in settings that need them [6, 12].

Furthermore, widespread applications of AI in healthcare delivery in unregulated contexts and by unregulated private providers, as is the case in many developing countries, could create challenges for government oversight of healthcare [8]. Any government that wants to play it safe might therefore likely discourage widespread AI integration into diagnostic imaging, particularly in the private sector.

3.5.3 Perceived transfer of human intervention

The ultimate goal of AI technologies should be to empower patients to assume control of their own evolving healthcare needs. However, inappropriate application of AI could result in situations where decisions hitherto taken by diagnostic imaging professionals and patients will be transferred to machines [8]. Consequently, this will undermine human autonomy, as neither how AI technologies arrive at decisions will be understood nor will humans be able to negotiate with the technology to reach a shared decision. Humans will therefore likely lose autonomy over control of healthcare systems and clinical decisions [4] in fully automated scenarios. There are already fears of loss of critical thinking skills and expertise [8, 12, 39, 64, 116, 117], as well as decreased patient consultation [4] in advanced AI application settings. These perceptions, though not fully practical, could hinder the widespread acceptance of AI systems in diagnostic imaging.

3.5.4 Standardization of operational language

One of the threats to widespread implementation of AI in diagnostic imaging units is the issue of merging the largely hand-written records using local terminologies with electronic AI systems [6]. Building a standard language corpus for each setting to suit their purpose will require substantial amount of effort, recompilation and interpretation of well-developed local terminologies into standardized medical terminologies [6, 148]. This will be both difficult to do and time-consuming especially in resource-poor countries. In practice, input data for AI systems do not satisfy the condition of semantic interoperability and are therefore not directly 'machine-readable' [28] incorporating capabilities of harmonized local terminologies.

3.5.5 Poor investment in infrastructure

For AI to thrive in diagnostic imaging applications in the developing world, conscious efforts must be made to address issues of technical competence in AI, provision of stable electrical power, sufficient internet bandwidth, and other IT related equipment to support AI penetration. This presents a high cost for adoption of AI in the short to long term.

Electrical power and the internet remain the two most critical factors that support successful implementation of AI. Meanwhile, it is estimated that in Africa, less than 30% of health facilities have access to reliable electricity [149], and internet penetration across the entire continent is estimated at only 39% [139]. Digital health infrastructure on which AI solutions can be built are also lacking [4, 150]. Even though there are already developed AI solutions off the shelf [139] which require less amount of time to deploy into the imaging subsector of the healthcare system, they still need to be customized to provide solutions to the contexts in which they will be applied. Again, technical competence to carry out this customization process is lacking in developing countries. These multifaceted issues need both time and significant funding by countries, whose economies are currently hard-hit by the COVID-19 pandemic and there are strict budgetary allocations to other competing demands in the economy.

3.5.6 Perceived job losses and redundancy

One of the pioneering publications debating the impact of AI on employment predicted that 47% of people working in the US in 2013 were 70% or more likely at risk of having their jobs automated within a decade [151]. Applying the same methodology in this study to developing countries (Nigeria, Ethiopia and South Africa included), the World Bank in its *World Development Report 2016: Digital Dividends* warned that there is rather a higher proportion of jobs likely to be automated in developing countries than was estimated for advanced countries, where many jobs had already been lost following the implementation of AI [152].

However, other schools of thought believe that instead of job losses, AI will rather lead to job polarization [153, 154], where job roles will be redistributed to favor more of those with high technical skills and competence who can adjust to this change. Even though the diagnostic imaging profession is responsive to technological changes, the pace at which developing countries can cope with this revolution may be limited by funding and infrastructural deficits. This will likely affect lower-rank professionals of diagnostic imaging who have been working in the healthcare system for ages, and yet may be considered redundant in the era of AI.

3.5.7 Lack of AI in the curriculum of diagnostic imaging training

Diagnostic imaging requires highly trained and qualified clinical experts who can carry out the demanding tasks of the profession [27]. However, these professionals are few in developing countries and where available, they are often concentrated in urban centers, including emigration of the most skilled among them to high-income countries [155, 156]. These dynamics negatively impact on the diagnostic imaging services in the developing world.

AI penetration will even worsen the situation for settings where technological competencies of professionals do not meet the implementation needs of AI applications. Lack of knowledge and understanding of the context in which AI can be applied will depend on availability of educational opportunities on the subject matter. However, the concept of AI is currently not incorporated into training of diagnostic imaging professionals, as there seem to be a disconnect between academic departments training IT and diagnostic imaging specialists on one hand, and imaging departments that need the skillset in AI on the other.

3.5.8 Security breaches

Large amounts of digital information either as input to, or outputs from, AI systems transmitted across the digital space need high-level of security against attacks. Both patients and care-givers may raise questions about how secure personal information may be, as with any other electronic system. There may be attacks against the dataset or the AI algorithm itself, in the forms of identification of persons from the data or manipulation of the program codes to produce an undesirable outcome, respectively [29].

3.6 Making a case for adoption of AI in diagnostic imaging

Disease burdens in developing countries are on the rise in recent times at a disproportionate rate compared to access to imaging services (these include both image acquisition and reporting). For example, 1 MRI scanner [157] and 1.9 radiologists [3] are available to every 1 million inhabitants in developing countries. This situation will, in no doubt, place a huge workload burden on the very few imaging equipment and diagnostic imaging professionals (including medical physicists, radiographers and radiologists) in this region. Clinical errors then will become a more common occurrence due to work-related stress and equipment breakdowns, with associated increased risk to patients. Meanwhile, diagnostic imaging is crucial for timely diagnosis and appropriate interventions. Unless these circumstances change (e.g., by deploying modern, AI-based imaging technologies to the region), it is very unlikely that most low and middle-income countries (LMICs) will meet the objectives of SDG 3, including universal health coverage by the estimated period of 2030/2035.

AI-based systems will make it possible for countries to be easily interconnected, generate large amounts of quality health data, and share ideas on improving the efficiencies of their imaging technologies over time. Out of this arrangement could emerge other digital technologies to help in disease surveillance at a global scale by public health agencies such as the WHO to monitor, identify and mitigate potential outbreaks or other public health emergencies [6]. Furthermore, the synthetic modality transfer capability of AI-based systems could make it possible for health facilities to save huge costs associated with the procurement of expensive imaging equipment and rely on inter-modality image conversions using only one appropriate equipment with such capability.

Given that LMICs are already benefiting from fast paced penetration of technological innovations, such as high-speed internet and computers, smartphones and watches, digitized health records, and AI-based imaging equipment, this is an opportunity for the diagnostic imaging community to embrace this revolution to expand access to imaging services, reduce diagnostic errors and patient waiting times, implement individualized care to improve efficiency and outcomes, and ensure an overall improved healthcare quality (see Appendix). However, training institutions in the relevant areas of AI technology must provide curriculum to technically equip professionals for this emerging role. Successful deployment of AI technologies in diagnostic imaging will empower practitioners to switch roles effectively to meet the workload demands in imaging units, and create more job opportunities in the diagnostic imaging pipeline.

3.7 Conclusion

Deployment of AI tools in diagnostic imaging is a timely event particularly for developing countries where imaging equipment and personnel are largely unavailable. Widespread implementation of this technology in these settings will however require addressing issues of data biases, system security and setting-specific applicability. Developing countries seeking to fully implement AI-based imaging technologies must learn from the experiences of advanced countries already benefiting from AI tools, and prepare their imaging workforce for AI applications, both in skill and policy direction through their national regulatory bodies and professional societies. Where applicable, AI tools should be developed by local expertise using human-centered design principles, including legal and ethical considerations. The process of AI implementation should be a collective responsibility of not only diagnostic imaging professionals and their professional societies, but should include computer programmers, equipment vendors, patients, ministries of health and information technology, lawyers, human rights advocates, academics in the relevant fields, and hospital managements. Through this wholistic approach, AI tools effectively integrated within diagnostic imaging processes will help propel most developing economies towards the achievement of the health-related targets in the SDGs, especially those related to the universal health coverage concept.

Appendix

Summary of SWOT analysis of AI in diagnostic imaging in the developing world.

Strengths	Weaknesses	Opportunities	Threats
– Enhanced diagnosis	– Need for big data	– Increased access to imaging services	– Input training data biases
– Disease monitoring and prediction	– Huge initial and maintenance costs	– Image evaluation and reporting	– Lack of accountability and regulatory frameworks
– Accurate tumor classification	– Inadequate technical expertise	– Synthetic modality transfer	– Perceived transfer of human intervention
– Efficient image acquisition	– Data ethical concerns	– Protocol optimization	– Standardization of operational language
– Image processing, segmentation and analysis	– Data security and accuracy	– Improved patient care	– Poor investment in infrastructure
	– Communal acceptability	– Audit of AI-based systems	– Perceived job losses and redundancy
		– Availability of support systems	– AI missing in diagnostic imaging training
			– Security breaches

References

1. Raso FA, Hilligoss H, Krishnamurthy V, Bavitz C, Kim L. Artificial intelligence & human rights: opportunities & risks. Harvard University, Cambridge, MA, US: Berkman Klein Center for Internet & Society; 2018:2018–6 pp.
2. Frija G, Blažić I, Frush DP, Hierath M, Kawooya M, Donoso-Bach L, et al. How to improve access to medical imaging in low-and middle-income countries? EClinical Med 2021;38:101034.
3. Hricak H, Abdel-Wahab M, Atun R, Lette MM, Paez D, Brink JA, et al. Medical imaging and nuclear medicine: a lancet oncology commission. Lancet Oncol 2021;22:e136–72.
4. Guo J, Li B. The application of medical artificial intelligence technology in rural areas of developing countries. Health Equity 2018;2:174–81.
5. Fale MI. Dr. Flynxz–A First Aid Mamdani-Sugeno-type fuzzy expert system for differential symptoms-based diagnosis. J King Saud Univ Comput Inf Sci 2020;34:1138–149.
6. Wahl B, Cossy-Gantner A, Germann S, Schwalbe NR. Artificial intelligence (AI) and global health: how can AI contribute to health in resource-poor settings? BMJ Global Health 2018;3:e000798.
7. Smith ML, Neupane S. Artificial intelligence and human development: toward a research agenda. In: White Paper. International Development Research Centre (IDRC); 2018. Available from: http://hdl.handle.net/10625/56949 [Accessed 31 May 2022].
8. World Health Organization (WHO). Ethics and governance of artificial intelligence for health: WHO guidance. Geneva, Switzerland: WHO; 2021.

9. Hamet P, Tremblay J. Artificial intelligence in medicine. Metabolism 2017;69:S36–40.
10. Lee EJ, Kim YH, Kim N, Kang DW. Deep into the brain: artificial intelligence in stroke imaging. J Stroke 2017;19:277–85.
11. Krittanawong C, Zhang H, Wang Z, Aydar M, Kitai T. Artificial intelligence in precision cardiovascular medicine. J Am Coll Cardiol 2017;69:2657–64.
12. Gwagwa A, Kraemer-Mbula E, Rizk N, Rutenberg I, De Beer J. Artificial Intelligence (AI) deployments in Africa: benefits, challenges and policy dimensions. Afr J Comput Ict 2020;26: 1–28.
13. Mitchell TM. Machine learning. New York City, United States: McGraw-Hill Higher Education; 1997.
14. Sutton RS, Barto AG. Introduction to reinforcement learning. Cambridge, Massachusetts, United States: MIT Press; 1998:1054 pp.
15. LeCun Y, Bengio Y, Hinton G. Deep learning. Nature 2015;521:436–44.
16. Poole DL, Mackworth AK. Artificial Intelligence: foundations of computational agents. Cambridge, United Kingdom: Cambridge University Press; 2010.
17. Goertzel B. Artificial general intelligence. Pennachin C, editor. New York: Springer; 2007.
18. Pennachin C, Goertzel B. Contemporary approaches to artificial general intelligence. In: Artificial general intelligence. Berlin, Heidelberg: Springer; 2007:1–30 pp.
19. Harfouche A, Saba P, Aoun G, Wamba SF. Guest editorial: cutting-edge technologies for the development of Asian countries. J Asia Bus Stud 2022;16:225–9.
20. De Dombal FT, Hartley JR, Sleeman DH. A computer-assisted system for learning clinical diagnosis. Lancet 1969;293:145–8.
21. Horrocks JC, McCann AP, Staniland JR, Leaper DJ, De Dombal FT. Computer-aided diagnosis: description of an adaptable system, and operational experience with 2,034 cases. Br Med J 1972; 2:5–9.
22. Shortliffe EH, editor. Computer-based medical consultations: MYCIN. New York: Elsevier; 2012.
23. Warner HR, Toronto AF, Veasey LG, Stephenson R. A mathematical approach to medical diagnosis: application to congenital heart disease. J Am Med Assoc 1961;177:177–83.
24. Miller DD, Brown EW. Artificial intelligence in medical practice: the question to the answer? Am J Med 2018;131:129–33.
25. Shafer GJ, Singh H, Thomas EJ, Thammasitboon S, Gautham KS. Frequency of diagnostic errors in the neonatal intensive care unit: a retrospective cohort study. J Perinatol 2022:1–7. https://doi.org/10.1038/s41372-022-01359-9.
26. Bera K, Schalper KA, Rimm DL, Velcheti V, Madabhushi A. Artificial intelligence in digital pathology—new tools for diagnosis and precision oncology. Nat Rev Clin Oncol 2019;16:703–15.
27. Lekadir K, Mutsvangwa T, Lazrak N, Zahir J, Cintas C, El Hassouni M, et al. From MICCAI to AFRICAI: African network for artificial intelligence in biomedical imaging. ICLR; 2022. Available from: https://pml4dc.github.io/iclr2022/pdf/PML4DC_ICLR2022_12.pdf [Accessed 30 May 2022].
28. Lambin P, Leijenaar RT, Deist TM, Peerlings J, De Jong EE, Van Timmeren J, et al. Radiomics: the bridge between medical imaging and personalized medicine. Nat Rev Clin Oncol 2017;14:749–62.
29. Kaissis GA, Makowski MR, Rückert D, Braren RF. Secure, privacy-preserving and federated machine learning in medical imaging. Nat Mach Intell 2020;2:305–11.
30. De Bruyn A, Viswanathan V, Beh YS, Brock JK, von Wangenheim F. Artificial intelligence and marketing: pitfalls and opportunities. J Interact Market 2020;51:91–105.
31. Sahiner B, Chan HP, Hadjiiski LM, Cascade PN, Kazerooni EA, Chughtai AR, et al. Effect of CAD on radiologists' detection of lung nodules on thoracic CT scans: analysis of an observer performance study by nodule size. Acad Radiol 2009;16:1518–30.
32. van Ginneken B, Hogeweg L, Prokop M. Computer-Aided diagnosis in chest radiography: beyond nodules. Eur J Radiol 2009;72:226–30.

33. Morton MJ, Whaley DH, Brandt KR, Amrami KK. Screening mammograms: interpretation with computer-aided detection–prospective evaluation. Radiology 2006;239:375–83.
34. Litjens G, Kooi T, Bejnordi BE, Setio AAA, Ciompi F, Ghafoorian M, et al. A survey on deep learning in medical image analysis. Med Image Anal 2017;42:60–88.
35. Nichols JA, Herbert Chan HW, Baker MAB. Machine learning: applications of artificial intelligence to imaging and diagnosis. Biophys Rev 2019;11:111–8.
36. Hardy M, Harvey H. Artificial intelligence in diagnostic imaging: impact on the radiography profession. Br J Radiol Suppl 2020;93:20190840.
37. Syed AB, Zoga AC. Artificial intelligence in radiology: current technology and future directions. In: Seminars in muscoskelosketal radiology. New York City, United States: Thieme Medical Publishers; 2018, vol. 22:540–5 pp.
38. Ahn SY, Chae KJ, Goo JM. The potential role of grid-like software in bedside chest radiography in improving image quality and dose reduction: an observer preference study. Korean J Radiol 2018; 19:526–33.
39. Wuni AR, Botwe BO, Akudjedu TN. Impact of artificial intelligence on clinical radiography practice: futuristic prospects in a low resource setting. Radiography 2021;27:S69–73.
40. Sarsby A. SWOT analysis. Lulu. com; 2016. Available from: https://books.google.com.gh/books?hl=en&lr=&id=Yrp3DQAAQBAJ&oi=fnd&pg=PA1&dq=Alan+sarsby+swot+analysis&ots=ODoeZuz2ZE&sig=unXr10T13IdbmibkxOFkZA4tfe8&redir_esc=y#v=onepage&q=Alan%20sarsby%20swot%20analysis&f=false [Accessed 31 May 2022].
41. Murry W. Strength, weakness, opportunity, and threat (SWOT) analysis. Investopedia; 2021. Available from: https://www.investopedia.com/terms/s/swot.asp [Accessed 31 May 2022].
42. Davenport TH, Hongsermeier TM, Kimberly Alba Mc, Cord KA. Using AI to improve electronic health records; 2018. Available from: https://hbr.org/2018/12/using-ai-to-improve-electronic-health-records [Accessed 25 May 2022].
43. Lin D, Lin H. Translating artificial intelligence into clinical practice. Ann Transl Med 2020;8:715.
44. Curtis C, Liu C, Bollerman TJ, Pianykh OS. Machine learning for predicting patient wait times and appointment delays. J Am Coll Radiol 2018;15:1310–6.
45. McNemar E. Adopting AI to improve patient outcomes, cost, savings, health equality. Health IT Analytics: Quality and governance news; 2022. Available from: https://healthitanalytics.com/news/adopting-ai-to-improve-patient-outcomes-cost-savings-health-equality [Accessed 20 May 2022].
46. Drexel University. Pros & cons of artificial intelligence in medicine; 2021. Available from: https://drexel.edu/cci/stories/artificial-intelligence-in-medicine-pros-and-cons/ [Accessed 23 May 2022].
47. Intrex Group. Artificial intelligence in radiology - use cases and trends. Available from: https://itrexgroup.com/blog/artificial-intelligence-in-radiology-use-cases-predictions/#header [Accessed 20 May 2022].
48. McNemar E. Top opportunities for artificial intelligence to improve cancer care. In: Health IT analytics: features. 2021 Nov 29. Available from: https://healthitanalytics.com/features/top-opportunities-for-artificial-intelligence-to-improve-cancer-care [Accessed 24 May 2022].
49. Pesapane F, Codari M, Sardanelli F. Artificial intelligence in medical imaging: threat or opportunity? Radiologists again at the forefront of innovation in medicine. Eur Radiol Exp 2018;2: 1–10.
50. Chetty S, Venter D, Speelman A. Determining the need for after-hours diagnostic radiological reporting in emergency departments at public hospitals in South Africa: perceptions of emergency physicians in KwaZulu-Natal. J Med Imag Radiat Sci 2020;51:470–9.
51. McNemar E. Benefits of artificial intelligence to radiology workflows. Health IT analytics: News.; 2019. Available from: https://healthitanalytics.com/news/benefits-of-artificial-intelligence-to-radiology-workflows [Accessed 20 May 2022].

52. Hamdi Y, Abdeljaoued-Tej I, Zatchi AA, Abdelhak S, Boubaker S, Brown JS, et al. Cancer in africa: the untold story. Front Oncol 2021;11:650117.
53. Marti-Bonmati L, Koh DM, Riklund K, Bobowicz M, Roussakis Y, Vilanova JC, et al. Considerations for artificial intelligence clinical impact in oncologic imaging: an AI4HI position paper. Insights Imaging 2022;13:1–11.
54. Coppola F, Faggioni L, Gabelloni M, De Vietro F, Mendola V, Cattabriga A, et al. Human, all too human? An all-around appraisal of the "Artificial Intelligence Revolution" in medical imaging. Front Psychol 2021;12:710982.
55. Ahuja AS. The impact of artificial intelligence in medicine on the future role of the physician. PeerJ 2019;7:e7702.
56. Do HM, Spear LG, Nikpanah M, Mirmomen SM, Machado LB, Toscano AP, et al. Augmented radiologist workflow improves report value and saves time: a potential model for implementation of artificial intelligence. Acad Radiol 2020;27:96–105.
57. Gore JC. Artificial intelligence in medical imaging. Magn Reson Imag 2020;68:A1–4.
58. Kulkarni S, Jha S. Artificial intelligence, radiology, and tuberculosis: a review. Acad Radiol 2020; 27:71–5.
59. Iezzi R, Goldberg SN, Merlino B, Posa A, Valentini V, Manfredi R. Artificial intelligence in interventional radiology: a literature review and future perspectives. J Oncol 2019;2019:6153041.
60. Nensa F, Demircioglu A, Rischpler C. Artificial intelligence in nuclear medicine. J Nucl Med 2019; 60:29S–37.
61. Gregory J, Welliver S, Chong J. Top 10 reviewer critiques of radiology artificial intelligence (AI) articles: qualitative thematic analysis of reviewer critiques of machine learning/deep learning manuscripts submitted to JMRI. J Magn Reson Imag 2020;52:248–54.
62. Belfiore MP, Urraro F, Grassi R, Giacobbe G, Patelli G, Cappabianca S, et al. Artificial intelligence to codify lung CT in Covid-19 patients. La Radiologia Med 2020;125:500–4.
63. Kamiński MF, Hassan C, Bisschops R, Pohl J, Pellisé M, Dekker E, et al. Advanced imaging for detection and differentiation of colorectal neoplasia: European Society of Gastrointestinal Endoscopy (ESGE) Guideline. Endoscopy 2014;46:435–57.
64. Antwi WK, Akudjedu TN, Botwe BO. Artificial intelligence in medical imaging practice in Africa: a qualitative content analysis study of radiographers' perspectives. Insights Imaging 2021;12:1–9.
65. Mollura DJ, Culp MP, Pollack E, Battino G, Scheel JR, Mango VL, et al. Artificial intelligence in low- and middle-income countries: innovating global health radiology. Radiology 2020;297:513–20.
66. Carrillo Larco R, Tudor Car L, Pearson-Stuttard J, Panch T, Miranda JJ, Atun R. Machine learning health-related applications in low-income and middle-income countries: a scoping review protocol. BMJ Open 2020;10:e035983.
67. Mollura DJ, Lugossy AM. RAD-AID: fostering opportunities to impact global health with technology. Appl Radiol 2021;50:36–7.
68. Dzobo K, Adotey S, Thomford NE, Dzobo W. Integrating artificial and human intelligence: a partnership for responsible innovation in biomedical engineering and medicine. OMICS A J Integr Biol 2020;24:247–63.
69. Gong B, Nugent JP, Guest W, Parker W, Chang PJ, Khosa F, et al. Influence of artificial intelligence on Canadian medical students' preference for radiology specialty: a national survey study. Acad Radiol 2019;26:566–77.
70. Savadjiev P, Chong J, Dohan A, Vakalopoulou M, Reinhold C, Paragios N, et al. Demystification of AI-driven medical image interpretation: past, present and future. Eur Radiol 2019;29:1616–24.
71. van Hoek J, Huber A, Leichtle A, Härmä K, Hilt D, von Tengg-Kobligk H, et al. A survey on the future of radiology among radiologists, medical students and surgeons: students and surgeons tend to be more skeptical about artificial intelligence and radiologists may fear that other disciplines take over. Eur J Radiol 2019;121:108742.

72. Patuzzi J. Big data, AI look set to come under scrutiny at ECR 2018; 2017. Available from: https://www.auntminnieeurope.com/index.aspx?sec=rca&sub=ecr_2018&pag=dis&ItemID=614795 [Accessed 24 May 2022].
73. Casey B, Yee KM, Ridley EL, Forrest W, Kim A. Top 5 trends from RSNA 2017 in Chicago; 2017. Available from: https://www.auntminnie.com/index.aspx?sec=rca&sub=rsna_2017&pag=dis&ItemID=119393 [Accessed 25 May 2022].
74. Ward P, Ridley E, Forrest W, Moan R. Top 5 trends from ECR 2018 in Vienna; 2018. Available from: https://www.auntminnie.com/index.aspx?sec=rca&sub=ecr_2018&pag=dis&ItemID=120195 [Accessed 25 May 2022].
75. Waller J, O'Connor A, Rafaat E, Amireh A, Dempsey J, Martin C, et al. Applications and challenges of artificial intelligence in diagnostic and interventional radiology. Pol J Radiol 2022;87:113–7.
76. Ngwa W, Addai BW, Adewole I, Ainsworth V, Alaro J, Alatise OI, et al. Cancer in sub-saharan africa: a lancet oncology commission. Lancet Oncol 2022;S1470–2045:00720–8.
77. International G. Agency for Research on Cancer, World Health Organization; 2021. Available from: https://gco.iarc.fr [Accessed 25 May 2022].
78. Chang YW, An JK, Choi N, Ko KH, Kim KH, Han K, et al. Artificial intelligence for breast cancer screening in mammography (AI-STREAM): a prospective multicenter study design in Korea using AI-based CADe/x. J Breast Cancer 2022;25:57.
79. Lunit. AI-assisted radiologists can detect more breast cancer with reduced false-positive recall. Available from: https://www.prnewswire.com/news-releases/ai-assisted-radiologists-can-detect-more-breast-cancer-with-reduced-false-positive-recall-301001971.html [Accessed 20 May 2022].
80. Yee KM. How AI can help improve hip fracture diagnosis. AuntMinnieEurope.com. Available from: https://www.auntminnieeurope.com/index.aspx?sec=ser&sub=def&pag=dis&ItemID= 619096 [Accessed 22 May 2022].
81. Cheng CT, Ho TY, Lee TY, Chang CC, Chou CC, Chen CC, et al. Application of a deep learning algorithm for detection and visualization of hip fractures on plain pelvic radiographs. Eur Radiol 2019;29:5469–77.
82. Han DH. FDA approves AI algorithm that helps detect wrist fracture. MPR; 2018. Available from: https://www.empr.com/home/news/fda-approves-ai-algorithm-that-helps-detect-wrist-fractures/ [Accessed 20 May 2022].
83. de la Fuente Garcia S, Ritchie CW, Luz S. Artificial intelligence, speech, and language processing approaches to monitoring Alzheimer's disease: a systematic review. J Alzheim Dis 2020;78: 1547–74.
84. MENA's TH. First free diagnostic tool for COVID-19 using AI in medical images. In: Omnia health; 2020. Available from: https://insights.omnia-health.com/radiology/menas-first-free-diagnostic-tool-covid-19-using-ai-medical-images [Accessed 22 May 2022].
85. Philips Foundation. Philips Foundation deploys AI software in South Africa to detect and monitor COVID-19 using chest x-rays. 2021 Mar 25. Available from: https://www.philips-foundation.com/a-w/articles/CAD4COVID.html [Accessed 25 May 2022].
86. Khoury K. Deep AI imaging diagnostics help doctors prioritise care. In: Springwise: health and wellbeing; 2022. Available from: https://www.springwise.com/innovation/health-wellbeing/deep-ai-scans-xrays-for-signs-of-disease [Accessed 24 May 2022].
87. Imaging Technology News. Oxipit partners with healthCre Konnect to bring AI diagnostics to Africa. News: Artificial intelligence. Available from: https://www.itnonline.com/content/oxipit-partners-healthcare-konnect-bring-ai-diagnostics-africa [Accessed 26 May 2022].
88. Laghmari S. Artificial intelligence, a key tool to improve the African health system. In: Infomineo: African health system. Available from: https://infomineo.com/artificial-intelligence-a-key-tool-to-improve-the-african-health-system/ [Accessed 23 May 2022].

89. Eliyatkın N, Yalçın E, Zengel B, Aktaş S, Vardar E. Molecular classification of breast carcinoma: from traditional, old-fashioned way to a new age, and a new way. Eur J Breast Health 2015;11:59.
90. Fox M. Brain cancer is now the leading cancer killer of kids. Health news; 2016. Available from: https://www.nbcnews.com/health/health-news/brain-cancer-now-leading-cancer-killer-kids-n649411 [Accessed 22 May 2022].
91. McGinley L. Brain cancer replaces leukemia as the leading cause of cancer deaths in kids. The Washington Post: Health; 2016. Available from: https://www.washingtonpost.com/news/to-your-health/wp/2016/09/16/brain-cancer-replaces-leukemia-as-the-leading-cause-of-cancer-deaths-in-kids/ [Accessed 22 May 2022].
92. World Health Organization. Childhood cancer. Available from: https://www.who.int/news-room/fact-sheets/detail/cancer-in-children [Accessed 23 May 2022].
93. McNemar E. Artificial intelligence bolsters colorectal cancer detection. In: Health IT analytics: tools and strategies news. 2021 Nov 4. Available from: https://healthitanalytics.com/news/artificial-intelligence-bolsters-colorectal-cancer-detection [Accessed 24 May 2022].
94. Krupinski EA. Current perspectives in medical image perception. Atten Percept Psycho 2010;72:1205–17.
95. Renard F, Guedria S, Palma ND, Vuillerme N. Variability and reproducibility in deep learning for medical image segmentation. Sci Rep 2020;10:1–6.
96. Chen R, Wang M, Lai Y. Analysis of the role and robustness of artificial intelligence in commodity image recognition under deep learning neural network. Plos One 2020;15:e0235783.
97. Yip SS, Parmar C, Kim J, Huynh E, Mak RH, Aerts HJ. Impact of experimental design on PET radiomics in predicting somatic mutation status. Eur J Radiol 2017;97:8–15.
98. Sutton EJ, Huang EP, Drukker K, Burnside ES, Li H, Net JM, et al. Breast MRI radiomics: comparison of computer-and human-extracted imaging phenotypes. Eur Radiol Exp 2017;1:1–0.
99. King AD, Chow KK, Yu KH, Mo FK, Yeung DK, Yuan J, et al. Head and neck squamous cell carcinoma: diagnostic performance of diffusion-weighted MR imaging for the prediction of treatment response. Radiology 2013;266:531–8.
100. Fusco R, Di Marzo M, Sansone C, Sansone M, Petrillo A. Breast DCE-MRI: lesion classification using dynamic and morphological features by means of a multiple classifier system. Eur Radiol Exp 2017;1:1–7.
101. Alhajeri M, Shah SG. Limitations in and solutions for improving the functionality of picture archiving and communication system: an exploratory study of PACS professionals' perspectives. J Digit Imag 2019;32:54–67.
102. Mahajan A, Vaidya T, Gupta A, Rane S, Gupta S. Artificial intelligence in healthcare in developing nations: the beginning of a transformative journey. Cancer Res Treat 2019;2:182.
103. El Saghir NS, Anderson BO, Gralow J, Lopes G, Shulman LN, Moukadem HA, et al. Impact of merit-based immigration policies on brain drain from low-and middle-income countries. JCO Glob Oncol 2020;6:185–9.
104. Mittelstadt BD, Floridi L. The ethics of big data: current and foreseeable issues in biomedical contexts. Sci Eng Ethics 2016;22:303–41.
105. Geis JR, Brady AP, Wu CC, Spencer J, Ranschaert E, Jaremko JL, et al. Ethics of artificial intelligence in radiology: summary of the joint European and North American multisociety statement. Can Assoc Radiol J 2019;70:329–34.
106. Drew BJ, Harris P, Zegre-Hemsey JK, Mammone T, Schindler D, Salas-Boni R, et al. Insights into the problem of alarm fatigue with physiologic monitor devices: a comprehensive observational study of consecutive intensive care unit patients. PLoS One 2014;9:e110274.
107. Rajkomar A, Hardt M, Howell MD, Corrado G, Chin HM. Ensuring fairness in machine learning to advance health equity. Ann Intern Med 2018;169:866–72.

108. Char DS, Shah NH, Magnus D. Implementing machine learning in health care - addressing ethical challenges. N Engl J Med 2018;378:981–3.
109. Dawson NV, Arkes HR. Systematic errors in medical decision making: judgment limitations. J Gen Intern Med 1987;2:183–7.
110. Belard A, Buchman T, Forsberg J, Potter KB, Dente JC, Kirk A, et al. Precision diagnosis: a view of the clinical decision support systems (CDSS) landscape through the lens of critical care. J Clin Monit Comput 2017;31:261–71.
111. Abdullah R, Fakieh B. Health care employees' perceptions of the use of artificial intelligence applications: survey study. J Med Internet Res 2020;22:e17620.
112. Bughin J, McCarthy B, Chui M. A survey of 3,000 executives reveals how businesses succeed with AI. Harv Bus Rev 2017. https://hbr.org/2017/08/a-survey-of-3000-executives-reveals-how-businesses-succeed-with-ai.
113. Topol EJ. High-performance medicine: the convergence of human and artificial intelligence. Nat Med 2019;25:44–56.
114. French J, Chen L. Preparing for artificial intelligence: systems-level implications for the medical imaging and radiation therapy professions. J Med Imag Radiat Sci 2019;50:S20–3.
115. Waljee AK, Weinheimer-Haus EM, Abubakar A, Ngugi AK, Siwo GH, Kwakye G, et al. Artificial intelligence and machine learning for early detection and diagnosis of colorectal cancer in sub-Saharan Africa. Gut 2022;71:1259–65.
116. Botwe BO, Akudjedu TN, Antwi WK, Rockson P, Mkoloma SS, Balogun EO, et al. The integration of artificial intelligence in medical imaging practice: perspectives of African radiographers. Radiography 2021;27:861–6.
117. Abuzaid MM, Elshami W, McConnell J, Tekin HO. An extensive survey of radiographers from the Middle East and India on artificial intelligence integration in radiology practice. Health Technol 2021;11:1045–50.
118. Hosny A, Parmar C, Quackenbush J, Schwartz LH, Aerts HJ. Artificial intelligence in radiology. Nat Rev Cancer 2018;18:500–10.
119. The Cancer Workforce Plan. Phase 1: Delivering the cancer strategy to 2021; 2017. Available from: https://www.hee.nhs.uk/sites/default/files/documents/Cancer%20Workforce%20Plan%20phase%201%20-%20Delivering%20the%20cancer%20strategy%20to%202021.pdf [Accessed 29 May 2022].
120. Jha S, Topol EJ. Adapting to artificial intelligence: radiologists and pathologists as information specialists. J Am Med Assoc 2016;316:2353–4.
121. Nie D, Cao X, Gao Y, Wang L, Shen D. Estimating CT image from MRI data using 3D fully convolutional networks. In: Deep learning and data labeling for medical applications. Cham: Springer; 2016:170–8 pp.
122. Chaibi H, Nourine R. New pseudo-CT generation approach from magnetic resonance imaging using a local texture descriptor. J Biomed Phys Eng 2018;8:53.
123. Teeuwisse W, Geleijns J, Veldkamp W. An inter-hospital comparison of patient dose based on clinical indications. Eur Radiol 2007;17:1795–805.
124. Foley SJ, McEntee MF, Rainford LA. Establishment of CT diagnostic reference levels in Ireland. Br J Radiol Suppl 2012;85:1390–7.
125. McFadden SL, Hughes CM, Winder RJ. Variation in radiographic protocols in paediatric interventional cardiology. J Radiol Prot 2013;33:313.
126. Sammy IA, Chatha H, Bouamra O, Fragoso-Iñiguez M, Lecky F, Edwards A. The use of whole-body computed tomography in major trauma: variations in practice in UK trauma hospitals. Emerg Med J 2017;34:647–52.
127. Liu J, Zarshenas A, Qadir A, Wei Z, Yang L, Fajardo L, et al. Radiation dose reduction in digital breast tomosynthesis (DBT) by means of deep-learning-based supervised image processing.

In: Medical imag 2018: image process international society for optics and photonics; 2018, vol 10574:105740F p.
128. Humphries T, Si D, Coulter S, Simms M, Xing R. Comparison of deep learning approaches to low dose CT using low intensity and sparse view data. In: Medical imaging 2019: physics of medical imaging. SPIE; 2019, vol 10948:1048–54 pp.
129. Ahn C, Heo C, Kim JH. Combined low-dose simulation and deep learning for CT denoising: application in ultra-low-dose chest CT. In: International forum on medical imaging in Asia 2019. SPIE; 2019, vol 11050:52–6 pp.
130. Wang S, Su Z, Ying L, Peng X, Zhu S, Liang F, et al. Accelerating magnetic resonance imaging via deep learning. In: IEEE 13th international symposium on biomedical imaging (ISBI). IEEE; 2016: 514–7 pp.
131. Wu L, Cheng JZ, Li S, Lei B, Wang T, Ni D. FUIQA: fetal ultrasound image quality assessment with deep convolutional networks. IEEE Trans Cybern 2017;47:1336–49.
132. Looney P, Stevenson GN, Nicolaides KH, Plasencia W, Molloholli M, Natsis S, et al. Fully automated, real-time 3D ultrasound segmentation to estimate first trimester placental volume using deep learning. JCI Insight 2018;3:e120178.
133. Kuo CC, Chang CM, Liu KT, Lin WK, Chiang HY, Chung CW, et al. Automation of the kidney function prediction and classification through ultrasound-based kidney imaging using deep learning. NPJ Digital Medicine 2019;2:1–9.
134. Liu F, Jang H, Kijowski R, Bradshaw T, McMillan AB. Deep learning MR imaging-based attenuation correction for PET/MR imaging. Radiology 2018;286:676–84.
135. Esses SJ, Lu X, Zhao T, Shanbhogue K, Dane B, Bruno M, et al. Automated image quality evaluation of T2-weighted liver MRI utilizing deep learning architecture. J Magn Reson Imag 2018;47:723–8.
136. Department of Health. The ionising radiation (medical exposure) regulations; 2017. Available from: http://www.legislation.gov.uk/uksi/2017/1322/contents/made [Accessed 30 May 2022].
137. Kim DH, Wit H, Thurston M. Artificial intelligence in the diagnosis of Parkinson's disease from ioflupane-123 single photon emission computed tomography dopamine transporter scans using transfer learning. Nucl Med Commun 2018;39:887–93.
138. Choi H, Ha S, Im HJ, Paek SH, Lee DS. Refining diagnosis of Parkinson's disease with deep learning-based interpretation of dopamine transporter imaging. Neuroimage: Clinica 2017;16: 586–94.
139. Owoyemi A, Owoyemi J, Osiyemi A, Boyd A. Artificial intelligence for healthcare in Africa. Frontiers in Digital Health 2020;2:6.
140. Council of European Union.Regulation (EU) 2017/745 of the European parliament and of the council of 5 April 2017 on medical devices. Available from: https://eur-lex.europa.eu/legal-content/EN/TXT/PDF/?uri=CELEX:32017R0745 [Accessed 30 May 2022].
141. Lakhani P, Prater AB, Hutson RK, Andriole KP, Dreyer KJ, Morey J, et al. Machine learning in radiology: applications beyond image interpretation. J Am Coll Radiol 2018;15:350–9.
142. Holst C, Sukums F, Radovanovic D, Ngowi B, Noll J, Winkler AS. Sub-Saharan Africa—the new breeding ground for global digital health. The Lancet Digital Health 2020;2:e160–2.
143. United Nations. Resource guide on artificial intelligence (AI) strategies; June 2021. Available from: https://sdgs.un.org/sites/default/files/2021-06/Resource%20Guide%20on%20AI%20Strategies_June%202021.pdf [Accessed 30 May 2022].
144. NIH awards nearly $75M to catalyze data science research in Africa; 2021. Available from: https://www.nih.gov/news-events/news-releases/nih-awards-nearly-75m-catalyze-data-science-research-africa [Accessed 30 May 2022].
145. Parker RK, Mwachiro MM, Ranketi SS, Mogambi FC, Topazian HM, White RE. Curative surgery improves survival for colorectal cancer in rural Kenya. World J Surg 2020;44:30–6.

146. Sheikhtaheri A, Sadoughi F, Hashemi Dehaghi Z. Developing and using expert systems and neural networks in medicine: a review on benefits and challenges. J Med Syst 2014;38:1–6.
147. Committee on Data (CODATA). Data sharing principles in developing countries: the Nairobi data sharing principles. Paris, France: International Science Council; 2014.
148. World Health Organization. Electronic health records: manual for developing countries. Manila, the Philippines: WHO Regional Office for the Western Pacific; 2006.
149. Adair-Rohani H, Zukor K, Bonjour S, Wilburn S, Kuesel AC, Hebert R, et al. Limited electricity access in health facilities of sub-Saharan Africa: a systematic review of data on electricity access, sources, and reliability. Glob Health: Science and Practice 2013;1:249–61.
150. Odekunle FF, Odekunle RO, Shankar S. Why sub-Saharan Africa lags in electronic health record adoption and possible strategies to increase its adoption in this region. Int J Health Sci 2017;11: 59–64.
151. Frey CB, Osborne MA. The future of employment: how susceptible are jobs to computerisation? Technol Forecast Soc Change 2017;114:254–80.
152. World Bank Group. World development report 2016: digital dividends. Washington, D.C., US: World Bank Publications; 2016.
153. Lee KF. AI superpowers: China, Silicon Valley, and the new world order. Boston, MA, US: Houghton Mifflin Harcourt; 2018.
154. Autor DH, Levy F, Murnane RJ. The skill content of recent technological change: An empirical exploration. Q J Econ 2003;118:1279–333.
155. Naicker S, Eastwood JB, Plange-Rhule J, Tutt RC. Shortage of healthcare workers in sub-Saharan Africa: a nephrological perspective. Clin Nephrol 2010;74:S129–33.
156. Duvivier RJ, Burch VC, Boulet JR. A comparison of physician emigration from Africa to the United States of America between 2005 and 2015. Hum Resour Health 2017;15:1–2.
157. Ogbole GI, Adeyomoye AO, Badu-Peprah A, Mensah Y, Nzeh DA. Survey of magnetic resonance imaging availability in West Africa. The Pan African Medical Journal 2018;30:240.

Srimanta Maji* and Akshaya K. Sahu

4 Modeling, simulation and mixing time calculation of stirred tank for nanofluids using partially-averaged Navier–Stokes (PANS) $k_u - \epsilon_u$ turbulence model

Abstract: The present study deals with the numerical simulation of stirred tank in the presence of nanofluids to see the effect of different volume fraction (ϕ) of nanoparticles on the behaviour of flow characteristics and for the calculation of mixing time in the entire tank. The flow is assumed to be steady, axisymmetric, two dimensional and incompressible. For the simulation of flow inside the vessel, partially-averaged Navier–Stokes (PANS) $k_u - \epsilon_u$ turbulence model is used. Control volume method has been taken to descretize the governing equation along with power-law schemes. Further, semi-implicit method for pressure-linked equations revised (SIMPLER) algorithm and line-by-line tri-diagonal matrix algorithm (TDMA) have been taken to obtain the solution. The objective is to investigate the influence of ϕ on the characteristic flow variables and to calculate mixing time for different ϕ of nanofluids and for different values of f_k, PANS model parameter. It is noted that with the increase in ϕ, mixing time has also been increased and it increases very fast for PANS $k_u - \epsilon_u$ model with the increase in filter width f_k.

Keywords: mixing time calculation; nanofluids; PANS $k_u - \epsilon_u$ model; stirred tank simulation.

4.1 Introduction

The mixing of fluids at micro- and nano-scale are the most important operations in the development of wastewater treatment, pharmaceutical, chemical and other industrial processes and mixing time is one of the most efficient parameter for characterizing the mixing of fluids in agitated tanks. Many technologies and different techniques (conductivity method [1,2], thermocouple method [3], etc.) have been used for calculating the mixing time in agitated tank for different liquids. However, these techniques are for calculating local mixing time. These types of measurements are varying in

***Corresponding author: Srimanta Maji,** Department of Mathematics, Institute of Chemical Technology, Marathwada, Jalna, India, E-mail: s.maji@marj.ictmumbai.edu.in
Akshaya K. Sahu, Department of Mathematics, Institute of Chemical Technology, Mumbai, India, E-mail: ak.sahu@ictmumbai.edu.in

As per De Gruyter's policy this article has previously been published in the journal Physical Sciences Reviews. Please cite as: S. Maji and A. K. Sahu "Modeling, simulation and mixing time calculation of stirred tank for nanofluids using partially-averaged Navier–Stokes (PANS) $k_u - \epsilon_u$ turbulence model" *Physical Sciences Reviews* [Online] 2022. DOI: 10.1515/psr-2022-0126 | https://doi.org/10.1515/9783110913361-004

different position and can not be used for the calculation of mixing time for the whole tank. With the rapid development in the field of computational fluid dynamics (CFD), mixing processes in agitated tanks can be calculated globally [4]. The most commonly used turbulence models for simulation and calculation the mixing time inside the stirred vessel are the Reynolds- average Naiver–Stokes (RANS) models (standard $k-\epsilon$, RNG $k-\epsilon$ and Reynolds stress (RSM) models, etc. [5] due to its less computational cost. In recent years, CFD turbulent flow simulation for nanofluids has been used for different types of problems [6–10]. However, the study of simulation of stirred tank and mixing time calculation for nanofluids are not found in the literature till now and experimental data are not available for stirred tank simulation for nanofluids in which we can compare the predicted results for the present study.

Therefore, a theoretical study has been made on the simulation of axial type flow in cylindrical baffled stirred vessel to analyze the flow characteristics for nanofluids and to calculate mixing time inside the tank using PANS $k_u-\epsilon_u$ turbulence model. In the next section, the mathematical formulation for the turbulence model which we have proposed for nanofluids has been discussed. Section 4.3 describes the boundary conditions for the modeled equations. In the next section, numerical method has been discussed for the present problem. Section 4.5 includes the results and discussion and the conclusions of the present work have been described in Section 4.6.

4.2 Mathematical formulation

In the study of turbulent flow simulation in the stirred tank, nanofluid has been taken in place of conventional fluid with the assumption that (i) nanofluids are regarded as Newtonian fluids, (ii) the based fluid contains uniformly sized and shaped nanoparticles that are evenly distributed, (iii) nano size particles and the base fluid are thought to exist in thermal equilibrium. The physical system which is taken for the study contains a cylindrical vessel with four equispaced baffles of width T/10 (T is diameter of the vessel with T = 500 mm) which are placed equally around the perimeter along with pitch bladed downflow turbine (PTD) as impeller which is of axial type having diameter 167 mm. The location of impeller is at H/3 from the bottom of vessel, H is the height of the liquid (H = T) and the center of the impeller plane is taken as origin of the physical system (Figure 4.1) [11]. In this study, the flow is assumed to be steady, axisymmetric, two dimensional and incompressible. Then, the non-dimensional form of the governing equations of turbulent flow for nanofluids can be expressed in the following compact mathematical form as:

$$\frac{1}{r}\frac{\partial}{\partial r}(ur\Theta)+\frac{\partial}{\partial z}(v\Theta)=\frac{1}{r}\frac{\partial}{\partial r}\left[r(\Gamma_{\text{eff}})_{nf}\frac{\partial\Theta}{\partial r}\right]+\frac{\partial}{\partial z}\left[(\Gamma_{\text{eff}})_{nf}\frac{\partial\Theta}{\partial z}\right]+(S)_{nf}^{\Theta} \tag{4.1}$$

where, Θ represents a general notation for the flow variables $(u,\ v,\ k_u,\ \epsilon_u)$. The expression for $(\Gamma_{\text{eff}})_{nf}$ and $(S)_{nf}^{\Theta}$ are given in Table 4.1. Here, the variables u, v, p, k_u, ϵ_u, r, and z are non-dimensional and are given as: $u=\frac{\overline{U}}{U_{\text{tip}}}$, $v=\frac{\overline{V}}{U_{\text{tip}}}$, $p=\frac{P}{\rho U_{\text{tip}}^2}$, $k_u=\frac{k}{U_{\text{tip}}^2}$,

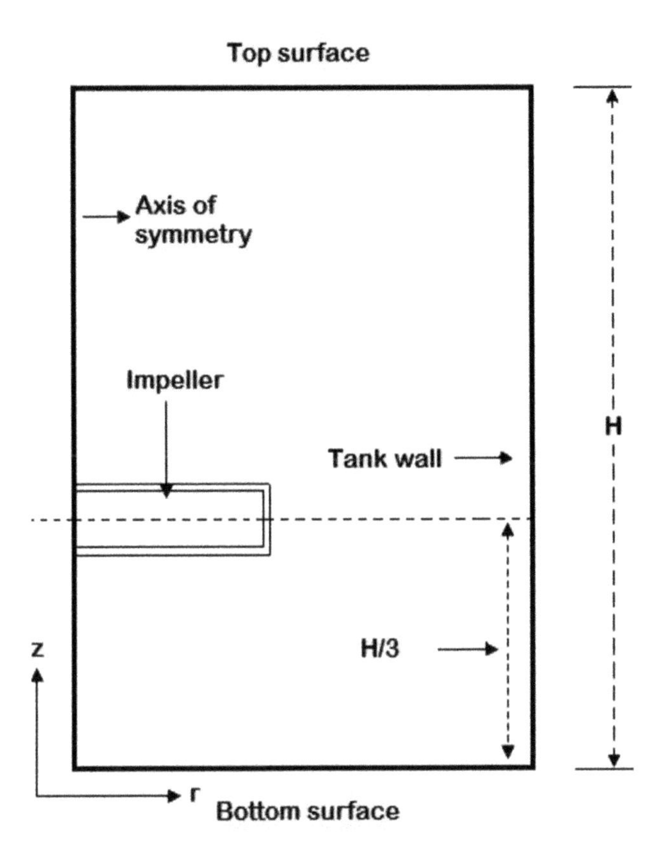

Figure 4.1: Solution domain for the present work.

$\epsilon_u = \frac{\epsilon R}{U_{\text{tip}}^3}$, $r = \frac{\bar{r}}{R}$ and $z = \frac{\bar{z}}{R}$, where U_{tip} is the velocity of impeller tip and R is the radius of the cylindrical vessel.

The equation (4.1) with Table 4.1 represents the governing equations for all the flow characteristic for nanofluids. Further, the equation for non-dimensional concentration

Table 4.1: Viscosity $(\Gamma_{\text{eff}})_{nf}$ and source terms S_{nf}^{Θ} for nanofluids for equation (4.1).

Θ	$(\Gamma_{\text{eff}})_{nf}$	S_{nf}^{Θ}
1	0	0
U	$\mu_{nf} + \mu_t$	$C_\mu\left[\frac{1}{r}\frac{\partial}{\partial r}\left(r\Gamma_{\text{eff}}\right)_{nf}\frac{\partial u}{\partial r}\right) + \frac{\partial}{\partial z}\left(\Gamma_{\text{eff}}\right)_{nf}\frac{\partial v}{\partial r}\right)\Big] - \frac{1}{\Phi}\left(\frac{\partial p}{\partial r} + \frac{2}{3}\frac{\partial k_u}{\partial r}\right)$
V	$\mu_{nf} + \mu_t$	$C_\mu\left[\frac{1}{r}\frac{\partial}{\partial r}\left(r\Gamma_{\text{eff}}\right)_{nf}\frac{\partial u}{\partial z}\right) + \frac{\partial}{\partial z}\left(\Gamma_{\text{eff}}\right)_{nf}\frac{\partial v}{\partial z}\right)\Big] - \frac{1}{\Phi}\left(\frac{\partial p}{\partial z} + \frac{2}{3}\frac{\partial k_u}{\partial z}\right)$
k_u	$\mu_{nf} + \frac{\mu_t}{\sigma_{k_u}}$	$G - \frac{1}{\Phi}\epsilon_u$
ϵ_u	$\mu_{nf} + \frac{\mu_t}{\sigma_{\epsilon_u}}$	$\frac{\epsilon_u}{k_u}\left(C_{1\epsilon}^{*}G - \frac{1}{\Phi}C_{2\epsilon}^{*}\epsilon_u\right)$

$G = \frac{1}{\Phi}C_\mu\mu_t\left[2\left[\left(\frac{\partial u}{\partial r}\right)^2 + \left(\frac{u}{r}\right)^2 + \left(\frac{\partial v}{\partial z}\right)^2\right] + \left(\frac{\partial u}{\partial z} + \frac{\partial v}{\partial r}\right)^2\right]$ $\mu_{nf} = \frac{\bar{\mu}_f}{C_\mu(1-\phi)^{2.5}Re}$ $Re = \frac{\rho_f U_{\text{tip}} R}{\mu_f}$, $\mu_t = \frac{1}{\Phi}\frac{k_u^2}{\epsilon_u}$ where $\Phi = (1 - \phi) + \phi\frac{\rho_s}{\rho_f}$.

has been incorporated for calculating the mixing time for stirred tank is given as follows,

$$\frac{\partial C}{\partial t}+\frac{1}{r}\frac{\partial}{\partial r}(urC)+\frac{\partial}{\partial z}(vC)=\frac{1}{r}\frac{\partial}{\partial r}\left[r(\Gamma_{\text{eff}})_{nf}\frac{\partial C}{\partial r}\right]+\frac{\partial}{\partial z}\left[(\Gamma_{\text{eff}})_{nf}\frac{\partial C}{\partial z}\right] \tag{4.2}$$

where, the non-dimensional concentration, $C=\frac{\overline{C}}{\overline{C}_{\text{avg}}}$, $\overline{C}_{\text{avg}}$ being the initial average concentration value and non-dimensional time, $t=\frac{\overline{t}}{R/U_{\text{tip}}}$. Here, $(\Gamma_{\text{eff}})_{nf}$ has been defined as mentioned in Table 4.1 and the source term $(S^{\Theta})_{nf}=0$.

After obtaining the converged solutions for all the flow characteristics related to equation (4.1), the concentration equation has been solved by using these results.

4.3 Boundary conditions

The boundary conditions for this physical problem have been defined as:

At $z=0$; $u=0$, $v=0$, $k_u=0$, $\epsilon_u=0$, $\partial C/\partial z=0$.

At $z=2$; $v=0$, $\partial u/\partial z=0$, $\partial k_u/\partial z=0$, $\partial \epsilon_u/\partial z=0$, $\partial C/\partial z=0$.

At $r=0$; $u=0$, $\partial v/\partial r=0$, $\partial k_u/\partial r=0$, $\partial \epsilon_u/\partial r=0$, $\partial C/\partial r=0$.

At $r=1$; $u=0$, $v=0$, $k_u=0$, $\epsilon_u=0$, $\partial C/\partial r=0$.

For numerical computation, boundary condition for no slip in the turbulent region are replaced by logarithmic velocity profile. The experimental data for u, v and k_u in the impeller region are given in the location of 30 mm below the impeller. Local dissipation rate in the region are calculated by the formula $\epsilon_u=\frac{1}{\Phi}C_\mu^{\frac{3}{4}}k_u^{\frac{3}{2}}/0.07l$ (l is the diameter of the vessel). The dissipation rate in the tank is obtained by $\epsilon_u=C_\mu^{\frac{3}{4}}k_u^{\frac{3}{2}}/\Phi\kappa y_p$. ($y_p$ stands for the shortest distance between the wall and nearest node point).

4.4 Numerical method

The governing equation (4.1) has been discretized by using control volume method. In the study [11], we compared the results for both upwind and power-law schemes, it showed that the prediction for power law scheme is better. Therefore in this study, we have discretized the convective part using power law scheme and used central difference scheme of second order for discretization of diffusive term. Then, a set of algebraic equations has been obtained in terms of Θ for all the flow variables as follows:

$$a_P\,\Theta_P=a_E\,\Theta_E+a_W\,\Theta_W+a_N\,\Theta_N+a_S\,\Theta_S+\overline{S}_{nf}^{\Theta} \tag{4.3}$$

here,

$$a_P=a_W+a_E+a_N+a_S+(F_n-F_s)+(F_e-F_w) \tag{4.4}$$

$$\begin{aligned}
a_E &= D_e A(|P_e|) + \max[-F_e, 0], \text{ and } P_e = \frac{F_e}{D_e} \\
a_W &= D_w A(|P_w|) + \max[F_w, 0] \text{ and } P_w = \frac{F_w}{D_w} \\
a_N &= D_n A(|P_n|) + \max[-F_n, 0] \text{ and } P_n = \frac{F_n}{D_n} \\
a_S &= D_s A(|P_s|) + \max[F_s, 0] \text{ and } P_s = \frac{F_s}{D_s}
\end{aligned} \tag{4.5}$$

where,

$$\begin{aligned}
F_e &= u_e r_e \Delta z \text{ and } D_e = (C_\mu)_{nf} \frac{r_e \Gamma_e \Delta z}{\Delta r} \\
F_w &= u_w r_e \Delta z \text{ and } D_w = (C_\mu)_{nf} \frac{r_w \Gamma_w \Delta z}{\Delta r} \\
F_n &= v_n r_P \Delta r \text{ and } D_n = (C_\mu)_{nf} \frac{r_P \Gamma_n \Delta r}{\Delta z} \\
F_s &= v_s r_P \Delta r \text{ and } D_s = (C_\mu)_{nf} \frac{r_P \Gamma_s \Delta r}{\Delta z}
\end{aligned} \tag{4.6}$$

and

$$\overline{S_{nf}^{\Theta}} = \left(S_{nf}^{\Theta}\right)_P r_P \Delta r \Delta z \tag{4.7}$$

The term S_{nf}^{Θ} is representing the source term for all the flow charecteristics which has been obtained from Table 4.2. The discretised source term $\overline{S_{nf}^{\Theta}}$ can be written as in equation (4.7). Further to obtain the matrix to be diagonally dominant matrix, the equation (4.7) can be written as

$$\overline{S_{nf}^{\Theta}} = (C_\mu)_{nf} \; (S_C^{\Theta} + S_P^{\Theta} \Theta_P) r_P \Delta r \Delta z \tag{4.8}$$

Table 4.2: Mixing time for different volume fraction ϕ for Ag-water nanofluid by PANS $k_u - \epsilon_u$ model.

	Mixing time		
Φ	**PANS $k_u - \epsilon_u$ model**	f_k	**PANS $k_u - \epsilon_u$ model**
0.0	79.058	0.2	12.03006
0.02	82.7093	0.5	27.140526
0.05	87.9603	0.8	62.29555
0.08	92.28130	1.0	78.2483006
0.1	93.5315		

where S_C term is positive and S_p is negative. As the variables k, ϵ, k_u and ϵ_u are positive, the source terms are positive for them. Here,

$$(C_\mu)_{nf} = \frac{C_\mu}{(1-\phi)^{2.5}\left((1-\phi)+\phi\frac{\rho_s}{\rho_f}\right)}$$

The equation (4.3) can be written in generalised form for all grid points as

$$\begin{aligned}(\overline{a_P})_{ij}(\Theta_P)_{ij} = {} & (a_E)_{ij}(\Theta_E)_{i+1j} + (a_W)_{ij}(\Theta_W)_{i-1j} + (a_N)_{ij}(\Theta_E)_{ij+1} \\ & + (a_S)_{ij}(\Theta_W)_{ij-1} + (C_\mu)_{nf}(S_C)_{ij} r_P \Delta r \Delta z\end{aligned} \tag{4.9}$$

where $(\overline{a_P})_{ij} = (a_P)_{ij} + (C_\mu)_{nf}(S_P)_{ij} r_P \Delta r \Delta z$. Further, to satisfy the continuity equation, we have taken SIMPLER algorithm [12] to obtain both pressure and pressure correction equation. It is clear from table (4.1) that for $\Theta = 1$, $\Gamma_{\text{eff}} = 0$ and $\overline{S}^{\Theta}_{nf} = 0$ and it represents the equation of continuity. The discretized continuity equation relative to pressure control volume can be written as

$$[(r_e u_e - r_w u_w)\Delta z + (v_n - v_s) r_P \Delta r] = 0 \tag{4.10}$$

According to the SIMPLER algorithm, we start the process by guessing the velocity fields u, v and then calculate the coefficients for momentum equations. Hence to calculate u_e, v_n by the formula,

$$u_e = \frac{\sum a_{nb} u_{nb} + \overline{S}^{u}_{nf}}{a_e} + d_e(p_P - p_E) \tag{4.11}$$

$$v_n = \frac{\sum a_{nb} v_{nb} + \overline{S}^{v}_{nf}}{a_n} + d_n(p_P - p_N) \tag{4.12}$$

The pseudo-velocities $\hat{u}_e$ and $\hat{v}_n$ has been defined as

$$\hat{u}_e = \frac{\sum a_{nb} u_{nb} + \overline{S}^{u}_{nf}}{a_e}, \hat{v}_n = \frac{\sum a_{nb} v_{nb} + \overline{S}^{v}_{nf}}{a_n}$$

to compose the neighbour velocities u_{nb} and contains no pressure term. The equation (4.11) and (4.12) can be written as,

$$u_e = \hat{u}_e + d_e(p_P - p_E) \tag{4.13}$$

$$v_n = \hat{v}_n + d_n(p_P - p_N) \tag{4.14}$$

Now, substituting the values of u_e, u_w, v_e and v_n in the discretised continuity equation (4.10), the pressure equation can be written as

$$A_P p_P = A_E p_E + A_W p_W + A_N p_N + A_S p_S + b \tag{4.15}$$

where,

$$
\begin{aligned}
A_E &= r_e d_e \Delta z \text{ and } d_e = \frac{r_P \Delta z}{A_P} \\
A_W &= r_w d_w \Delta z \text{ and } d_w = \frac{r_P \Delta z}{A_P} \\
A_N &= r_n d_n \Delta r \text{ and } d_n = \frac{r_P \Delta r}{A_P} \\
A_S &= r_s d_s \Delta r \text{ and } d_s = \frac{r_P \Delta r}{A_P}
\end{aligned} \tag{4.16}
$$

where,

$$
b = \left(r_w \widehat{u}_w - r_e \widehat{u}_e \right) \Delta z + \left(\widehat{v}_s - \widehat{v}_n \right) r_P \Delta r \tag{4.17}
$$

By treating the pressure field p as p^*, we obtain u^* and v^* to solve the momentum equations.

$$
a_e u_e^* = \sum a_{nb} u_{nb}^* + b + (p_P^* - p_E^*) A_e \tag{4.18}
$$

$$
a_n v_n^* = \sum a_{nb} v_{nb}^* + b + (p_P^* - p_N^*) A_n \tag{4.19}
$$

After discretizing the continuity equation we have got the equation for pressure correction p' as follows,

$$
A_P p_P' = A_E p_E' + A_W p_W' + A_N p_N' + A_S p_S' + b_1 \tag{4.20}
$$

where, A_E, A_W, A_N and A_S are calculate similar to the equations (4.10) and mass residue term

$$
b_1 = (r_w u_w^* - r_e u_e^*) \Delta z + (v_s^* - v_n^*) r_P \Delta r \tag{4.21}
$$

Then correct the velocity fields by the formula

$$
u_e = u_e^* + d_e (p_P' - p_E') \tag{4.22}
$$

$$
v_n = v_n^* + d_n (p_P' - p_N') \tag{4.23}
$$

And then all the variables u, v, k_u and ϵ_u are calculated by using the equation (4.9). To solve the set of algebraic equations line-by-line TDMA iterative method has been applied and the iterative process terminates when the mass residue exceeds 10^{-6} on each control volume. After solving the equations for (u, v, k_u, ϵ_u), we have computed the concentration equation. Hence, the discretised concentration equation can be represented as,

$$
\bar{a}_P C_P^{(\bar{t}+1)} = a_E C_E^{(\bar{t}+1)} + a_W C_W^{(\bar{t}+1)} + a_N C_N^{(\bar{t}+1)} + a_S C_S^{(\bar{t}+1)} + S_m^{(\bar{t})} \tag{4.24}
$$

here, $a_P = a_W + a_E + a_N + a_S + (F_n - F_s) + (F_e - F_w) + r_P \Delta r \Delta z / \Delta t$ and a_E, a_W, a_N, a_W are given by the equations (4.4) and (4.5) at time t. The source term $S_m^{(\bar{t})}$ is given by,

$$S_m^{(\bar{t})} = \frac{C_P{}^{\bar{t}} r_P \Delta r \Delta z}{\Delta t}$$

For the calculation of mixing time, at $t = 0$, a tracer input has been given in the vessel to see the evolution of concentration and it has been calculated by solving the concentration equation for turbulent flow and the mixing time has been taken when the tracer concentration reached within 99% of the final tracer concentration.

4.5 Results and discussion

This study has been made for the simulation of baffled, cylindrical stirred tank to see the effect of different volume fraction ϕ of nanofluid on the flow variables and for the calculation of mixing time. *Ag*-water nanofluid have been taken for the study. Figure 4.1 is displayed the domain of simulation for the present work and for numerical simulation, PANS $k_u - \epsilon_u$ turbulence model is used. For numerical simulation, 61×85 non-uniform grid size has been considered for the current numerical calculations since it provides a grid-free solution (Figure 4.2a,b)). The objective of the present study is to investigate theoretically the effect of different ϕ of nanofluids on the flow variables throughout the domain as the experimental data are not available for nanofluids for

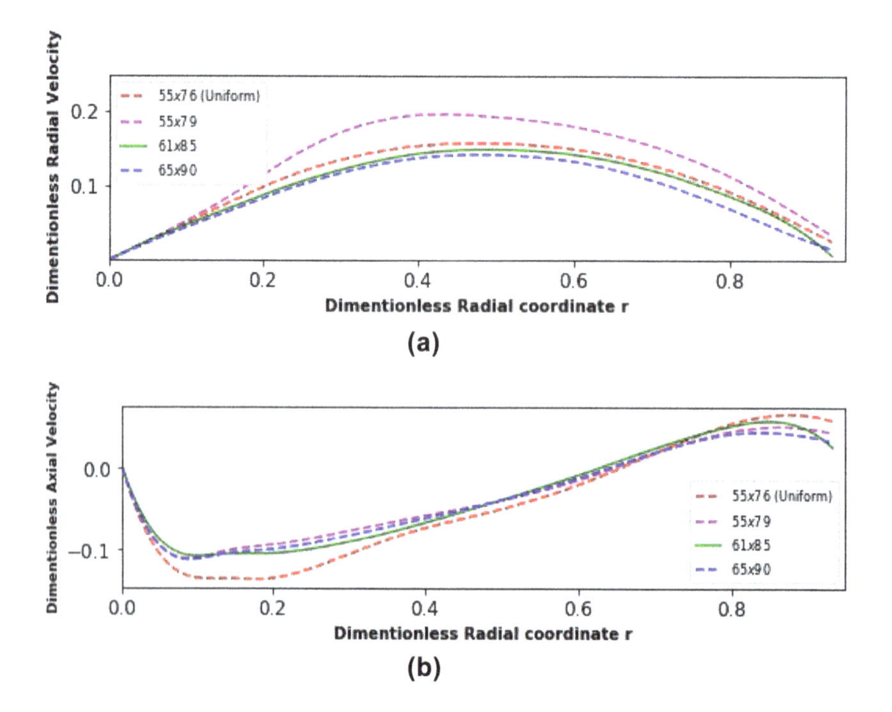

Figure 4.2: Grid independence test with different grid sizes for radial and axial velocity profiles.

simulating the stirred vessel. For the present simulation of stirred tank, PANS $k_u - \epsilon_u$ turbulence model is considered for a fixed value of $f_k = 1$. The numerical study of stirred tank simulation was made by Maji and Sahu [11] using PANS $k_u - \epsilon_u$ turbulence model for its various filter parameter f_k. Therefore, for a fixed f_k for PANS model, we have studied the effects of different ϕ of nanofluids on the flow characteristics. Further, mixing time calculations in the stirred vessel has been calculated using PANS $k_u - \epsilon_u$ model for different f_k. Also, for different ϕ mixing time has been calculated throughout the tank.

4.5.1 Radial velocity

Figure 4.3 presents the $k - \epsilon$ model predicted radial velocity profiles at two different axial level below the impeller location for different volume fraction ϕ of nanofluids. At $z = 0.18$, below the impeller, the radial velocity profiles for different ϕ has been represented by Figure 4.3a. It has been observed that near axis of symmetry the radial velocities are much weaker for all ϕ. This happens due to the flow simulation is of axial type. Then the velocities gradually increase and attend their maximum at $r = 0.45$ for all ϕ. Beyond $r = 0.45$ velocities are gradually decreased. It has been observed that for $\phi = 0.0$ velocity profile is much stiffer than other ϕ. At $z = 0.306$, just below the impeller, Figure 4.3b velocity profiles are not much stiffer for all ϕ as seen in Figure 4.2a. Except $\phi = 0.0$, the $k - \epsilon$ model predicted radial velocity profiles are almost same for all ϕ.

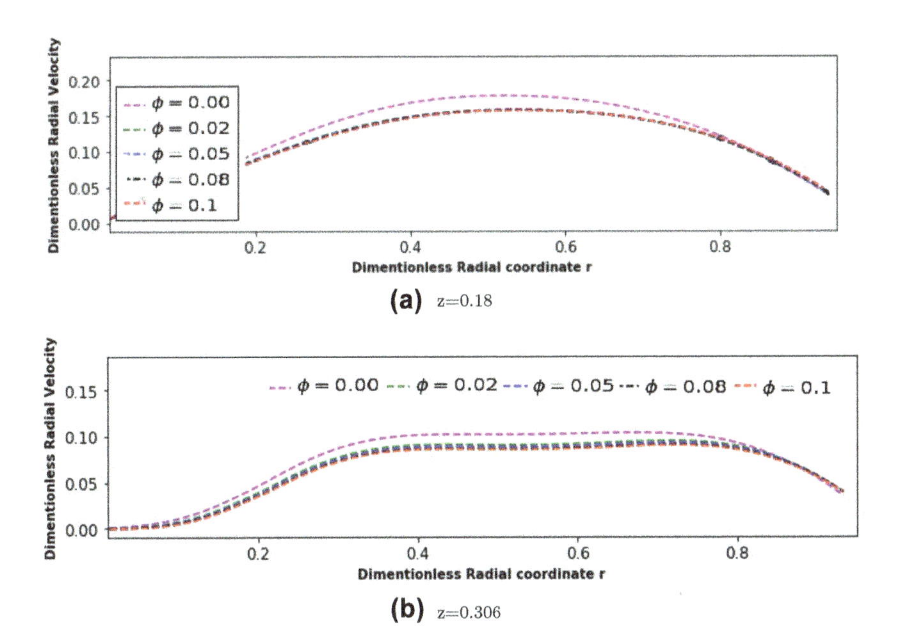

Figure 4.3: Radial velocity profiles for different ϕ at different z-location below the impeller.

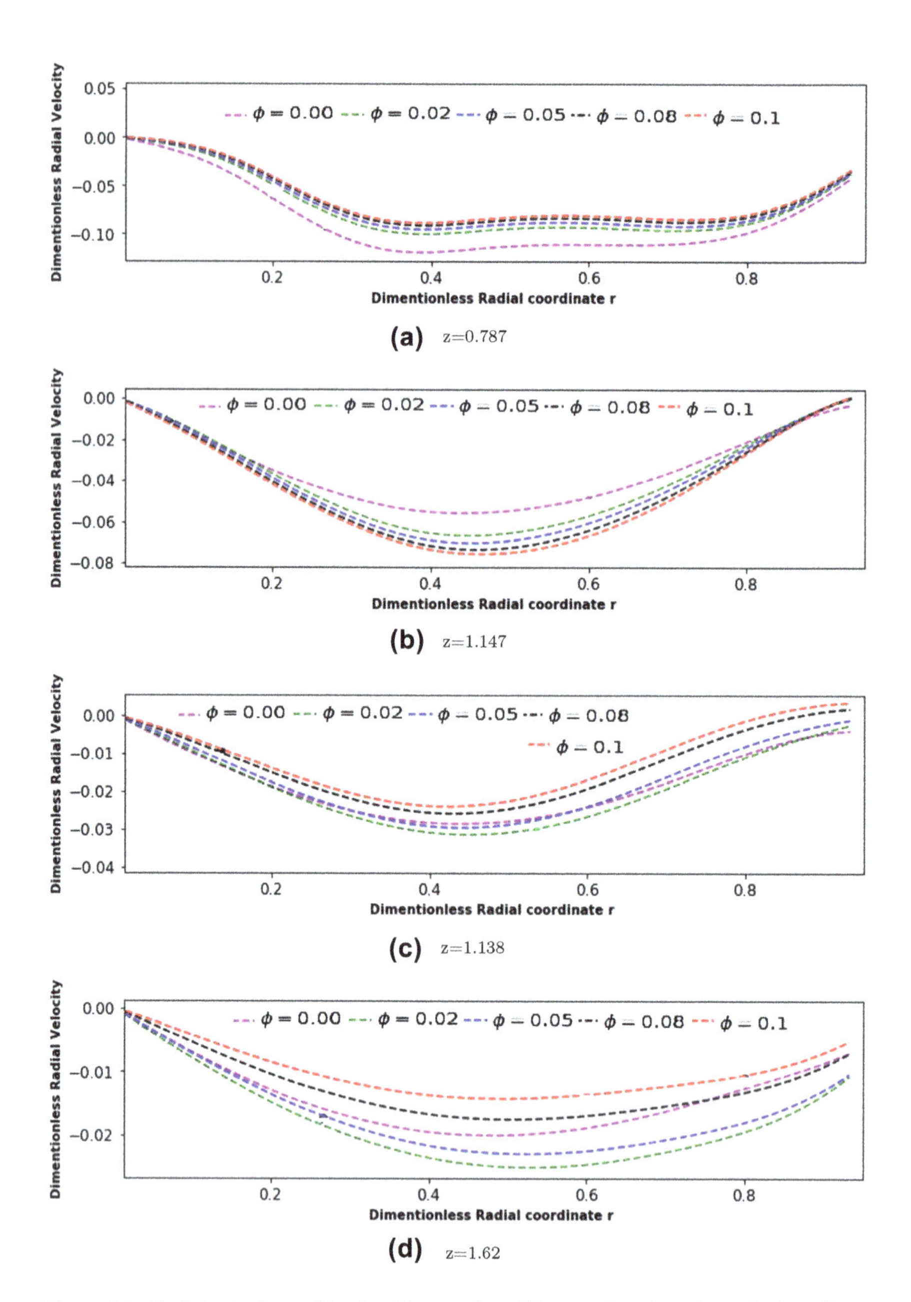

Figure 4.4: Radial velocity profiles for different ϕ at different z-location above the impeller.

Thus, below the impeller locations, there are no such variation in the radial velocity profiles with the increase in ϕ for nanofluids. The variation for different ϕ in radial velocity profiles have been observed above the impeller. Just above the impeller at $z = 0.787$, in the region $0.2 < r < 0.8$ (Figure 4.4a), and for the location $z = 1.147$, in the range

$0.2 < r < 0.8$ (Figure 4.4b), variations in the radial velocity profile observed for different ϕ. However, the deviation has been noted much for $\phi = 0.0$ from the other ϕ. Much away from the impeller, the variation in radial velocities have been observed throughout the radial direction for both the location $z = 1.38$ and $z = 1.62$ (Figure 4.4c,d)). For both the axial levels, the velocity profiles gradually increase from the axis of symmetry and then attend maximum in the neighbourhood of $r = 0.45$. It again decreases upto the near wall. Both the cases for $\phi = 0.02$ attends its maximum and $\phi = 0.1$ attends minimum near $r = 0.4$ for amongst all ϕ.

4.5.2 Axial velocity

The profile for axial velocities at different axial locations for the flow domain are plotted in the Figures 4.5 and 4.6. As the flow for axial flow impeller is downward type, thus, from the axis of symmetry, the velocity increases sharply downward and attends its maximum and then it decreases slowly. The same physical phenomena have been observed in our study also. In the Figure 4.5a,b), It's been noticed that the axial velocity profiles are much stiffer near the axis, then gradually decrease. For both the locations, the axial velocity attends its maximum near r = 0.2. Further, the axial velocity profile for the location z = 0.306 is much stiffer as compared to z = 0.18 as the flow is of axial type and z = 0.18 is nearer to the impeller. Then the velocity profiles changes its direction towards positive direction and becomes zero at $r = 0.8$. Beyond $r > 0.8$, it changes in the

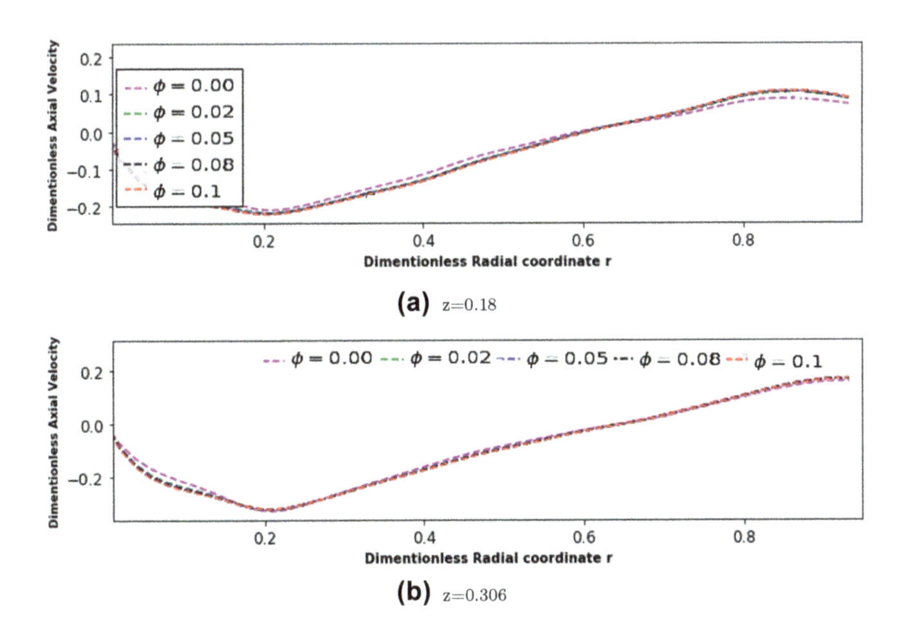

Figure 4.5: Axial velocity profiles for different ϕ at different z-location below the impeller.

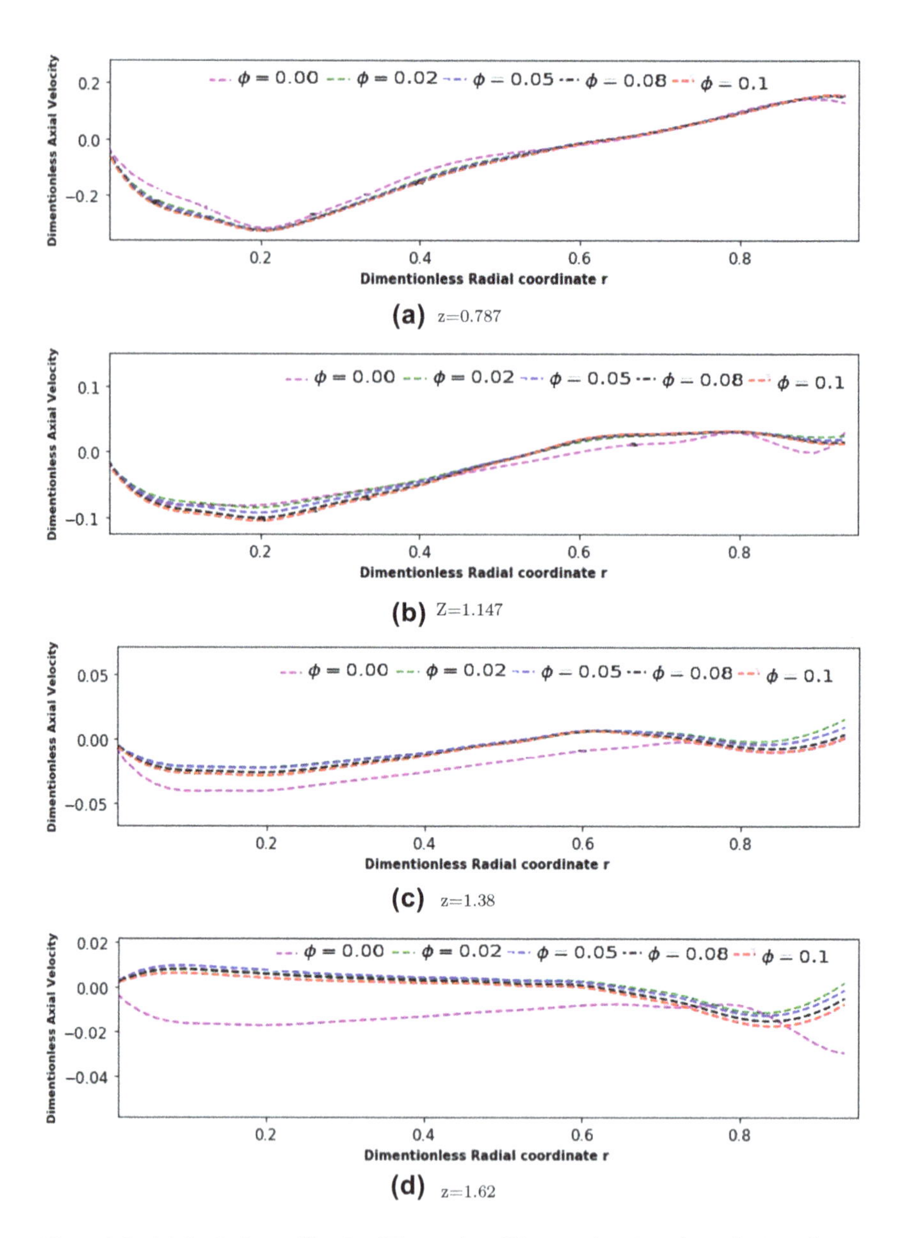

Figure 4.6: Axial velocity profiles for different ϕ at different z-location above the impeller.

positive direction for both the locations. It can be noted that the gradient are sharply visible between $0 < r < 0.2$ and $0.2 < r < 0.8$ in Figure 4.5b. However, there are not much variation noticed for the axial velocity profiles for different ϕ below the impeller locations. Upper the impeller position at z = 0.787 and z = 1.147, there are not so much variation has been noticed in the axial velocity profiles for different ϕ as shown in the

Figure 4.6a,b). However, the velocity profile is much flat at z = 1.147 and beyond $r > 0.45$ the variation in the axial velocity profile observes for $\phi = 0$ from the others. The variations in the velocity fields at the level of the axis $z = 1.147$ and $z = 1.38$ which are much away from the impeller have been displayed in the Figure 4.6c and Figure 4.6d. Here the velocity profiles are much flat and the variation in the axial velocity fields has been observed for $\phi = 0.0$ from other values of ϕ. This happens because near the free surface, the effect of the impeller on the velocity fields are much lesser.

4.5.3 Turbulent kinetic energy (TKE)

The predicted results of TKE for different volume fraction of nanoparticles ϕ of *Ag*-water nanofluid have been plotted in the Figure 4.7 and Figure 4.8 for different axial level. TKE prediction for different ϕ at $z = 0.18$ increases sharply from close to the axis upto $r = 0.2$ for each ϕ (Figure 4.7a). Then it decreases drastically beyond this point upto near wall. The prediction is maximum for $\phi = 0.1$ and minimum for $\phi = 0.0$ i.e., for conventional fluids. Therefore, it is clear from the predictions that with the increase in ϕ there is an increase in TKE. Further, it has been observed that the prediction for $\phi = 0$ is much lesser as compared to others $\phi(\neq 0)$, i.e., when the nanoparticles are included in the base fluid, the prediction for TKE increases. Hence it can be said that the TKE prediction increases with the increase in ϕ. Similar pattern for TKE prediction of $z = 0.18$

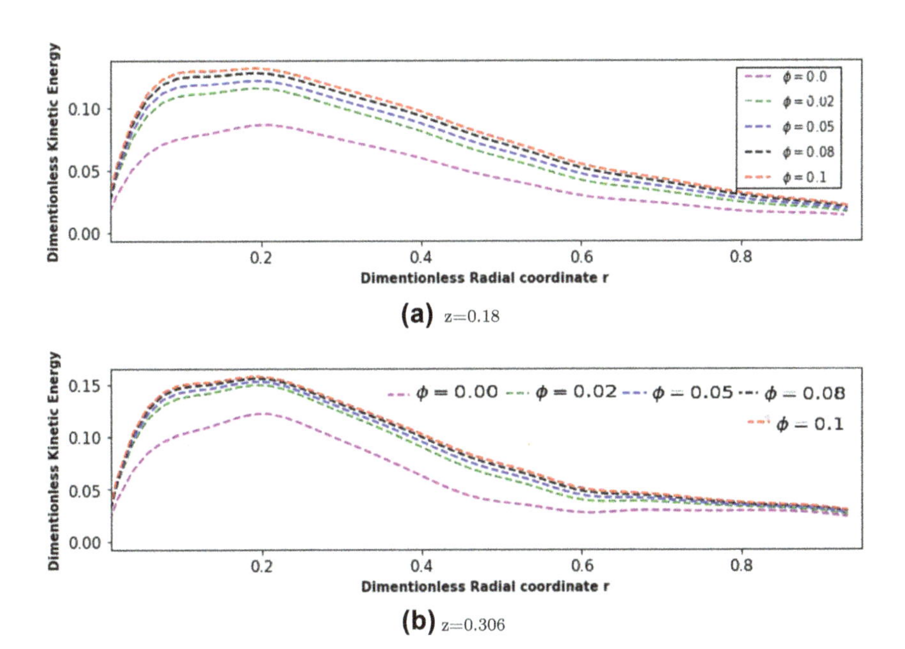

Figure 4.7: TKE profiles for different ϕ at some z-location below the impeller.

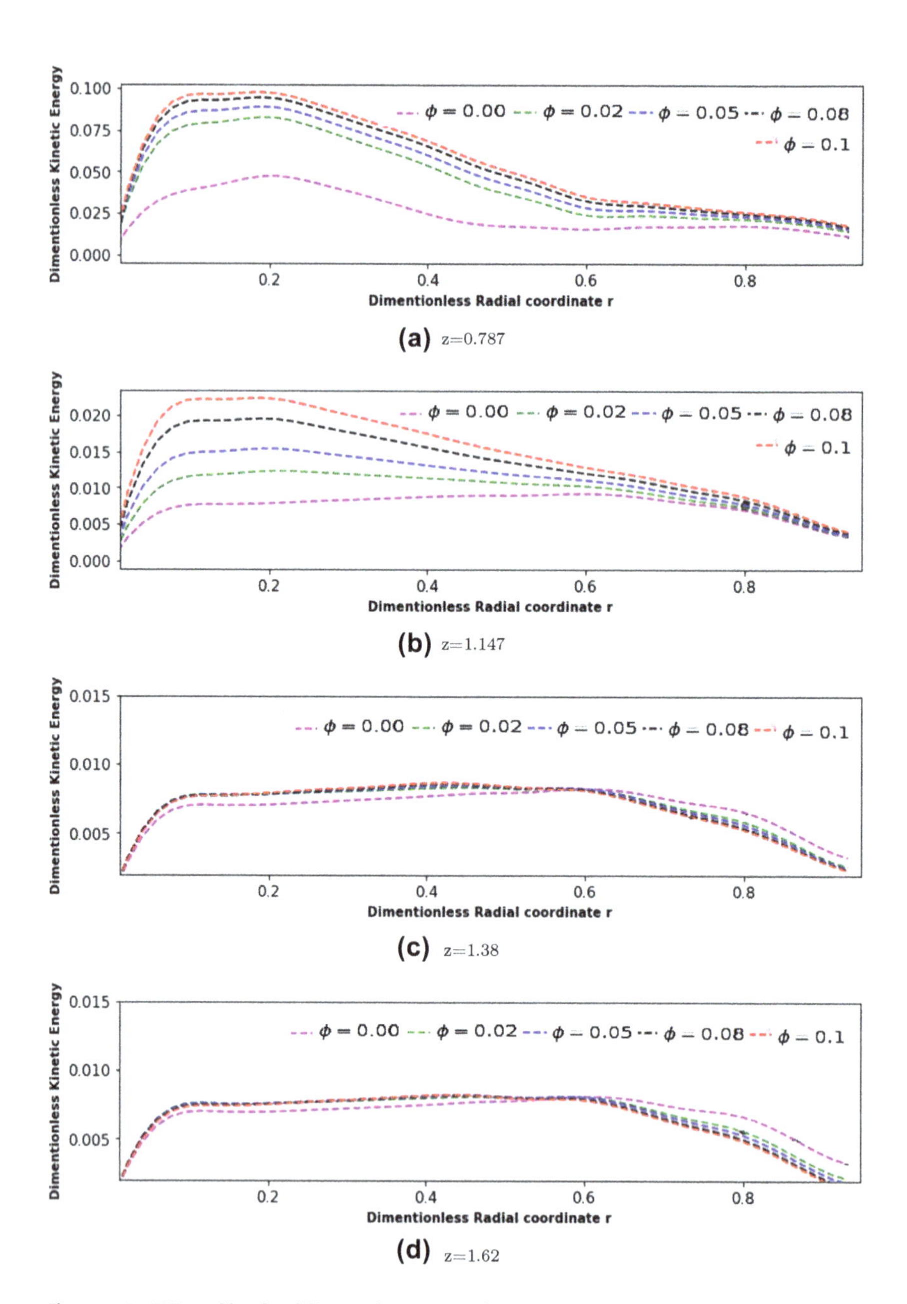

Figure 4.8: TKE profiles for different ϕ at some z-location above the impeller.

has been also seen at the level of the axis z = 0.306 (Figure 4.7b). Here the predictions are more for TKE for each ϕ. Further, the predictions for each $\phi(\neq 0)$ are more for TKE as compared to base fluid $\phi(=0)$ in the region $0.05 < r < 0.65$. This is because of the flow is of axial type and the location is near the impeller. At $z = 0.787$, just above the impeller,

Figure 4.8a gives the same pattern as above locations. However, the predictions for TKE for all ϕ's are less in this location and the prediction for $\phi = 0$ are much lesser than other $\phi(\neq 0)$. With the increase in distance from the impeller the prediction of TKE also reduces. At the location $z = 1.147$, the predictions of TKE for different ϕ are much lesser. The variation in the TKE profiles for different ϕ are much more. With the increase in ϕ, the TKE profile increase as shown in the Figure 4.8b. Much away and above at the level of the axis location at $z = 1.38$ (Figure 4.8c), the TKE predictions for all $\phi(\neq 0)$ are almost same throughout the radial direction. However, in the region $0.1 < r < 0.57$, the predicted TKE for $\phi = 0$ is lesser from the prediction of all $\phi(\neq 0)$ and beyond $r > 0.62$, i.e., predictions for TKE for $\phi = 0$ is higher from others ϕ. Almost same trends have been noticed at $z = 1.38$ (Figure 4.8d). However, the TKE predictions are little lesser at this location for all ϕ with the comparison to the location $z = 1.147$. Figure 4.9 has been displayed for the visualisation of flow field for different volume fraction ϕ of *Ag*-water nanofluid. There are no such impact is visible in the flow field for different ϕ as shown

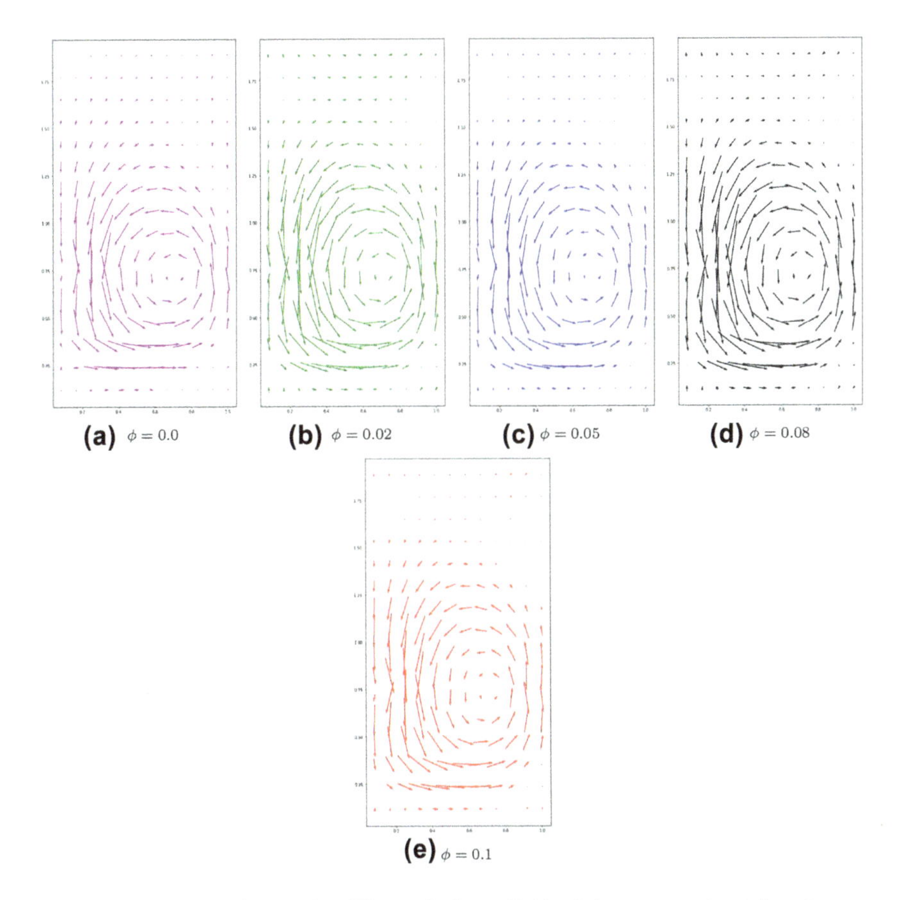

Figure 4.9: Flow visualisation for different ϕ of nanofluids of the computational domain.

in the Figure 4.9a–e. Further, it is observed that the flow is directed downwards after interaction with the impeller. Then, it reaches to the bottom wall and goes towards vertical wall. Thereafter, interaction with the vertical wall, the direction of flow fields become upward and near the surface, it changes its direction to downward due to loss of energy. Thus, the rotation of flow fields inside the stirred tank is of axial type and no significant effect of radial velocity has been observed in the impeller region as axial flow type impeller is used in the present study.

4.5.4 Mixing time calculation

The mixing time has been calculated for different ϕ of nanofluids. It has been observed that with increase in ϕ, the mixing time has also increased as shown in Table 4.2. For $\phi = 0$, there is an excellent agreement with the calculation of mixing time with the previous work [4]. Further, mixing time is calculated for various f_k for PANS $k_u - \epsilon_u$ model. It has been observed that with the increase in f_k, mixing time increases very fast for PANS $k_u - \epsilon_u$ model. This happens as the unresolved TKE increase with the increase in f_k (as $k_u = f_k\, k$). Hence, mixing time increases with increasing f_k.

4.6 Conclusions

In the study, a theoretical study has been made for the numerical simulation of stirred tank in the presence of nanofluid to see the behaviour of flow characteristics for different ϕ of nanofluids and then calculate the mixing time for different ϕ of nanofluids. Further, for different f_k of PANS $k_u - \epsilon_u$ model, mixing time has been calculated for the simulation of stirred vessel for nanofluids.

1. There is no significant variation has been observed in the radial velocity profiles below the impeller for different volume fraction ϕ of nanoparticle. Whereas, just above the impeller, in the radial direction upto $r > 0.4$ for $\phi = 0.02$ and away from the impeller for all ϕ throughout the radial location the variation in the radial velocity profiles has been observed.
2. Below and just above the impeller location, there is no such variation in the axial velocity profiles has been observed for different ϕ. Whereas, away from the impeller location, much variation in the axial velocity field has been observed for $\phi = 0$.
3. Sharp variation in the TKE profile has been observed through out the domain. However, near the impeller both below and above, the variations are more. Further it has been noticed that with increase in ϕ, the TKE profiles are also increased. Thus, with the inclusion of nanoparticles in the base fluid, the prediction for TKE increases. This imply that the TKE prediction increases with increase in ϕ.

4. In case of calculating the mixing time, it has been observed that with increase in ϕ, the mixing time has also increased. Further, with increase in f_k for PANS $k_u - \epsilon_u$ model, mixing time increases with increase in f_k.

References

1. Biggs RD. Mixing rates in stirred tanks. AIChE J 1963;9:636–40.
2. Raghav Rao KS, Joshi JB. Liquid phase mixing in mechanically agitated vessels. Chem Eng Commun 1988;74:1–25.
3. Rewatkar VB, Joshi JB. Effect of impeller design on liquid phase mixing in mechanically agitated reactors. Chem Eng Commun 1991;102:1–33.
4. Sahu AK, Kumar P, Patwardhan AW, Joshi JB. CFD modelling and mixing in stirred tanks. Chem Eng Sci 1999;54:2285–93.
5. Joshi JB, Nere NK, Rane CV, Murthy BN, Mathpati CS, Patwardhan AW, et al. CFD simulation of stirred tanks: comparison of turbulence models (Part II: axial flow impellers, multiple impellers and multiphase dispersions). Can J Chem Eng 2011;89:754–816.
6. Sheikholeslami M, Jafaryar M, Ali JA, Hamad SM, Divsalar A, Shafee A, et al. Simulation of turbulent flow of nanofluid due to existence of new effective turbulator involving entropy generation. J Mol Liq 2019;291:111283.
7. Davarnejad R, Jamshidzadeh M. CFD modeling of heat transfer performance of MgO-water nanofluid under turbulent flow. Eng Sci Technol Int J 2015;18:536–42.
8. Meriläinen A, Seppälä A, Saari K, Seitsonen J, Ruokolainen J, Puisto S, et al. Influence of particle size and shape on turbulent heat transfer characteristics and pressure losses in water-based nanofluids. Int J Heat Mass Tran 2013;61:439–48.
9. Boertz H, J. Baars A, Cieśliński JT, Smolen S. Numerical study of turbulent flow and heat transfer of nanofluids in pipes. Heat Tran Eng 2018;39:241–51.
10. Akbari OA, Toghraie D, Karimipour A. Numerical simulation of heat transfer and turbulent flow of water nanofluids copper oxide in rectangular microchannel with semi-attached rib. Adv Mech Eng 2016;8: 1687814016641016.
11. Maji S, Sahu AK. Stirred tank simulation using Partially-Averaged Navier-Stokes $k_u - \epsilon_u$ turbulence model. SN Appl Sci 2021;3:1–6.
12. Patankar SV. Numerical heat transfer and fluid flow. USA: Taylor & Francis; 1980.

Mayanglambam Maneeta Devi, Okram Mukherjee Singh* and Thokchom Prasanta Singh*

5 Synthesis of *N*-containing heterocycles in water

Abstract: An organic reaction with water as a medium has numerous benefits, like improvement in reactivities and selectivities, simple workup techniques, possibility of recycling the catalyst with milder reaction conditions and eco-friendly synthesis. Further, exploring of water as a reaction medium gives rise to unusual reactivities and selectivities, supplementing the organic chemist's necessity for reaction media. This review focus on the use of water for the synthesis of Nitrogen-containing heterocycles covering from 2011 to 2021.

Keywords: green chemistry; heterocycles; inexpensive; non-flammable; water.

5.1 Introduction

Nowadays, solvents are used in large quantities in the chemical industry as a reaction medium for producing per mass of final products. Thus, the choice of an appropriate solvent plays a key role from environment, welfare, economic, handling and separation technique of product points of view [1]. Keeping this concept in mind, water a potentially green and natural solvent possessing distinct physical and chemical properties along with the strong hydrogen bonding, and an extensive range of temperature to be in the liquid state [2]. The use of water as a reaction medium is an excellent option from the environment and economical perspective as it is a cheap, non-flammable and possibility of catalyst recycling with extensively accessible resources [3]. Thus, organic chemists are in constant hunt for new discoveries of biologically active compounds or replacement of organic solvents in various known reactions by water as it minimizes the release of toxic chemicals in the environment and easy workup process. Hence, numerous organic transformations have been carried out in water and this article will briefly review recent research progress concerning synthesizing various nitrogen-containing heterocycle compounds using water as a reaction medium.

***Corresponding authors: Okram Mukherjee Singh and Thokchom Prasanta Singh,** Chemistry Department, Manipur University, Canchipur-795003, Manipur, Imphal, India,
E-mail: ok_mukherjee@yahoo.co.in (O. M. Singh), prasantath@gmail.com (T. Prasanta Singh).
Mayanglambam Maneeta Devi, Chemistry Department, Manipur University, Canchipur-795003, Manipur, Imphal, India

As per De Gruyter's policy this article has previously been published in the journal Physical Sciences Reviews. Please cite as: M. M. Devi, O. M. Singh and T. Prasanta Singh "Synthesis of *N*-containing heterocycles in water" *Physical Sciences Reviews* [Online] 2022. DOI: 10.1515/psr-2022-0127 | https://doi.org/10.1515/9783110913361-005

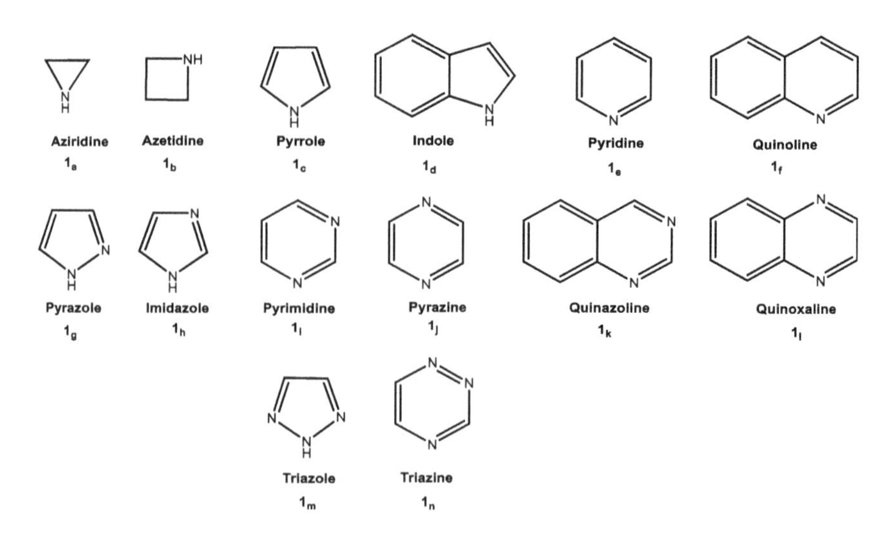

Figure 5.1: Different types of *N*-containing heterocycles.

5.2 Synthesis of *N*-containing heterocycles

The *N*-containing heterocycles are found in various alkaloids (both natural and synthetic) possessing wide ranges of biological properties. They are constituents of different bioactive molecules such as vitamins, antibiotics, nucleic acids, dyes and agrochemicals [4]. Therefore, synthesizing various *N*-containing heterocycles is of considerable interest, and motivated research is needed to develop new economic, efficient and selective synthetic strategies. Based on the number and position of a nitrogen atom(s) in three, four, five and six-membered rings of heterocycles (Figure 5.1), they are classified and discussed considering the recent literature reports to demonstrate the importance of water in organic synthesis and interpret the chemistry involves in the procedures.

5.2.1 Synthesis of aziridines

The synthesis of aziridines $\mathbf{1_a}$ through ring closure reaction initiated by the aza-Michael reaction using vinyl selenones **2** and 1° amines **3** in water at room temperature was established [5]. In this protocol, the aziridines synthesis in water was shorter in reaction duration as compared to other organic solvents (Scheme 5.1). The proposed reaction mechanism demonstrated that water plays both as an H-bond donor and acceptor with the **2** and **3**, respectively, thereby enhancing the electrophilic and nucleophilic characters to **2** and **3**. Further, the H-bond established among H_2O and the β-aminoselenone results in the cyclization step in this method.

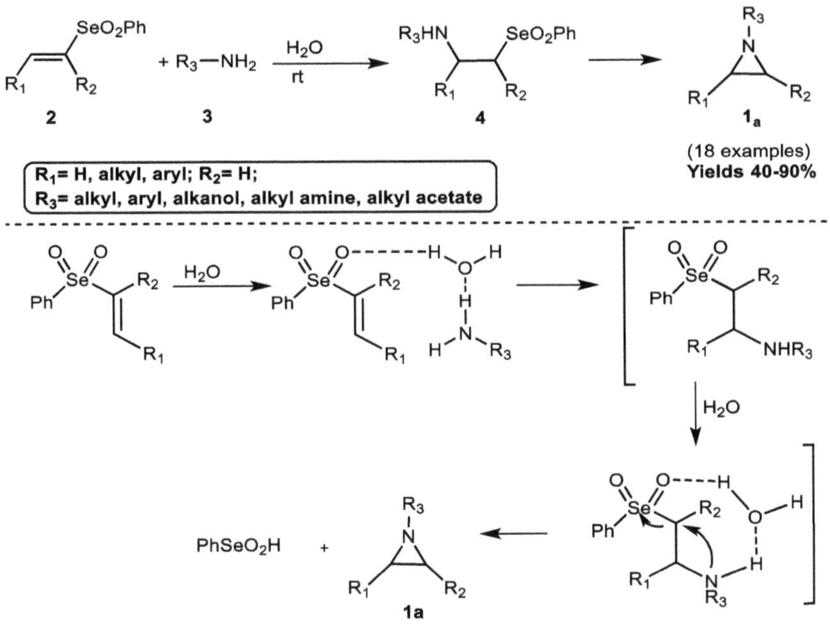

Scheme 5.1: The synthesis of aziridines in water.

The synthesis of the tosylaziridines 1_a using graphene oxide (GO) in the presence of I_2 in water was described utilizing *N*-(*p*-toluenesulfonyl)imino]phenyliodinane (PhI = NTs) **6** as the nitrene source [6]. The process consists of nitrene insertion into various styrenes **5**, and the substituents in the benzene ring have minimal effect on product outcome at room temperature. However, the substituents at the *p*-position of the benzene ring give the aziridines 1_a in good yields. This reaction has many advantages compare to the reported methods for synthesizing activated aziridines, such as green protocol and easy setup of reaction environment with higher yields of products (Scheme 5.2).

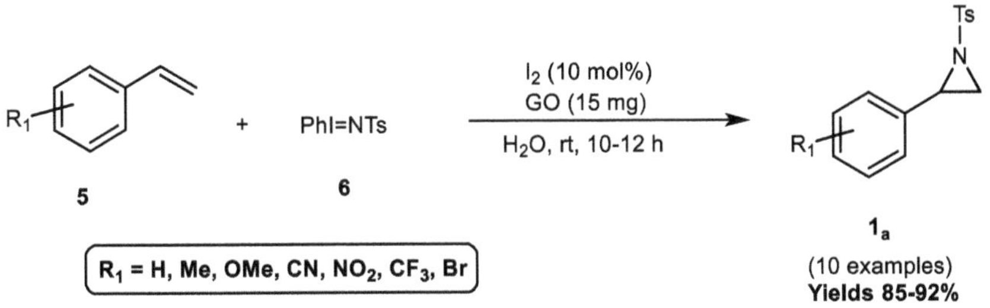

Scheme 5.2: The synthesis of the tosylaziridines from olefins.

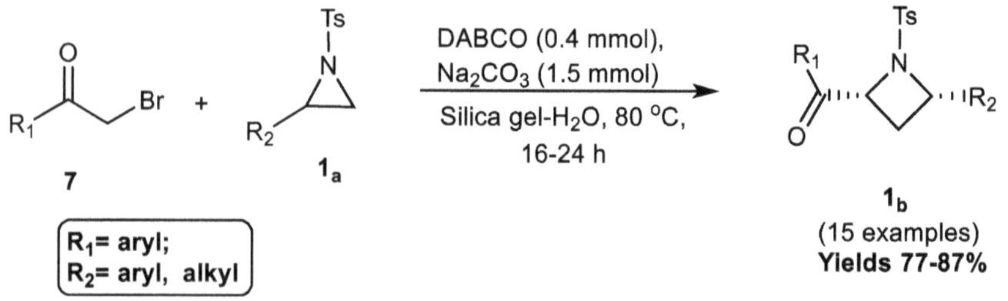

R₁= aryl; R₂= aryl, alkyl

DABCO (0.4 mmol), Na_2CO_3 (1.5 mmol)

Silica gel-H_2O, 80 °C, 16-24 h

1_b (15 examples) **Yields 77-87%**

Scheme 5.3: Synthesis of azetidines from aziridines.

5.2.2 Synthesis of azetidines

The synthesis of tosylazetidines $\mathbf{1_b}$ from *N*-tosylaziridines $\mathbf{1_a}$ via ring expansion using 1,4-diazabicyclo[2.2.2]octane (DABCO) with nitrogen ylides formed *in situ* from phenacyl bromides **7** in a silica gel-H_2O system was established [7]. This reaction could afford biologically active azetidines $\mathbf{1_b}$ in good yields through greener approaches with *cis*-stereoselectivities in a one-pot fashion (Scheme 5.3).

5.2.3 Synthesis of pyrroles

The exploration of an efficient synthetic method for the trisubstituted pyrroles $\mathbf{1_c}$ from enaminones **8** and *β*-bromonitrostyrenes **5'** was successfully demonstrated in water [8]. In this method, both *β*-bromonitrostyrenes and enaminones with diverse substitution patterns could be utilized providing the pyrroles in high yields (Scheme 5.4).

Green synthesis of the pyrroles $\mathbf{1_c}$ *via* the Paal-Knorr protocol using poly(ethylene glycol)-confined sulfonic acid (PEG-SO_3H) in water was demonstrated [9], in which the amines **3** condensed with 2,5-hexadione **9** at ambient temperature (Scheme 5.5). The reactions proceeded well enough with various amines having different substituents,

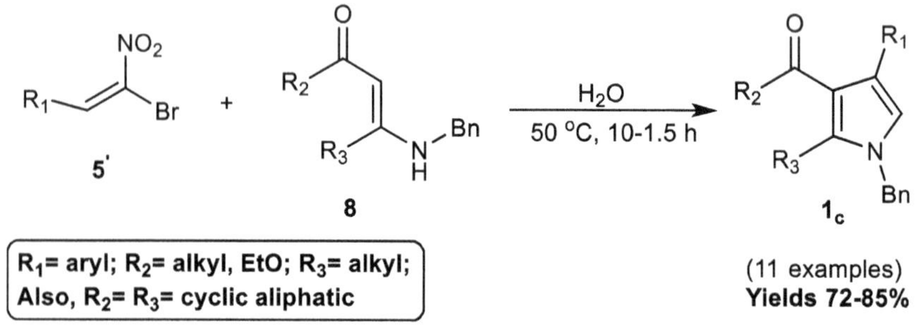

R_1= aryl; R_2= alkyl, EtO; R_3= alkyl; Also, R_2= R_3= cyclic aliphatic

H_2O, 50 °C, 10-1.5 h

1_c (11 examples) **Yields 72-85%**

Scheme 5.4: The synthesis of pyrroles in water.

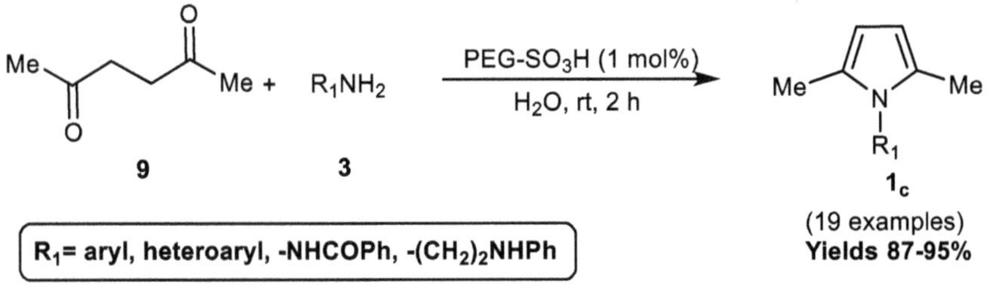

Scheme 5.5: The synthesis of pyrrole using 2,5-hexadione.

and isolation of the products was performed through simple filtration without needing for chromatography techniques.

The reaction between arylglyoxals **10**, 1,3-diketones **11** and enaminoketones **8** leading to the formation of polysubstituted tetraones **12** in water was performed, which was then further converted to polysubstituted pyrroles $\mathbf{1_c}/\mathbf{1_{c'}}/\mathbf{1_{c''}}$ using triethylamine (NEt_3) or NaOH in a water–ethanol mixture [10]. In this method, various arylglyoxals and enaminoketones with different substituents pattern were well tolerated to give the products (Scheme 5.6).

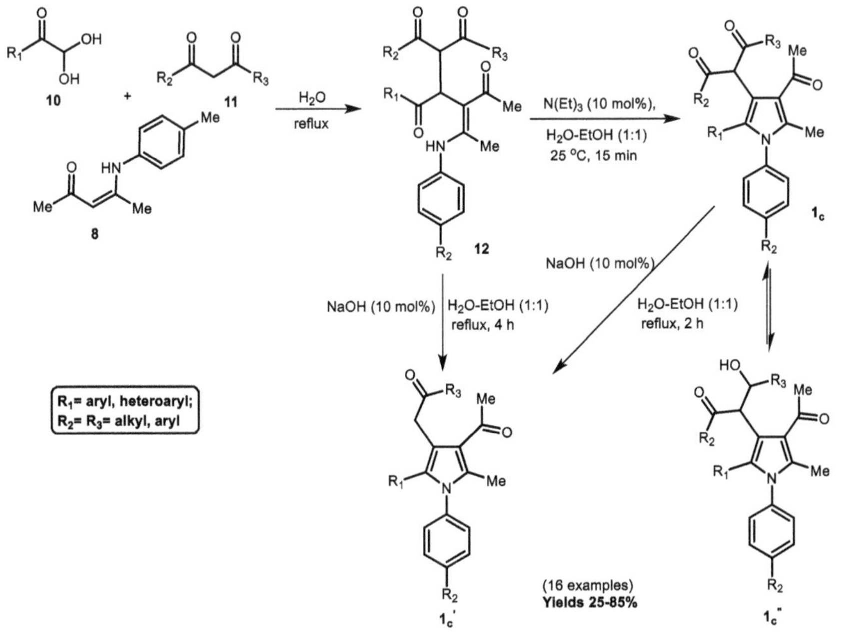

Scheme 5.6: The synthesis of poly-substituted pyrroles in water.

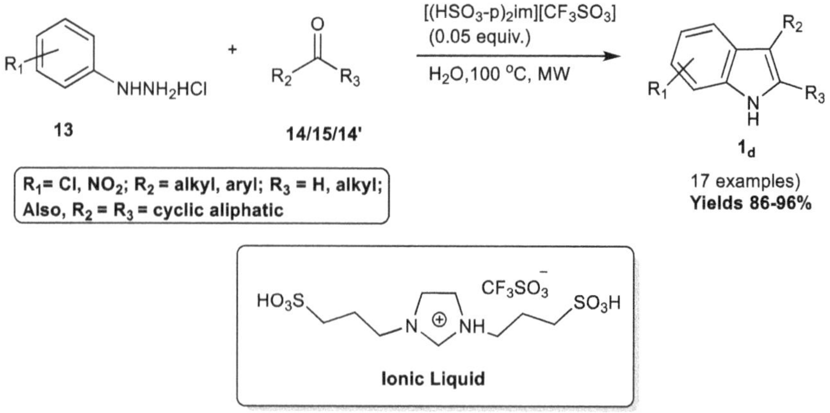

Scheme 5.7: Microwave-assisted fischer indoles synthesis in water.

5.2.4 Synthesis of indoles

One-pot Fischer indoles $\mathbf{1_d}$ synthesis using SO_3H-functionalized Brønsted acidic ionic liquids $[(HSO_3\text{-}p)_2im][CF_3SO_3]$ in water was reported without the need for organic solvents [11]. The indoles $\mathbf{1_d}$ were prepared from various carbonyl compounds **14/15/14'** and aromatic hydrazine hydrochlorides **13** in water under microwave irradiation providing excellent yields up to 96% (Scheme 5.7). This method involves simple separation of the product as the indole products were insoluble in water and the possibility of recycling the catalyst without losing its activity.

Polystyrene-stabilized Pt nanoparticles (PS-PtNPs) were used for the synthesis of indoles $\mathbf{1_d}$ in H_2O using aminophenethyl alcohol **16** under O_2 atmosphere through aerobic alcohol oxidation and subsequent cyclization [12]. The substituted indoles $\mathbf{1_d}$ were obtained in moderate yields. However, with additive tetra-*n*-butylammonium bromide (TBAB) high yields of indoles $\mathbf{1_d}$ up to 88% was obtained (Scheme 5.8). The 2-aminophenethyl alcohol with 4/6-substituted gives the corresponding indoles in good yields, signifying that substituents at these positions have minimal effects on the system. The nanocatalyst was recovered by simple decantation after every reaction and recycled; the transmission emission microscopes (TEM) image displayed that the size of the recycled NPs increased to some extent (3.6 « 0.6 nm) as compared to newly prepared NPs (2.0 « 0.3 nm).

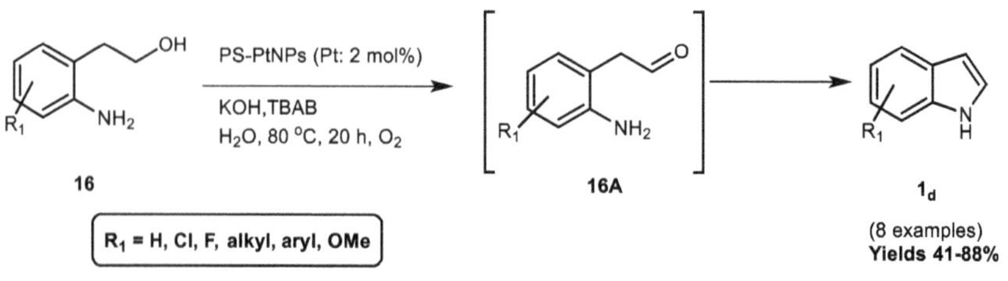

Scheme 5.8: Synthesis of indoles in water using Pt nanoparticles.

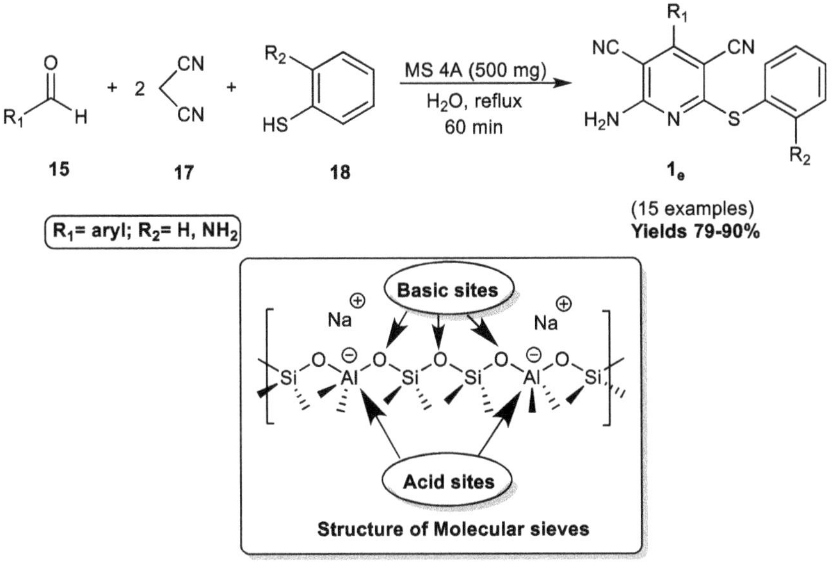

Scheme 5.9: Synthesis of poly substituted pyridines using molecular sieves in water.

5.2.5 Synthesis of pyridine

The synthesis of poly substituted pyridines **1e** [13] in tap/deionize water from aromatic aldehyde **15**, malononitrile **17** and thiol **18** using unmodified microporous molecular sieves (MS 4A) was reported. The catalytic activity of two different sites existing over the MS 4A has been verified in this mechanism. The Al^{3+} and O^{2-} sites over the vast surface area of MS 4A and aid of ions existing in tap water improves the basicity of O^{2-}. Further, the hydrophobic interactions established between the surface of MS 4A and organic compounds demonstrate the collective effects to complete the reaction in a short span of 40–120 min (Scheme 5.9). The reaction assisted by microwave irradiation further shortens the reaction duration to 30–70 min.

The synthesis of polysubstituted pyridines **1e** [14] from ketones **14**, cinnamaldehyde **15'**, and ammonium acetate **19** using Fe_3O_4 decorated multi-walled carbon

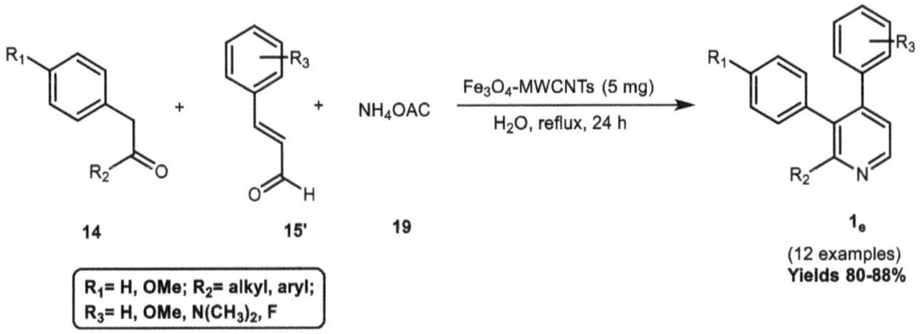

Scheme 5.10: Synthesis of pyridine derivatives using Fe_3O_4-(MWCNT) catalyst in water.

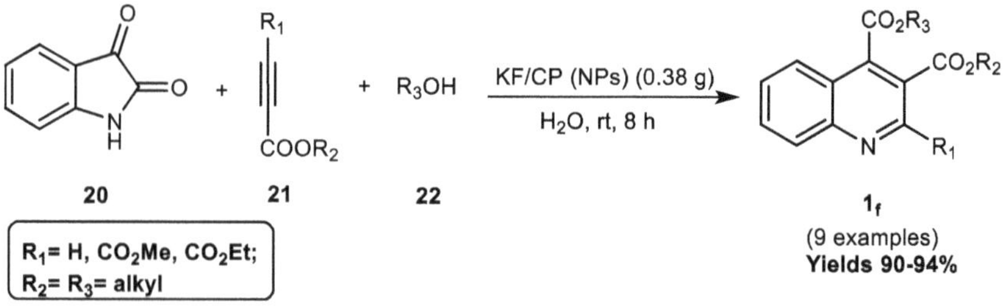

Scheme 5.11: Synthesis of quinolines in water using nanocatalyst.

nanotube (MWCNT) as an effective catalyst in water was established (Scheme 5.10). The nanocatalyst was recovered by an external magnet and recycled 5 times without losing any activity.

5.2.6 Synthesis of quinolines

An efficient synthesis of functionalized quinolines **1_f** [15] in water was carried out by the condensation of isatin **20**, acetylenes **21,** and aliphatic alcohols **22** in the presence of potassium fluoride impregnated clinoptilolite KF/CP (NPs) as a nanocatalyst at ambient temperature (Scheme 5.11). The ambient and cleaner reaction environments, usage of natural catalysts and high yields of the products make this method an exciting alternative to the other reported protocols.

Then, a new protocol for synthesizing substituted quinolines **1_f** from 2-aminoaryl alcohol **16′** and alcohol **22** in water was demonstrated [16] using Ir-catalyst (bidentate *N*-containing Ir-complexes), Scheme 5.12. In this protocol, various 2° alcohols having

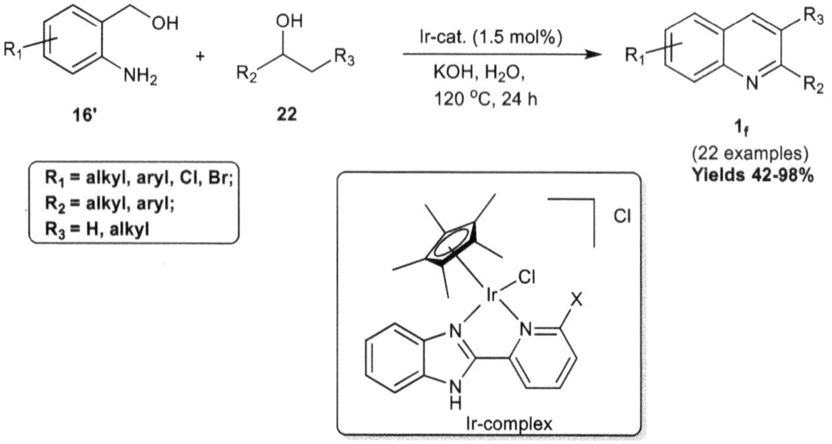

Scheme 5.12: Synthesis of quinolines using Ir-catalyst in water.

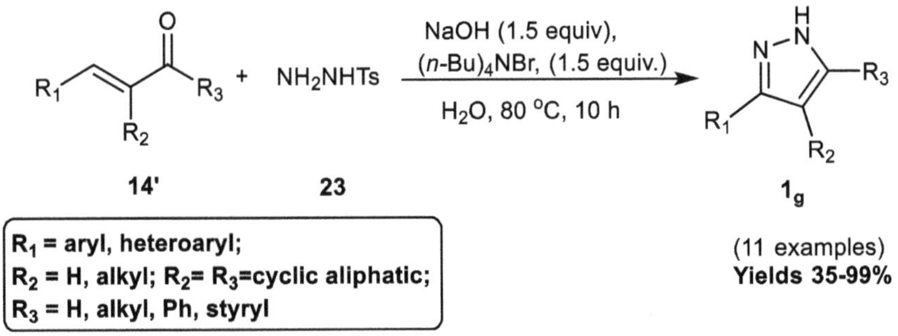

R_1–CH=C(R_2)–C(O)–R_3 **14'** + NH_2NHTs **23** → [NaOH (1.5 equiv), $(n\text{-}Bu)_4NBr$, (1.5 equiv.), H_2O, 80 °C, 10 h] → $\mathbf{1_g}$

R_1 = aryl, heteroaryl;
R_2 = H, alkyl; R_2= R_3=cyclic aliphatic;
R_3 = H, alkyl, Ph, styryl

(11 examples)
Yields 35-99%

Scheme 5.13: The synthesis of pyrazoles using carbonyl compounds.

different substituents in 4- and 3-position gave 78–98% yields of quinolines. However, 2° alcohols with substituents in the 2-position did not proceed to give the products in good yields.

5.3 Synthesis of two *N*-containing heterocycles

5.3.1 Synthesis of pyrazoles

An environmentally friendly process for synthesizing substituted 1*H*-pyrazoles $\mathbf{1_g}$ was established *via* a one-pot reaction of carbonyl compounds **14'** with tosyl hydrazide **23** in H_2O [17]. This method could tolerate with different functional groups and aromatic moiety having electronic substituents giving the 1*H*-pyrazoles in good yields (Scheme 5.13).

The synthesis of the pyrazoles $\mathbf{1_g}$ [18] from enaminones **8** and sulfonyl hydrazines **23'** in H_2O was established in the presence of I_2, *tert*-butyl hydroperoxide TBHP, and $NaHCO_3$ [18]. The aromatic sulfonyl hydrazines were well tolerated in this process providing good yields of the pyrazoles with various enaminones compared to alkyl-based **23'** (Scheme 5.14).

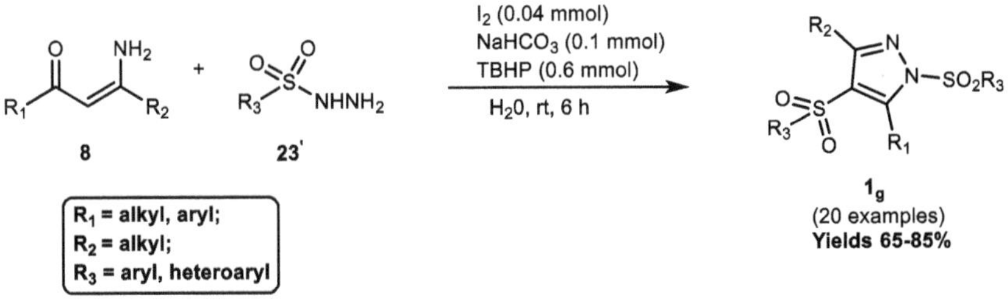

8 + $R_3SO_2NHNH_2$ **23'** → [I_2 (0.04 mmol), $NaHCO_3$ (0.1 mmol), TBHP (0.6 mmol), H_2O, rt, 6 h] → $\mathbf{1_g}$ (N–SO_2R_3)

R_1 = alkyl, aryl;
R_2 = alkyl;
R_3 = aryl, heteroaryl

(20 examples)
Yields 65-85%

Scheme 5.14: Synthesis of substituted pyrazoles in water.

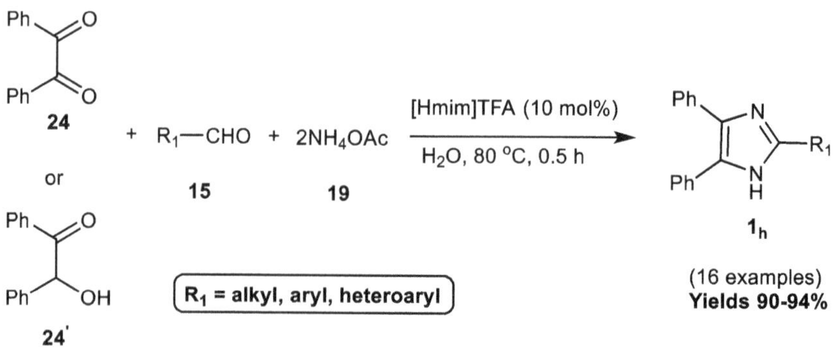

Ph, Ph, O, O, 24, or, Ph, Ph, O, OH, 24', + R_1—CHO + $2NH_4OAc$, 15, 19, [Hmim]TFA (10 mol%), H_2O, 80 °C, 0.5 h, Ph, Ph, N, N, H, R_1, 1_h, R_1 = alkyl, aryl, heteroaryl, (16 examples), **Yields 90-94%**

Scheme 5.15: Synthesis of imidazoles using 1-methylimidazolium trifluoroacetate ([Hmim]TFA).

5.3.2 Synthesis of imidazoles

The highly efficient method for the synthesis of imidazoles $\mathbf{1_h}$ [19] using 1-methylimidazolium trifluoroacetate ([Hmim]TFA) in water was demonstrated (Scheme 5.15). This one-pot reaction between benzil **24** or benzoin **24'**, NH_4OAc **19**, and aldehydes **15** was highly applicable, and greener approaches for the synthesis trisubstituted imidazoles $\mathbf{1_h}$ in good yields (90–94%).

Another protocol was also reported, in which benzil **24**, arylaldehydes **15** and NH_4OAc **19** were allowed to react in one-pot manner to give trisubstituted imidazoles $\mathbf{1_h}$ [20] in the presence of ZnO nanorods (ZnO NRs) (Scheme 5.16). In this method, both aryls and heteroaryl aldehydes participate in the reaction to give good yields of the desired imidazoles. However, the 3-and 2-substituted arylaldehydes' reaction duration is slightly longer compared to 4-substituted analogous, owing to steric hindrance in *2*- and *3*-substituted derivatives for the sluggish response.

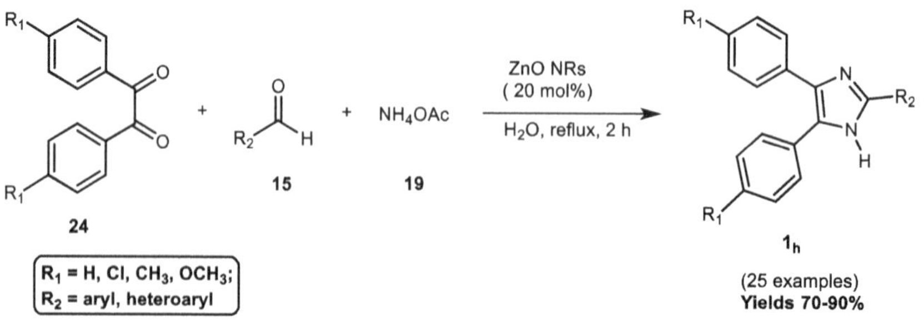

R_1, R_1, O, O, 24, + R_2, O, H, 15, + NH_4OAc, 19, ZnO NRs (20 mol%), H_2O, reflux, 2 h, R_1, R_1, N, N, H, R_2, 1_h, R_1 = H, Cl, CH_3, OCH_3; R_2 = aryl, heteroaryl, (25 examples), **Yields 70-90%**

Scheme 5.16: Synthesis of imidazoles using ZnO nanorods in H_2O.

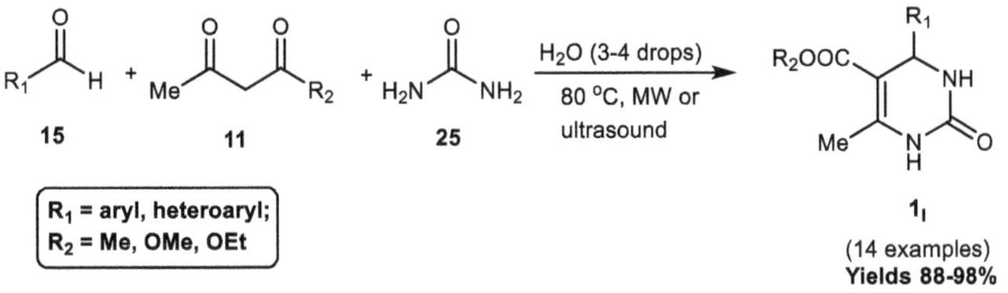

Scheme 5.17: Synthesis of 3,4-dihydropyrimidinones using water as a promoter.

5.3.3 Synthesis of pyrimidines

An efficient protocol for the synthesis of dihydropyrimidinones **1_l** [21] in H_2O was established (Scheme 5.17). The condensation of different aldehydes **15**, *β*-dicarbonyl compounds **11** and urea **25** took only 2 min in microwave irradiation conditions (88–98% yields) when compared to conventional heating and the use of ultrasound. This reaction proceeds under easy workup, improved yield, cheap and readily available catalyst, and H_2O in ambient temperature.

Similarly, the synthesis of novel pyrimidines **1_l** [22] *via* the reaction between arylaldehydes **15**, 2-aminobenzothiazole **26** and dimedone **11'** using thiamine

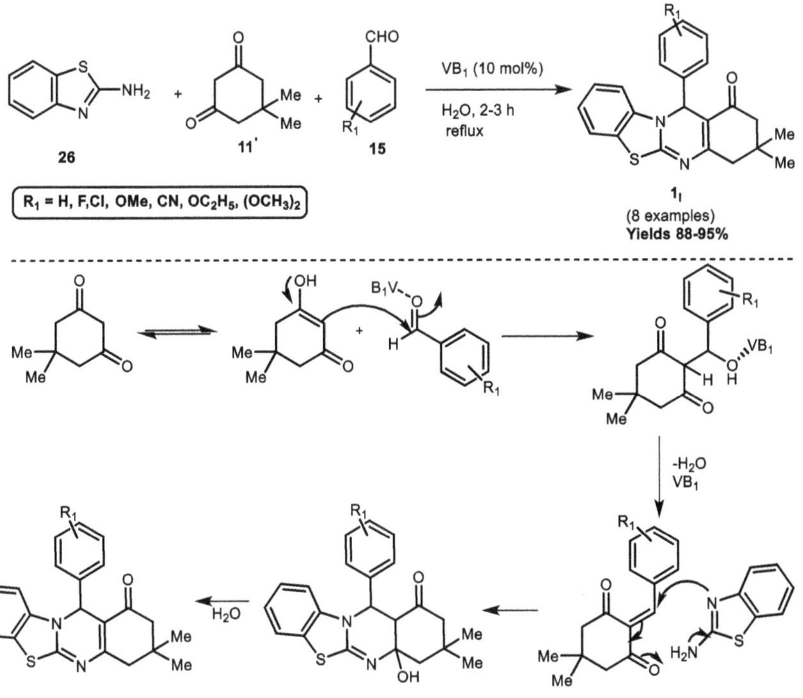

Scheme 5.18: Synthesis of pyrimidines using VB_1 catalyst in water.

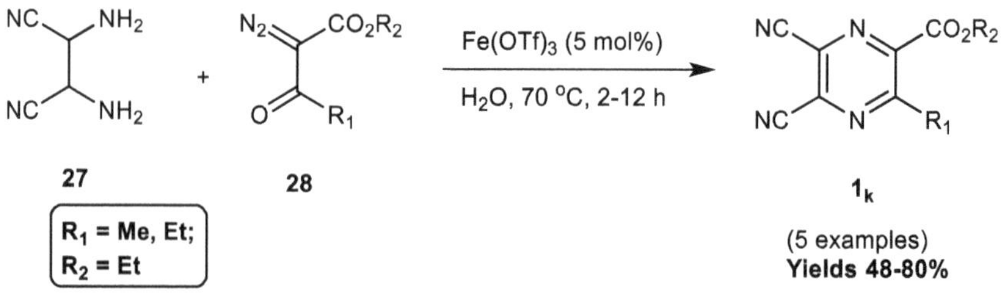

Scheme 5.19: Synthesis of pyrazine derivatives catalyzed by $Fe(OTf)_3$ in water.

hydrochloride (VB_1) as a catalyst in water was demonstrated (Scheme 5.18). The aldehydes with substituents such as Cl, F, and CN resulted in high yields of the product as compared to an aldehyde having OMe within a short time. The advantages of this methodology include good yields of pyrimidines in environmentally sustainable conditions, and easily recyclability of the catalyst. In the proposed mechanism, VB_1 acts as an efficient promoter for smooth Knoevenagel condensation between aldehyde and dimedone. Then, Michael addition followed by further cyclization intramolecularly to give pyrimidobenzothiazole with the elimination of the water molecule.

5.3.4 Synthesis of pyrazines

The one-pot synthesis of pyrazine $\mathbf{1_k}$ [23] *via* the tandem annulation reaction of diaminomaleonitrile **27** with diazo compounds **28**, affording the products in 48–80% yields, was established. The reaction was catalyzed by $Fe(OTf)_3$ and was heated at 70 °C for 2–12 h in H_2O, as shown in Scheme 5.19.

Then, the synthesis of pyrrolo fused pyrazines $\mathbf{1_k}$ in the presence of $Fe_3O_4@SiO_2$-OSO_3H as a reusable magnetic nanocatalyst was also described [24]. The one-pot multicomponent reaction of ethylenediamine **27'**, dialkylacetylenedicarboxylates **29**, and *β*-nitrostyrenes **5''** in water as a green solvent was completed within 2 h and isolated with high yields of the products (Scheme 5.20). The reaction in other solvents

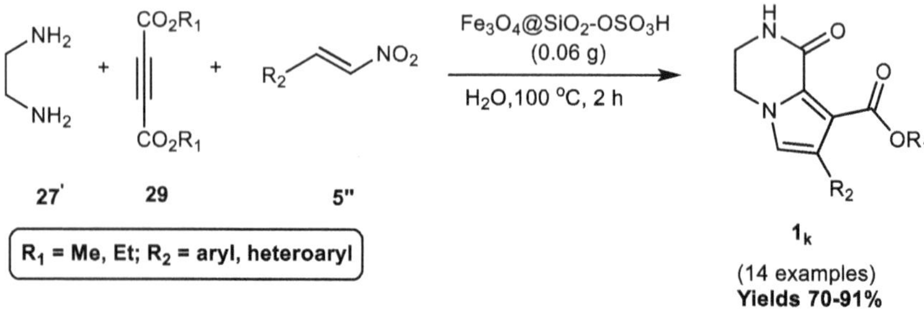

Scheme 5.20: The synthesis of pyrrolo fused pyrazines using Fe-nanocatalyst.

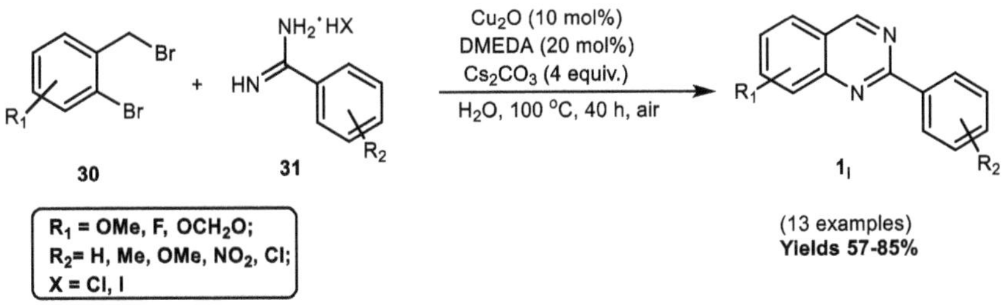

Scheme 5.21: Synthesis of substituted quinazolines.

such as PEG, DMF, EtOH, MeOH, MeCN and toluene were run and examined for 2 h; none of them was suitable as water for this protocol.

5.3.5 Synthesis of quinazolines

A method for the synthesis of quinazolines **1**$_{\mathbf{i}}$ [25] using the reaction between *ortho*-bromobenzyl bromides **30** and benzamidines **31** *via* Cu_2O-catalyst along with *N,N'*-dimethylethylenediamine (DMEDA) and Cs_2CO_3 as additives in H_2O was demonstrated (Scheme 5.21). The reactions in *i*PrOH and ethylene glycol (polar), DMF, DMSO and MeCN (polar aprotic solvents) and 3-xylene (nonpolar) were less effective compared to H_2O as solvent.

An efficient procedure for the synthesis of quinazolines **1**$_{\mathbf{i}}$ [26] from 2-aminobenzylamines **32** and α,α,α-trihalotoluenes **33** using NaOH and O_2 in H_2O at 100 °C was developed (Scheme 5.22). This protocol has several advantages like transition

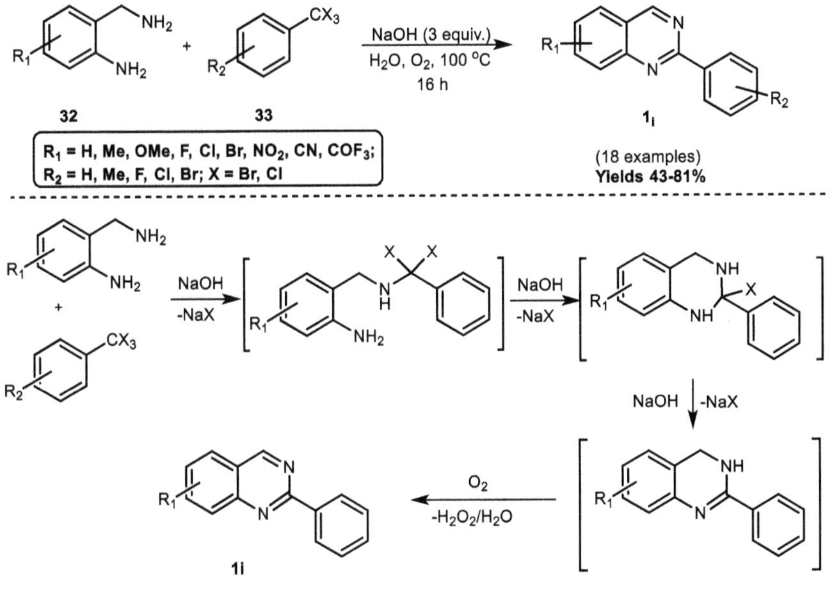

Scheme 5.22: Synthesis of quinazolines in water.

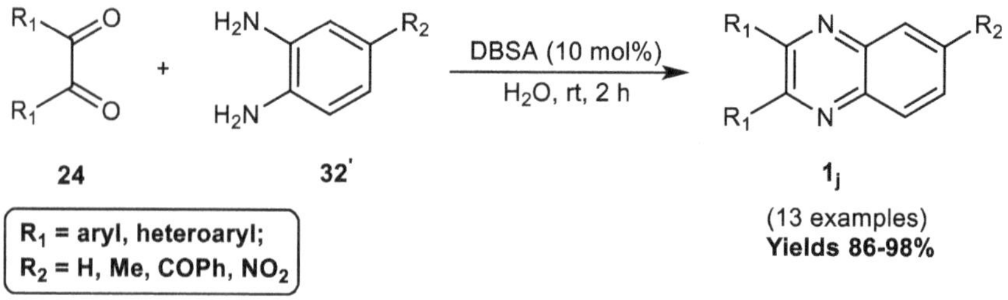

DBSA (10 mol%)
H_2O, rt, 2 h
24
32'
1_j
(13 examples)
Yields 86-98%
R_1 = aryl, heteroaryl;
R_2 = H, Me, COPh, NO_2

Scheme 5.23: Synthesis of quinoxalines in water.

metal-free, molecular oxygen as an oxidant, inexpensive base and water as a reaction medium. Further, the NaX (X = Cl, Br) was the only side product in the system, that could be removed by washing the crude reaction mixture. The reaction proceeds with the base-mediated intermolecular substitution between the substrates, again intramolecular substitution followed by elimination and oxidization by O_2 as the rate-determining step to furnish quinazolines.

5.3.6 Synthesis of quinoxalines

A green protocol for the synthesis of quinozaline $\mathbf{1_j}$ [27] from various 1,2-dicarbonyl **24** and *o*-phenylenediamine **32'** using *p*-dodecylbenzensulfonic acid (DBSA) as a catalyst in water was established under the mild condition at room temperature (Scheme 5.23). The consequence of other surfactants was also examined for this protocol; however, DBSA was the best applicable option for this method.

An efficient and straightforward methodology for the synthesis quinoxalines $\mathbf{1_j}$ in H_2O using bio-renewable alcohols **34** with various 1,2-diamines **32** or nitroamines **32'**

Ir-Cat (2.0 mol%)
KOH (1.5 equiv.)
H_2O, reflux, 24 h, air
32/32'
34
1_j
(21 examples)
Yields 62-92%
R_1 = H, Br, Me, OMe, Cl, Ph;
R_2 = NH_2, NO_2;
R_3 = H, alkyl, Ph;
R_4 = H, alkyl, Ph
Ir (III) Complex

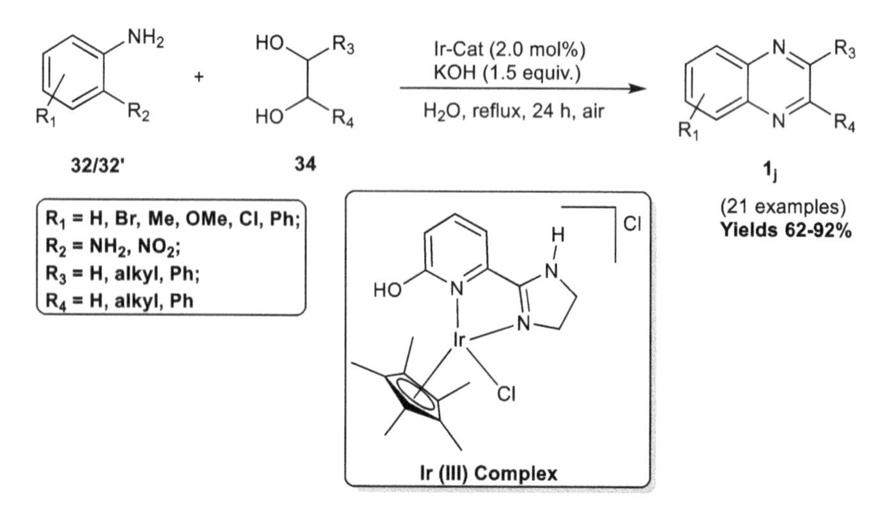

Scheme 5.24: The synthesis of various quinoxalines in water.

was developed [28]. In this method, Ir (III) complex was used as a catalyst with KOH giving good yields of the desired compounds (Scheme 5.24). Moreover, various symmetrical and unsymmetrical diol derivatives react well with benzene 1,2-diamine, giving good yields of the products. However, the reaction of unsymmetrical diamines and diols led to two types of regioisomers.

5.4 Synthesis of 3 *N*-containing heterocycles

5.4.1 Synthesis of triazoles

The one-pot multicomponent reaction for the synthesis of triazoles $\mathbf{1_m}$ [29] in the presence of copper nanoparticles on activated carbon (CuNPs/C) catalyst in a water medium was reported *via* 1,3-dipolar cycloaddition of organic halides **35**, NaN_3 **36** and alkynes **37** (Scheme 5.25). In this reaction, both the activated and deactivated alkyl

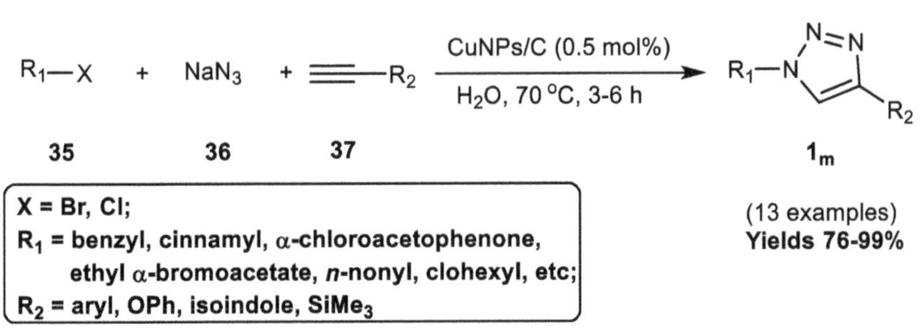

R_1—X (**35**) + NaN_3 (**36**) + ≡—R_2 (**37**) → [CuNPs/C (0.5 mol%), H_2O, 70 °C, 3-6 h] → **1m**

X = Br, Cl;
R_1 = benzyl, cinnamyl, α-chloroacetophenone, ethyl α-bromoacetate, *n*-nonyl, clohexyl, etc;
R_2 = aryl, OPh, isoindole, $SiMe_3$

(13 examples)
Yields 76-99%

Scheme 5.25: Synthesis of 1,2,3-triazole using CuNPs/C catalyst in water.

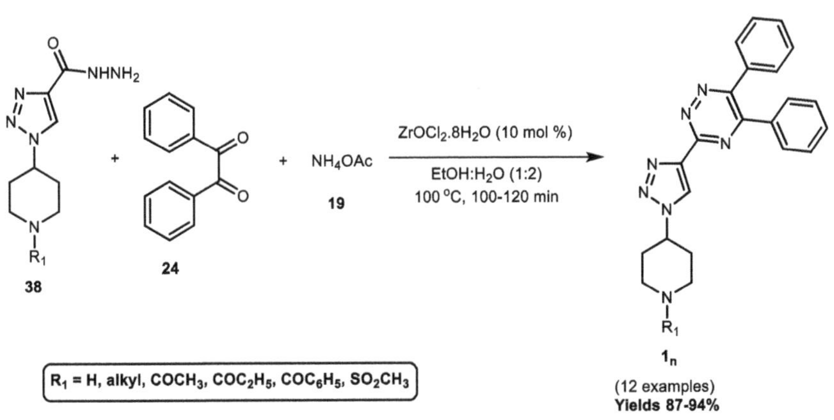

38 + **24** + NH_4OAc (**19**) → [$ZrOCl_2.8H_2O$ (10 mol %), EtOH:H_2O (1:2), 100 °C, 100-120 min] → **1n**

R_1 = H, alkyl, $COCH_3$, COC_2H_5, COC_6H_5, SO_2CH_3

(12 examples)
Yields 87-94%

Scheme 5.26: Synthesis of trisubstituted-triazines derivatives C_2H_5OH-H_2O medium.

halides can be used as the azide precursors. And it was adequate for other alkynes such as phenyl propargyl ether and *N*-propargylphthalimide.

5.4.2 Synthesis of triazines

The synthesis of novel-triazines $\mathbf{1}_n$ [30] using a unique hydrazide **38**, benzil **24** and ammonium acetate **19** was demonstrated using $ZrOCl_2.8H_2O$ as a catalyst in C_2H_5OH-H_2O medium (Scheme 5.26). The methodology could tolerate different hydrazides affording the products in the 87–94% range.

5.5 Conclusion

In this article, the recent advances in synthesizing *N*-containing heterocycles are being elaborated using water as a solvent. The *N*-containing heterocycles are helpful in designing for new therapeutic agents for improving physiological and pharmacological activity. The utilization of water as a reaction medium fits the constructive option from the eco-friendly and commercial perspective as it is a non-flammable, cheap, abundantly accessible resource and it is one of the most potent tools of green chemistry.

References

1. Wolfson A, Dlugy C, Shotland Y. Glycerol as a green solvent for high product yields and selectivities. Environ Chem Lett 2007;5:67–71.
2. María CP, María LM, Merichel P. Water as green extraction solvent: principles and reasons for its use. Curr Opin Green Sustain Chem 2017;5:31–6.
3. Joana FC, Sabine BR. Greener synthesis of nitrogen-containing heterocycles in water, peg, and bio-based solvents. Catalysts 2020;10:429.
4. Nagaraju K, Lalitha G, Suresh M, Kranthi KG, Sreekantha BJ. A review on recent advances in nitrogen-containing molecules and their biological applications. Molecules 2020;25:1909.
5. Silvia S, Francesca M, Francesca DV, Antonella C, Lorenzo T, Marcello T. One-pot synthesis of aziridines from vinyl selenones and variously functionalized primary amines. Tetrahedron 2010; 66:6851–7.
6. Prashant S, Suhasini M, Anjumala S, Manorama S, Vijai KR, Ankita R. First graphene oxide promoted metal-free nitrene insertion into olefins in water: towards facile synthesis of activated aziridines. RSC Adv 2017;7:48723–9.
7. Garima, Shrivastava VP, Yadav LDS. The first example of ring expansion of N-tosylaziridines to 2-aroyl-N-tosylazetidines with nitrogen ylides in an aqueous medium. Green Chem 2010;12: 1460–5.
8. Rueping M, Parra A. Fast, efficient, mild, and metal-free synthesis of pyrroles by domino reactions in water. Org Lett 2010;12:5281–3.

9. Abbas AJ, Sakineh A, Fatemah T. A green, chemoselective, and efficient protocol for paal-knorr pyrrole and bispyrrole synthesis using biodegradable polymeric catalyst PEG-SO3H in water. J Appl Polym Sci 2012;125:1339–45.
10. Mohammad AA, Frershteh NS, Marziyeh M. A green method for the synthesis of pyrrole derivatives using arylglyoxals, 1, 3-diketones and enaminoketones in water or water–ethanol mixture as solvent. Mol Divers 2020;24:1205–22.
11. Bai LL, Dan QX, Ai GZ. Novel SO3H-functionalized ionic liquids catalyzed a simple, green and efficient procedure for Fischer indole synthesis in water under microwave irradiation. J Fluor Chem 2012;144:45–50.
12. Atsushi O, Mao K, Kazuhiro T, Go H, Yasuhiro U, Tsutomu S, et al. Linear polystyrene-stabilized Pt nanoparticles catalyzed indole synthesis in water via aerobic alcohol oxidation. Chem Lett 2016; 45:758–76.
13. Shinde PV, Labde VB, Shingate BB, Shingare MS. Application of unmodified microporous molecular sieves for the synthesis of poly functionalized pyridine derivatives in water. J Mol Catal Chem 2011;336:100–5.
14. Basavegowda N, Mishra K, Lee YR. Fe3O4-decorated MWCNTs as an efficient and sustainable heterogeneous nanocatalyst for the synthesis of polyfunctionalised pyridines in water. Mater Technol 2019;34:558–69.
15. Sajjadi-Ghotbabadi H, Javanshir S, Rostami-Charati F. Nano KF/Clinoptilolite: an effective heterogeneous base nanocatalyst for synthesis of substituted quinolines in water. Catal Lett 2016; 146:338–44.
16. Maji M, Chakrabarti K, Panja D, Kundu S. Sustainable synthesis of N-heterocycles in water using alcohols following the double dehydrogenation strategy. J Catal 2019;373:93–102.
17. Wen J, Fu Y, Zhang RY, Zhang J, Chen SY, Yu XQ. A simple and efficient synthesis of pyrazoles in water. Tetrahedron 2011;67:9618–21.
18. Wang G, Wei L, Wan JP. Domino C-H sulfonylation and pyrazole annulation for fully substituted pyrazole synthesis in water using hydrophilic enaminones. J Org Chem 2019;84:2984–90.
19. MaGee DI, Bahramnejad B, Dabiri M. Highly efficient and eco-friendly synthesis of 2-alkyl and 2-aryl-4, 5-diphenyl-1H-imidazoles under mild conditions. Tetrahedron Lett 2013;54:2591–4.
20. Nikoofar K, Haghighi M, Lashanizadegan M, Ahmadvand Z. ZnO nanorods: efficient and reusable catalysts for the synthesis of substituted imidazoles in water. J Taibah Univ Sci 2015;9:570–8.
21. Tamuli KJ, Dutta D, Nath S, Bordoloi M. A greener and facile synthesis of imidazole and dihydropyrimidine derivatives under solvent-free condition using nature-derived catalyst. ChemistrySelect 2017;2:7787–91.
22. Sethiya A, Soni J, Manhas A, Jha PC, Agarwal S. Green and highly efficient MCR strategy for the synthesis of pyrimidine analogs in water via C–C and C–N bond formation and docking studies. Res Chem Intermed 2021;47:4477–96.
23. Pandit RP, Kim SH, Lee YR. Iron-catalyzed annulation of 1, 2-diamines and diazodicarbonyls for diverse and polyfunctionalized quinoxalines, pyrazines, and benzoquinoxalines in water. Adv Synth Catal 2016;358:3586–99.
24. Rostami H, Shiri L. One-pot multicomponent synthesis of pyrrolo[1, 2-a]pyrazines in water catalyzed by Fe3O4@SiO2-OSO3H. ChemistrySelect 2018;3:13487–92.
25. Malakar CC, Baskakova A, Conrad J, Beifuss U. Copper-catalyzed synthesis of quinazolines in water starting from o-bromobenzylbromides and benzamidines. Chem Eur J 2012;18:8882–5.
26. Chatterjee T, Kim DI, Cho EJ. Base-Promoted synthesis of 2-aryl quinazolines from 2-aminobenzylamines in water. J Org Chem 2018;83:7423–30.
27. Kolvari E, Zolfigol MA, Peiravi M. Green synthesis of quinoxaline derivatives using p-dodecylbenzensulfonic acid as a surfactant-type Bronsted acid catalyst in water. Green Chem Lett Rev 2012;5:155–9.

28. Chakrabarti K, Maji M, Kundu S. Cooperative iridium complex catalyzed synthesis of quinoxalines, benzimidazoles and quinazolines in water. Green Chem 2019;21:1999–2004.
29. Alonso F, Moglie Y, Radivoy G, Yus M. Multicomponent synthesis of 1, 2, 3-triazoles in water catalyzed by copper nanoparticles on activated carbon. Adv Synth Catal 2010;352:3208–14.
30. Jaiprakash NS, Devanand BS. One pot synthesis and SAR of some novel 3-substituted 5, 6-diphenyl-1, 2, 4-triazines as antifungal agents. Bioorg Med Chem Lett 2010;20:742–5.

Kongbrailatpam Gayatri Sharma*

6 Inorganic nanoparticles promoted synthesis of heterocycles

Abstract: The application of inorganic nanoparticles as nanocatalyst for synthesizing of nitrogen containing heterocycles are reviewed. While an inclusive summary of the various catalysts utilized in the synthesis of heterocycles is demonstrated with limited focus on the preparation or characterization of the catalyst. The review is being summarized into different sections based on the size and the number of *N*-atoms in the cyclic compounds.

Keywords: green chemistry; nanoparticles; organic synthesis; reusability.

6.1 Introduction

Nanotechnology is an emerging technology involving the synthesis of nanoparticles (NPs) with one of the dimensions ranging from 1 to 100 nm and its application [1]. They display a unique chemical, physical and biological activities compared to their atomic/molecular/bulk materials, attributed to their comparatively greater surface area to the volume, increased reactivity and greater mechanical strength [2–4]. A NP could be either a zero-dimensional (0D) in which the length, breadth and height are fixed at a single point, like nano dots or 1D where it possesses only one parameter, such as graphene and 2D where it has length and breadth, for example C-nanotubes. Further, it can be 3D where it has all the parameters such as length, breadth and height, Au nanoparticles as an example. They can be of different shapes such as spherical, tubular, cylindrical, conical, spiral, flat, hollow core, etc. or irregular shapes and differing from 1 to 100 nm in range. They can have different physicochemical properties, i.e., can have uniform or irregular surface variations and can exist as crystalline and amorphous with single or multi crystal solids either loosely bonded or clustered [5]. Now-a-days, various physico–chemical methods are being used for synthesizing the NPs, in such a way to improve the properties of NPs in cost efficiency ways.

The NPs are normally divided into the organic, inorganic and C-based. Inorganic NPs consist of metal and metal oxide; all the metals can be synthesized into their NPs. The frequently used metals for NP synthesis are Al, Co, Cu, Au, Fe, Pb, Ag and Zn. The metal oxide-based NPs are synthesized to modify the properties of their respective metal-based NPs, for example Fe NPs instantly oxidises to Fe_2O_3 in presence of O_2 to

***Corresponding author: Kongbrailatpam Gayatri Sharma**, Chemistry Department, Oriental College (Autonomous), Imphal, 795001, Manipur, India, E-mail: gayatrish83@gmail.com

As per De Gruyter's policy this article has previously been published in the journal Physical Sciences Reviews. Please cite as: K. G. Sharma "Inorganic nanoparticles promoted synthesis of heterocycles" *Physical Sciences Reviews* [Online] 2022. DOI: 10.1515/psr-2022-0129 | https://doi.org/10.1515/9783110913361-006

increases its reactivity compared to Fe NPs [6]. The common examples are Al_2O_3, Fe_2O_3, Fe_3O_4, SiO_2, TiO_2 and ZnO.

6.1.1 As catalysis

NPs comprise greater surface area offering them a desirable component of catalysts, resulting in a faster reaction time [7]. They can be easily separated and recycled without losing its catalytic activity. The use of Pt NPs in the automobile catalytic converters is one of the significant applications as little amount of Pt is needed with considerable reduce in the cost and enhancing performance. The NPs as catalysts are used in numerous organic syntheses, such as, oxidation of CO in aqueous solutions, hydrogenation of alkenes and hydrosilylation of olefins in organic medium. In this article, focus on inorganic nanocatalysts for the synthesis of *N*-containing heterocycles is summarised.

6.2 Synthesis of *N*-containing heterocycles

6.2.1 Synthesis of 3- & 4-membered rings with 1*N*-atom

A green methodology for the synthesis of aziridines **2** using Fe_3O_4-dopamine-Cu catalyst was explored [8]. These magnetic NPs were adapted with dopamine and used as a support to coordinate Cu metal. The reaction used olefin **1** as starting material and *N*-(*p*-toluenesulfonyl)imino]phenyliodinane (PhI = NTs) as nitrene source with microwave irradiation. The advantages of this method involve easy recyclability of the catalyst and reusability for 5 cycles without significant loss of yields, Scheme 6.1.

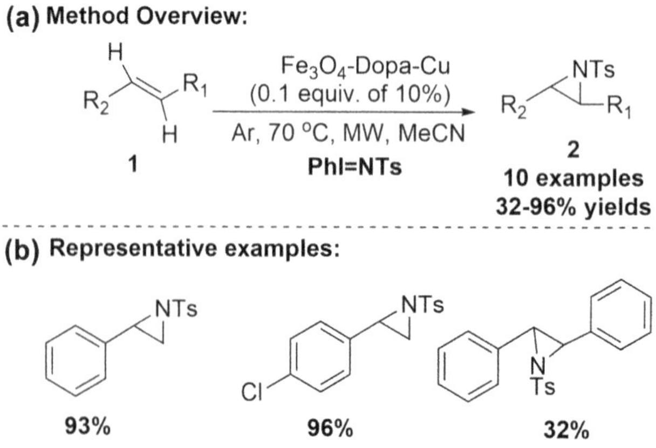

(a) Method Overview:

Fe_3O_4-Dopa-Cu (0.1 equiv. of 10%), Ar, 70 °C, MW, MeCN, PhI=NTs

1 → **2**, **10 examples**, **32-96% yields**

(b) Representative examples:

93% · 96% · 32%

Scheme 6.1: Magnetic nanoparticle catalysed azirines synthesis.

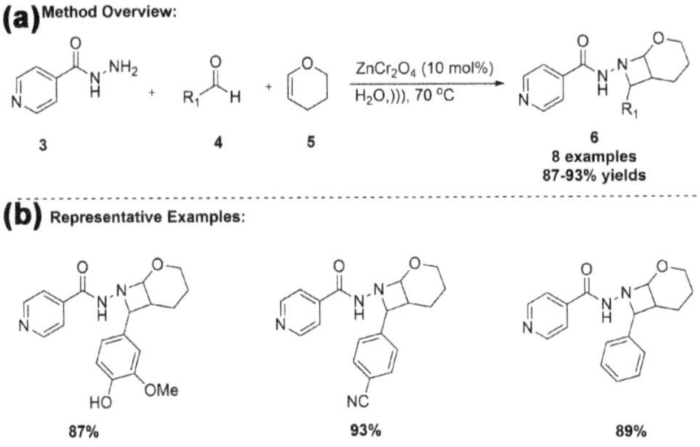

Scheme 6.2: Synthesis of substituted azetidines using $ZnCr_2O_4$NPs.

The synthesis of azetidines **6** in a one pot multicomponent fashion using $ZnCr_2O_4$NPs under ultrasonic irradiation was reported [9]. This $(2\pi + 2\pi)$ cycloaddition of isoniazid **3**, arylaldehydes **4** and 3,4-dihydro-2*H*-pyran **5** give azetidines in the excellent yields with just 10 mol% of catalyst at 70 °C of ultrasonic bath, Scheme 6.2.

6.2.2 Synthesis of 5-membered rings containing 1*N*-atom

In 2014, the Paal–Knorr reaction for the synthesis of *N*-substituted pyrroles by sulfamic acid-functionalized magnetic Fe_3O_4 NPs (MNPs/DAG-SO_3H) was established [10]. The general route for the fabrication of NPs is shown in Scheme 6.3. The X-ray diffraction (XRD) and transmission electron microscopy (TEM) techniques were used to characterize the newly synthesis MNPs. The average size of 20 nm having spherical

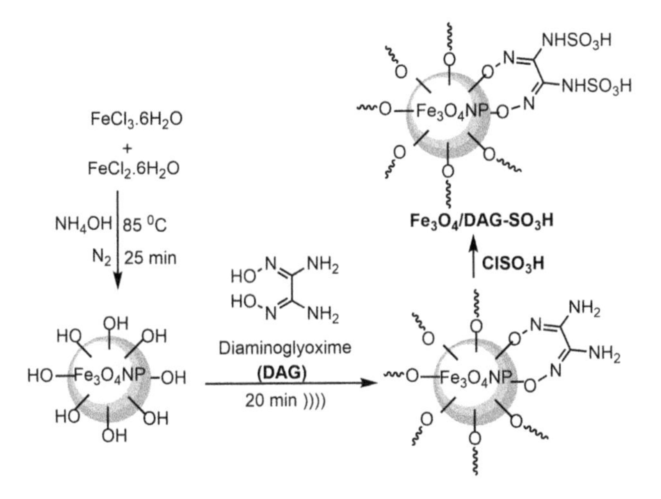

Scheme 6.3: Synthesis of sulfamic acid-functionalized magnetic Fe_3O_4NPs.

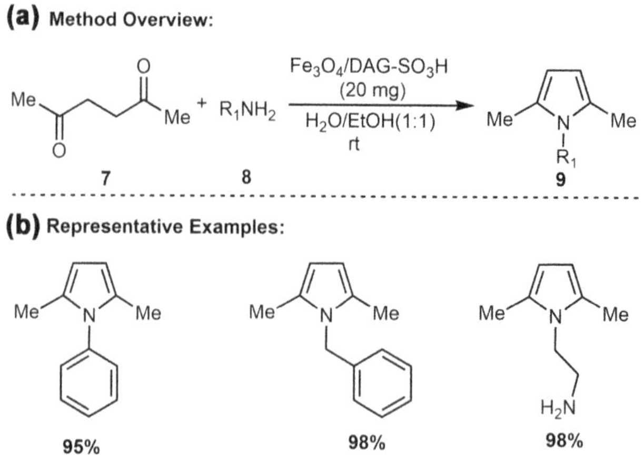

Scheme 6.4: Synthesis of pyrroles using sulfamic acid-functionalized magnetic Fe_3O_4NPs.

structure was confirmed by TEM. This method utilized the reaction between 1° amines (1 mmol) and hexan-2,5-dione (1 mmol) in presence of 20 mg of MNPs/DAG-SO_3H in H_2O–EtOH at 25 °C is shown in Scheme 6.4.

A one-pot method for the synthesis of pyrroles **12** using $NiFe_2O_4$ magnetic nanomaterials (NMNs) as catalyst [11] was reported. The reaction pot comprises of nitromethane **10**, aromatic aldehydes **4**, 1,3-dicarbonyl compounds **11** and 1° amines **8** in presence of $NiFe_2O_4$ catalyst at 100 °C giving the best results (Scheme 6.5). In this reaction, various anilines produce the desired pyrroles in high yields (80–96%) and the

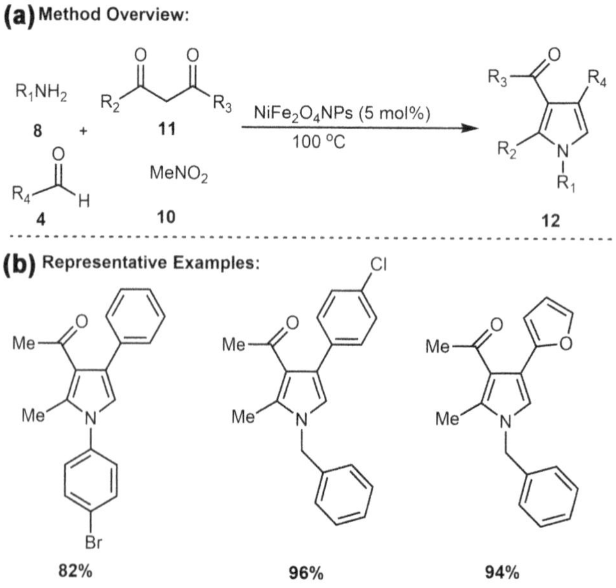

Scheme 6.5: Synthesis of tetra-substituted pyrroles using $NiFe_2O_4$ magnetic nanomaterials.

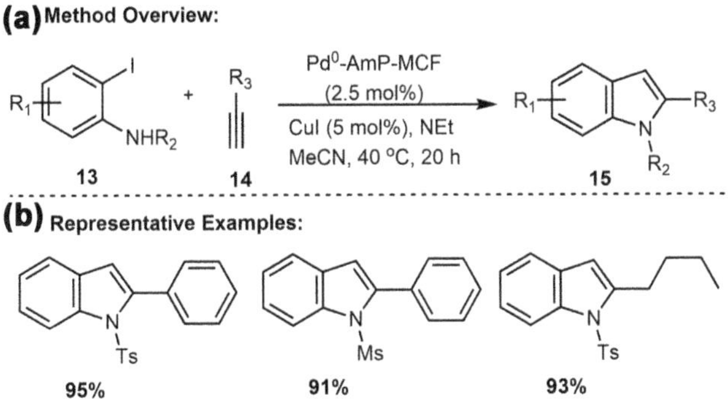

(a) Method Overview:

Pd^0-AmP-MCF (2.5 mol%)
CuI (5 mol%), NEt
MeCN, 40 °C, 20 h

(b) Representative Examples:

95% 91% 93%

Scheme 6.6: PdNPs for the synthesis of indoles.

substituents of arylaldehydes have no effect to the reaction outcome. For example, furan-2-carbaldehyde could participate in this reaction effectively giving 94% yield.

The synthesis of indole derivatives **15** was demonstrated using PdNPs supported on a siliceous mesocellular foam (Pd^0-AmP-MCF) [12]. The reaction between protected 2-iodoanilines **13** and alkynes **14** in the presence of the Pd NPs gave 2-substituted indoles with excellent yields, the NPs could be reused 5 times without substantial loss of activity, as shown in Scheme 6.6.

Then, an operationally simple protocol for the Co_2Rh_2 NPs/charcoal-catalyzed tandem cyclization of (2-nitroaryl)acetonitriles **16** to indoles **15′** was reported in 2016 [13]. This tandem reaction could tolerate various substituents of the **16** under the mild conditions of 1 atm H_2 and ambient temperature with reasonable to good yields of the indoles prepared, Scheme 6.7.

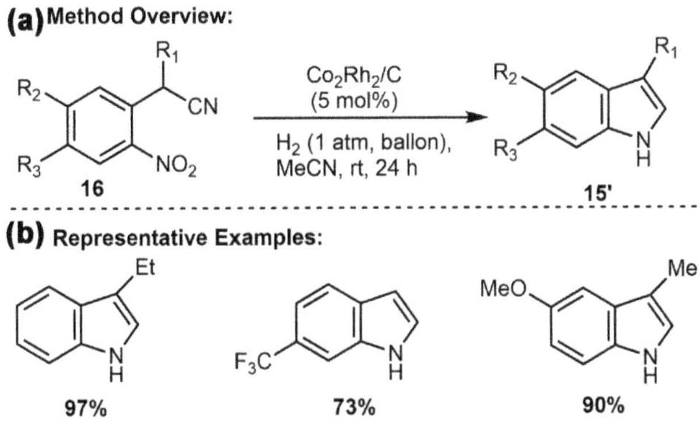

(a) Method Overview:

Co_2Rh_2/C (5 mol%)
H_2 (1 atm, ballon), MeCN, rt, 24 h

(b) Representative Examples:

97% 73% 90%

Scheme 6.7: The synthesis of indoles from (2-nitroaryl)acetonitriles using Co_2Rh_2NPs.

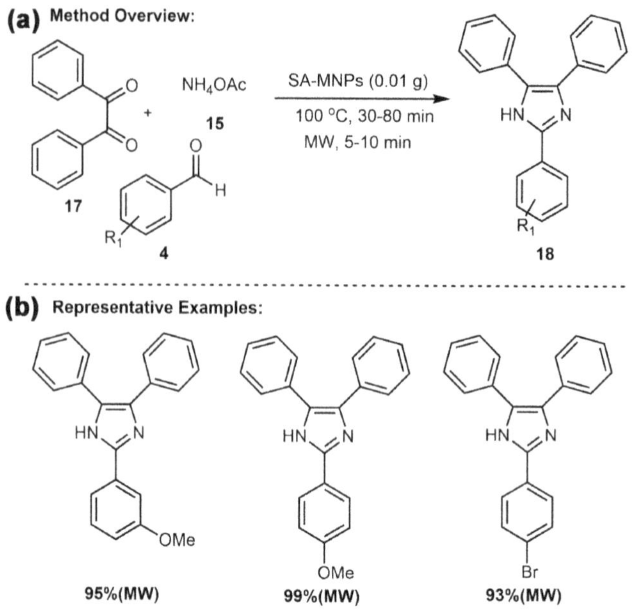

Scheme 6.8: The synthesis of imidazoles using Fe_3O_4NPs.

6.2.3 Synthesis of 5-membered rings containing 2*N*-atoms

The trisubstituted imidazoles **18** was synthesis by reacting 1,2-diketones **17**, arylaldehydes **4** and NH_4OAc **15** using sulphamic acid functionalized magnetic Fe_3O_4NPs (SA-MNPs) under microwave irradiation [14], Scheme 6.8. In this protocol, the NPs were separated using a magnet of 2000 Gs. The size of the particles and morphology of NPs were confirmed using TEM micrographs and the average diameter of both the pristine and SA-functionalized NPs core was around 18 nm which was in agreement with the XRD pattern observed.

In the meanwhile, an efficient and environmentally friendly synthesis of 1,2,4,5-tetrasubstituted imidazoles **18′** using 1,2-diketones **17**, arylaldehydes **4**, NH_4OAc **15** and 1° amines **8** in one pot model using microwave irradiation without any solvent was also explored [15]. The reactions using the ionic liquid (1-methyl-3-(3-trimethoxysilylpropyl) imidazolium chloride) immobilized on superparamagnetic Fe_3O_4 NPs (IL-MNPs) under the conventional heating conditions gave less yields along with longer reaction time compare to the microwave-assisted reactions, as shown in Scheme 6.9. This reaction conditions make this condensation leading to rapid access to products with safe operation and low pollution.

The synthesis of the pyrazols **21** by reaction of various phenyl hydrazines/semicarbazide/thiosemicarbazide **20** with 1,3-ketoester **19** at 25 °C in water was established using ZnO nanocatalyst [16], as shown in Scheme 6.10. The structure of the ZnO was

Scheme 6.9: The synthesis of 1,2,4,5-tetrasubstituted imidazoles using Fe_3O_4NPs.

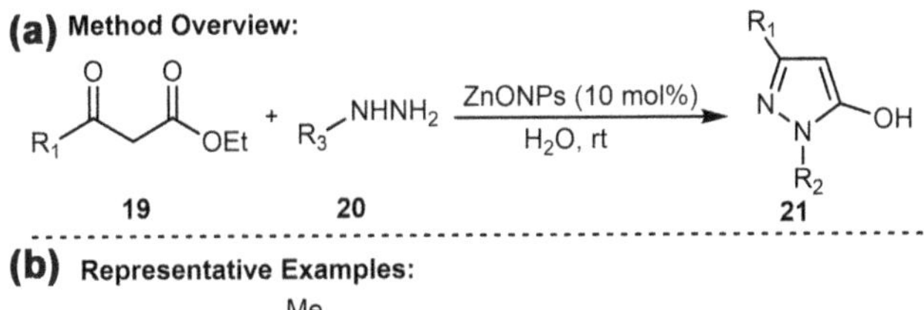

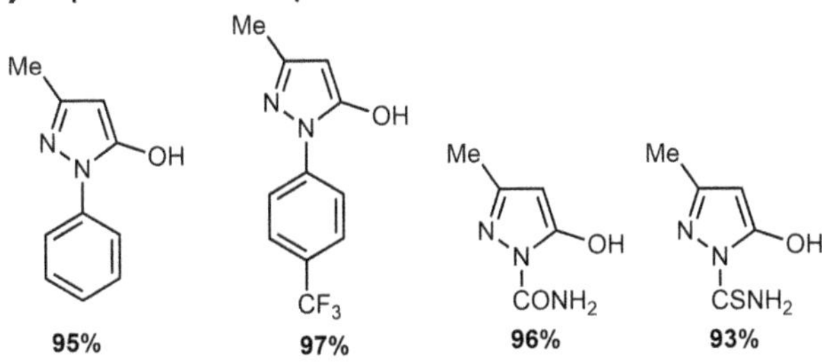

Scheme 6.10: The synthesis of pyrazols using ZnONPs.

around 30–50 nm as studied by XRD analysis and collaborated by TEM image of the particle, which was found to be in 30 nm. The reason for longer reaction time of 45 min for thiosemicarbazide compare to 4 min of semicarbazide was attributed to electron withdrawing nature of (C=S) in thiosemicarbazide.

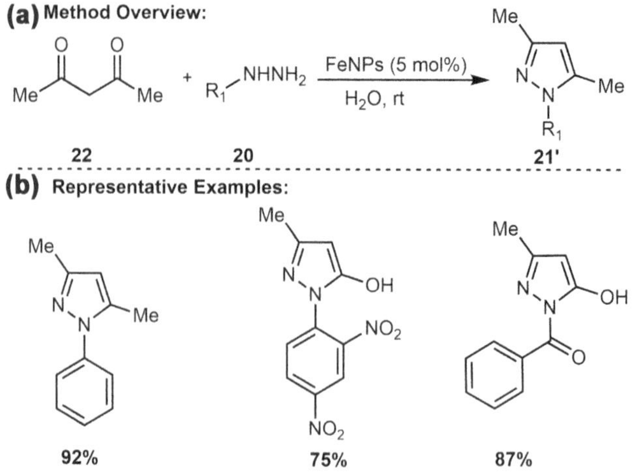

(a) Method Overview:

Me, Me, 22 + R_1–$NHNH_2$ 20 → FeNPs (5 mol%), H_2O, rt → 21'

(b) Representative Examples:

92% 75% 87%

Scheme 6.11: The synthesis of pyrazoles using FeNPs.

The catalytic application of Fe NPs using aqueous leaves extract of *Boswellia serrata* [17] was explored. The FeNPs was employed for the synthesis of pyrazole derivatives **21′** by reacting various phenyl hydrazine/hydrazide **20** with acetyl acetone **22** giving moderate to excellent yields (75–92%) in water at ambient conditions, Scheme 6.11.

6.2.4 Synthesis of 5-membered rings containing 3*N*-atoms

An efficient method for the synthesis of triazoles **24** from azides **23** and alkynes **14** in presence of the Cu NPs supported on nanocellulose (CuNPs/NC) was reported in 2017 [18]. The HRTEM images indicated the average particle size to be 6–7 nm in range. In this cycloaddition, both aromatic as well as alkyl terminal alkynes react with the various azides readily to give the corresponding products in excellent yields, Scheme 6.12.

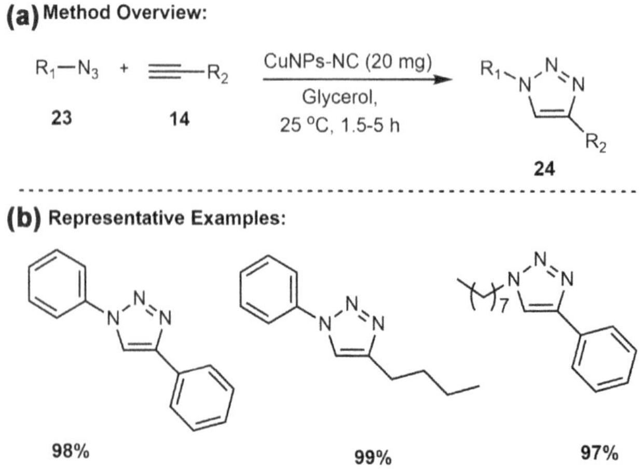

(a) Method Overview:

R_1–N_3 23 + ≡–R_2 14 → CuNPs-NC (20 mg), Glycerol, 25 °C, 1.5-5 h → 24

(b) Representative Examples:

98% 99% 97%

Scheme 6.12: The synthesis of triazoles using CuNPs.

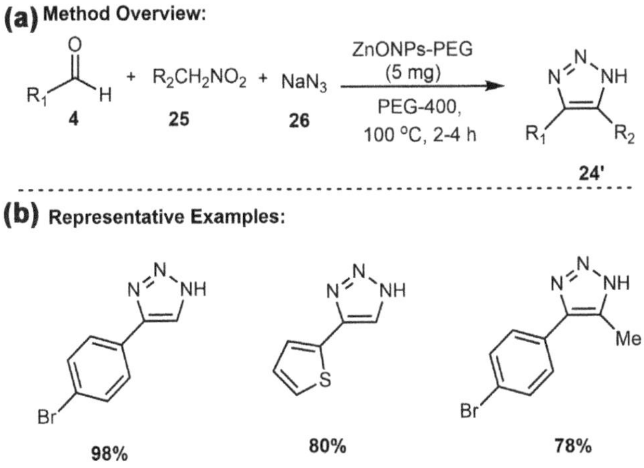

(a) Method Overview:

R_1CHO **4** + $R_2CH_2NO_2$ **25** + NaN_3 **26** → (ZnONPs-PEG (5 mg), PEG-400, 100 °C, 2-4 h) → **24'**

(b) Representative Examples:

98% 80% 78%

Scheme 6.13: The synthesis of triazoles using ZnONPs.

The ZnO nanocatalyst was used for the synthesis of tetrasubstituted-triazoles **24′** from aldehydes **4**, nitroalkane **25** and sodium azide **26** in a one-pot model [19]. The ZnONPs with controlled size of 15–25 nm was prepared using a sonochemical strategy. The protocol afforded good yields of the NH-triazoles irrespective of various substituents present in aldehydes, however, aromatic aldehydes having halogen and CN as substituents gave excellent yields of the triazoles, Scheme 6.13. The catalysis used in this protocol along with green solvent polyethylene glycol (PEG) 400 offers eco-friendly method to the prevailing approaches using organic solvents.

6.2.5 Synthesis of 6-membered rings containing 1*N*-atom

The one-pot synthesis of trisubstituted pyridines **29** in good to excellent yields from benzyl alcohols **27**, acetophenone **28** and NH_4OAc **15** was explored [20] under microwave irradiation in the presence of γ-MnO_2NPs, as shown in Scheme 6.14.

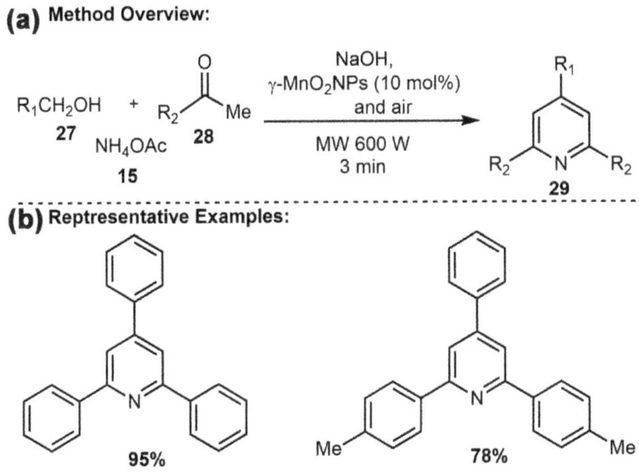

(a) Method Overview:

R_1CH_2OH **27** + R_2COMe **28** + NH_4OAc **15** → (NaOH, γ-MnO_2NPs (10 mol%) and air, MW 600 W, 3 min) → **29**

(b) Reptresentative Examples:

95% 78%

Scheme 6.14: The synthesis of pyridines using γ-MnO_2NPs.

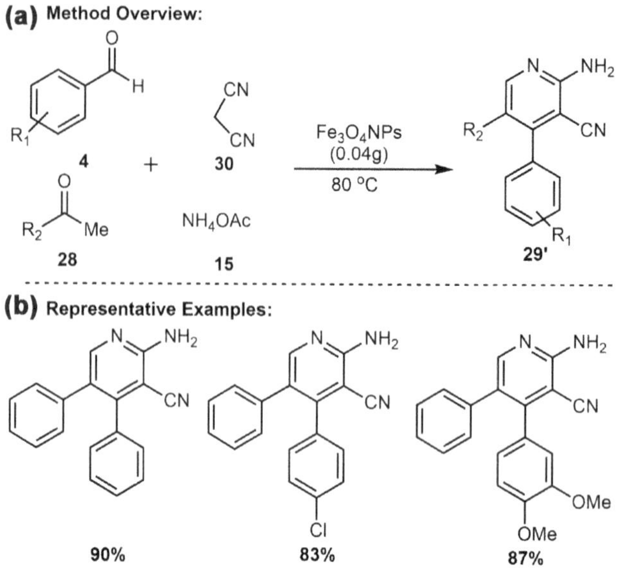

Scheme 6.15: Synthesis of tetra-substituted pyridines using Fe_3O_4NPs.

The synthesis of poly-functionalized pyridines **29′** *via* a one-pot reaction of benzaldehyde **4**, acetophenone **28**, NH_4OAc **15** and malononitrile **30** in the presence of Fe_3O_4NPs (MNPs) was demonstrated [21]. The catalyst was re-used at least 4 times without any notable change of catalytic efficiency in this reaction, as shown in Scheme 6.15.

A library of acridinedione derivatives **32** were synthesized from the condensation of dimedone **31**, arylaldehydes **4** and several arylamines **8** using monodisperse Pt NPs supported with reduced graphene oxide (PtNPs@rGO) as a catalyst [22]. The high yields of the products was attributed to the structure and small particle size, along with high Pt (0) contents (%) of rGO furnishing the Pt NPs, Scheme 6.16.

The synthesis of polysubstituted quinolines **33** from the cyclization of various anilines **8** and arylalkyl/alkyl aldehydes **4′** using SiO_2-supported AuNPs was successfully demonstrated [23], Scheme 6.17. In this protocol, the AuNPs/SiO_2 acts as Lewis acid and the Au NPs along with O_2 aided the oxidative conversion of 1,2-dihydroquinolines to different quinolines with the product yields up to 95%.

Then, the reaction of 1,3-carbonyl compounds **11/11′** and 2-aminobenzophenone **34** giving the quinolines **33'/33"** in H_2O was demonstrated [24] using Co(III) salen complex immobilized on Co ferrite-silica NPs ($CoFe_2O_4$@SiO_2@Co(III) salen complex). The catalyst was prepared as shown in Scheme 6.18 and studied by TEM, SEM-EDX, vibrating-sample magnetometer (VSM), FT-IR and thermogravimetric analysis (TGA) analyses. The grain size of the (i) $CoFe_2O_4$, (ii)$CoFe_2O_4$@SiO_2 and $CoFe_2O_4$@SiO_2@Co(III) salen NPs have a narrow distribution of sizes ranging from 18–30 nm as revealed from TEM. Thus, two set of quinolines were synthesized using the same reaction conditions, Scheme 6.19.

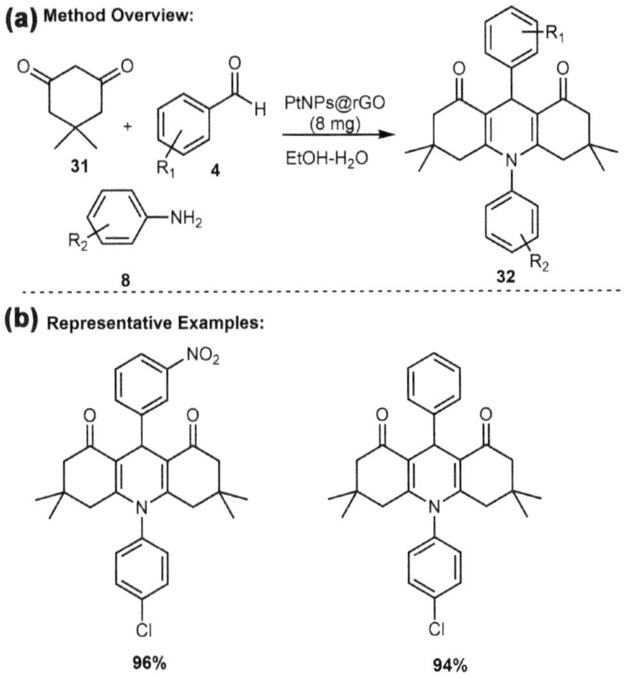

Scheme 6.16: Synthesis of acridines using PtNPs.

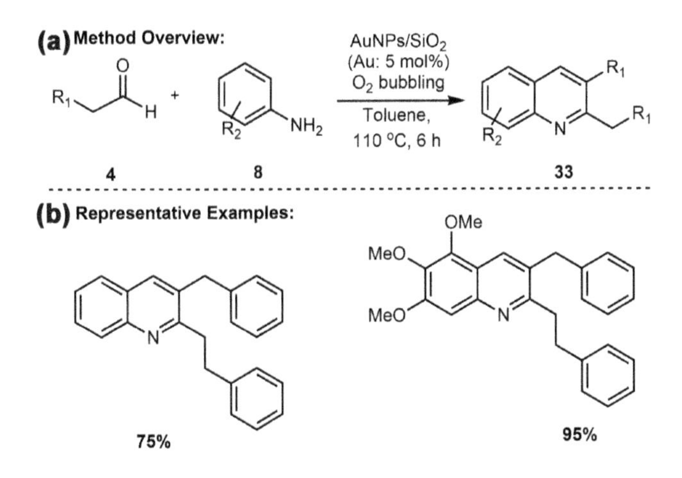

Scheme 6.17: Synthesis of polysubstituted quinolines using $AuNPs/SiO_2$.

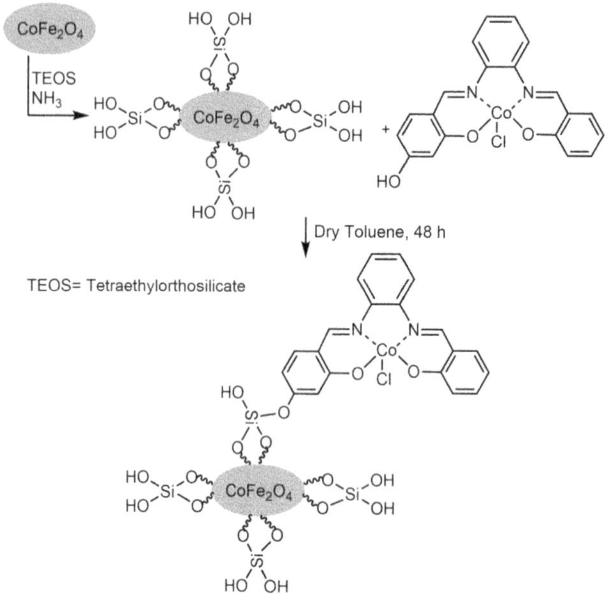

$CoFe_2O_4$
TEOS
NH_3
HO OH
Si
HO Si OH
$CoFe_2O_4$
+
N Co N
O Cl O
HO
Dry Toluene, 48 h
TEOS= Tetraethylorthosilicate

Scheme 6.18: Preparation of Co(III) salen complex.

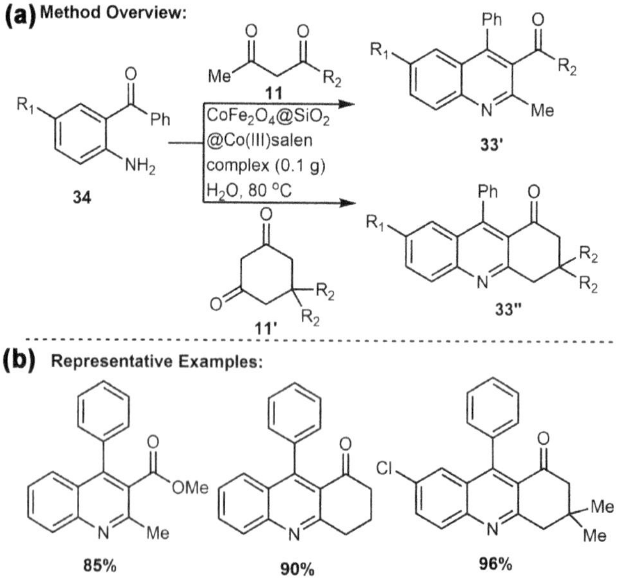

(a) Method Overview:
R_1
Ph
NH_2
34
Me
R_2
11
$CoFe_2O_4@SiO_2$
@Co(III)salen
complex (0.1 g)
H_2O, 80 °C
11'
33'
33''
(b) Representative Examples:
OMe
Me
85%
90%
Cl
Me
Me
96%

Scheme 6.19: Synthesis of quinolines using $CoFe_2O_4@SiO_2@Co$ (III) salen complex.

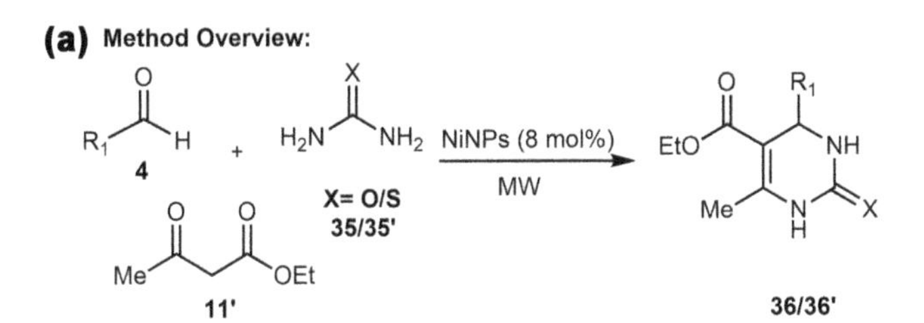

(b) Representative Examples:

98%

80%

95%

Scheme 6.20: The synthesis of dihydropyrimidinones using NiNPs.

6.2.6 Synthesis of 6-membered rings containing 2*N*-atoms

The Ni NPs catalyzed synthesis of dihydropyrimidinone derivatives from aryl/heteroaryl/alkylaldehydes **4**, urea/thiourea **35/35′** and ethyl acetoacetate **11′** under microwave irradiation was explored [25]. In this reaction, aliphatic aldehydes gave less yields compare to both aryl as well as heteroarylaldehydes with diverse substituents at different position, Scheme 6.20.

The synthesis of pyrazines from α-diketones **37** and ammonium formate **38** using xanthphos supported Ru NPs as catalyst [26] was explored, Scheme 6.21. The NPs stabilized with the xanthphos was analysed by TEM and the size of the NPs was < 2–3 nm.

(a) Method Overview:

37 + 38 → 39 (RuNPs@Xanthphos (1.0 mol%), DMF, 85 °C, 1 h)

(b) Representative Examples:

86%

61%

Scheme 6.21: Synthesis of substituted pyrazines catalyzed by RuNPs.

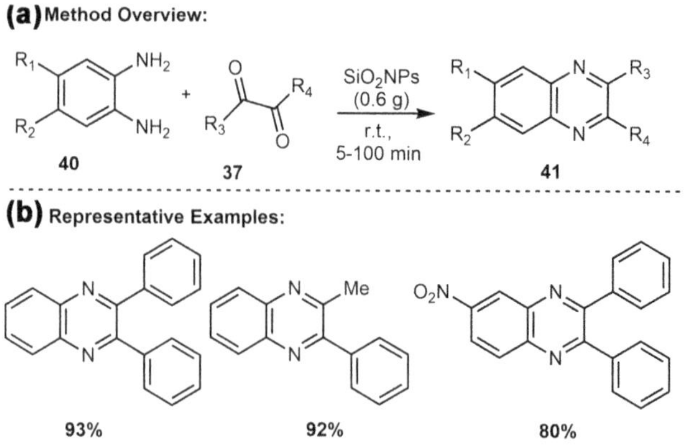

Scheme 6.22: Synthesis of quinoxalines using SiO_2NPs.

In 2012, an eco-friendly method for the synthesis of quinoxalines **41** using silica NPs *via* reaction of 1,2-diamines **40** with 1,2-diketones **37** at ambient temperature was reported [27], affording high yields of products. It was observed that electron-donating substituents had no substantial outcome on the reaction results; however, electron-withdrawing substituents reduced the yields with longer reaction duration, Scheme 6.22.

The synthesis of quinazolines **42** and quinazolinones **42′** using ultrafine Ni_2PNPs supported on N,P-codoped biomass-derived porous carbon (Ni_2P@NPC-800) through oxidative coupling of various alcohols **27** with diamines **40′** or 2-aminobenzamides **40''** using atmospheric air as the oxidant [28] was demonstrated, Scheme 6.23. The TEM images of Ni_2P@NPC-800 reveal that the ultrafine Ni_2PNPs were homogenously spread having narrow size distribution (3.2 ± 0.7 nm) on the graphitic carbon material.

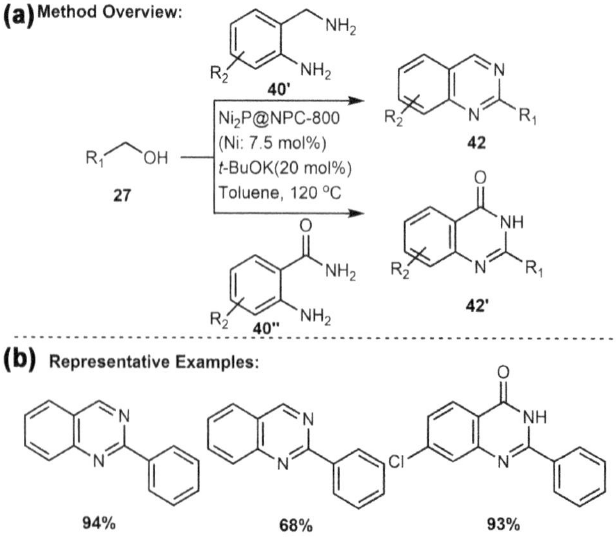

Scheme 6.23: Synthesis of quinazolines and quinazolinones using Ni_2P@NPC.

6.2.7 Synthesis of 7-membered rings

In 2021, the synthesis of azepines **45** *via* multicomponent reaction of isatins **43**, activated alkynes (**14** and **14′**), alkyl bromides **44** and NH_4OAc **15** using ionic liquid (IL) 1-octhyl-3-methyl imidazolium bromide ([OMIM]Br) at room temperature was demonstrated [29]. This IL acts as a stabilizer in the presence of Fe_3O_4/TiO_2/multi walled carbon nanotubes (MWCNTs) magnetic nanocomposite (MNCs) as an effective catalyst giving the azepines in high yields, Scheme 6.24. The crystal identity of Fe_3O_4/TiO_2 and Fe_3O_4/TiO_2/MWCNT MNCs were identified using XRD measurements. The morphology of the Fe_3O_4/TiO_2/MWCNT MNCs were also investigated by TEM.

The synthesis of diazepine derivatives **46** in a one-pot multicomponent process involving different 1,2-diamines **40** and terminal alkynes **14** in the presence of silica-supported iron oxide (Fe_3O_4/SiO_2) nanoparticles (S-MMNPs) in ethanol at ambient temperature [30] was performed, as shown in Scheme 6.25. The TEM images displayed monodispersed spherical-shaped NPs in which the Fe_3O_4 NPs were supported on silica.

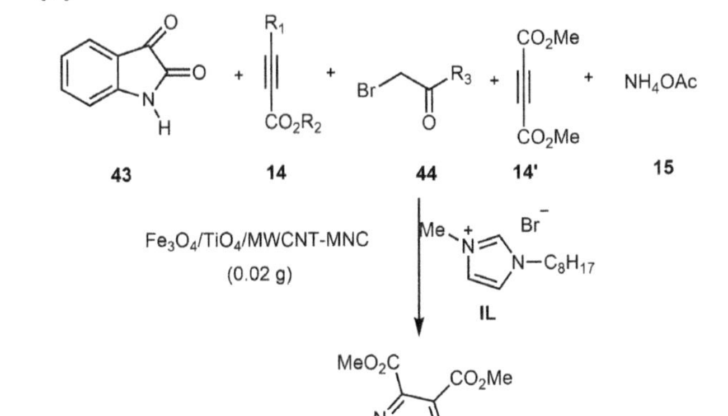

Scheme 6.24: Synthesis of azepines using Fe_3O_4 nanocomposite.

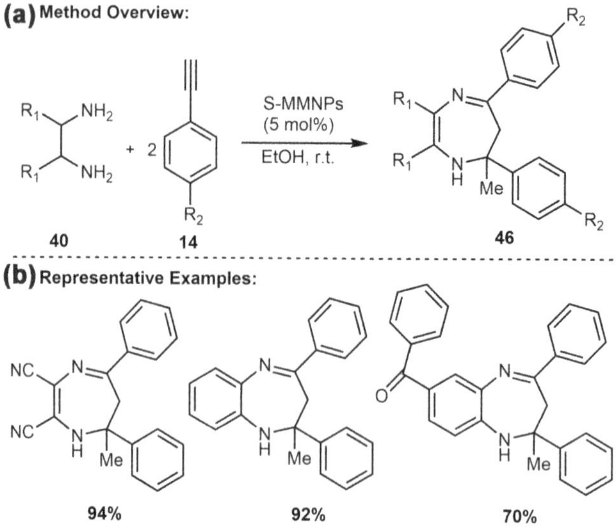

Scheme 6.25: Synthesis of diazepines using Fe_3O_4/SiO_2NPs.

The synthesis of the benzodiazepines **47** using ZnS NPs *via* reaction of *o*-phenylenediamine **40** and dimedone **31** with different arylaldehydes **4** in refluxing ethanol [31] was reported. The morphology of ZnS NPs was studied by SEM and TEM, and were found to be accumulated into about 20–30 nm spherical structures. The reaction was feasible in both aprotic and protic solvents such as EtOH, MeOH, H_2O, MeCN, $CHCl_3$ and *n*-hexane, however, ethanol was found to be the best for this reaction. In this one-pot multicomponent reaction, a wide range of 2-, 3- and 4-substituted aromatic aldehydes react with the other two substrates to give products in excellent yields, Scheme 6.26.

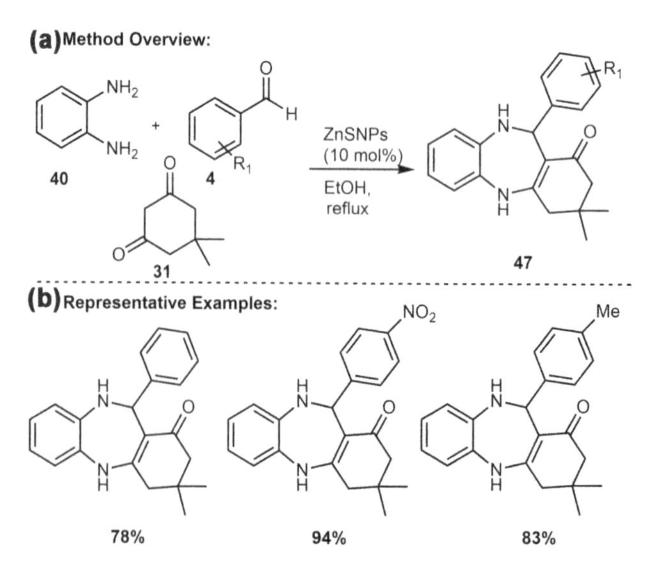

Scheme 6.26: Synthesis of benzodiazepines using ZnSNPs.

6.3 Conclusion

There is an ever-increasing demand for designing newer methods for the synthesis of *N*-heterocycles amongst the organic chemists. In this prospect, the synthesis using inorganic nanoparticles as nanocatalyst offer advantages in terms of activity, selectivity, efficiency and reusability. They can be reused by simply recovering by magnet or filtration without noticeable change in its catalytic activity, still there is need to explore as well as improve and to synthesize new NPs with more properties. This chapter provides an inclusive understanding on organic reactions which are catalyzed using environmentally friendly inorganic nanoparticles.

References

1. Kaushik N, Thakkar MS, Snehit S, Mhatre MS, Rasesh Y, Parikh MS. Biological synthesis of metallic nanoparticles. Nanomed Nanotechnol Biol Med 2010;6:257–62.
2. Kubik KB, Sugisaka M. From molecular biology to nanotechnology and nanomedicine. Biosystems 2002;65:123–38.
3. Daniel MC, Astruc D. Gold nanoparticles: assembly, supramolecular chemistry, quantum-size-related properties, and applications toward biology, catalysis, and nanotechnology. Chem Rev 2004;104:293–346.
4. Cho EJ, Holback H, Liu KC, Abouelmagd SA, Park J, Yeo Y. Nanoparticle characterization: state of the art, challenges, and emerging technologies. Mol Pharm 2013;10:2093–110.
5. Machado S, Pacheco JG, Nouws HPA, Albergaria JT, Delerue-Matos C. Characterization of green zero-valent iron nanoparticles produced with tree leaf extracts. Sci Total Environ 2015;533:76–81.
6. Tai CY, Tai CT, Chang MH, Liu HS. Synthesis of magnesium hydroxide and oxide nanoparticles using a spinning disk reactor. Ind Eng Chem Res 2007;46:5536–41.
7. Crooks RM, Zhao M, Sun L, Chechik V, Yeung LK. Dendrimer-encapsulated metal nanoparticles: synthesis, characterization, and applications to catalysis. Acc Chem Res 2001;34:181–90.
8. Khodadadi MR, Pourceau G, Becuwe M, Wadouachi A, Toumieux S. First sustainable aziridination of olefins using recyclable copper immobilized magnetic nanoparticles. Synlett 2019;30:563–6.
9. Bangale S, Jondhale V, Pansare D, Chavan P. Reusable $ZnCr_2O_4$ nano catalyzed one pot three-component cycloaddition reaction for synthesis of azetidine derivatives under ultrasound irradiation. Polycycl Aromat Comp 2021;42:6398–410.
10. Veisi H, Mohammadi P, Gholami J. Sulfamic acid heterogenized on functionalized magnetic Fe_3O_4 nanoparticles with diaminoglyoxime as a green, efficient and reusable catalyst for one-pot synthesis of substituted pyrroles in aqueous phase. Appl Organomet Chem 2014;28:868–73.
11. Moghaddam FM, Foroushani BK, Rezvani HR. Nickel ferrite nanoparticles: an efficient and reusable nanocatalyst for a neat, one-pot and four component synthesis of pyrroles. RSC Adv 2015;5: 18092–6.
12. Bruneau A, Gustafson KPJ, Yuan N, Tai CW, Persson I, Zou X, et al. Synthesis of benzofurans and indoles from terminal alkynes and iodoaromatics catalyzed by recyclable palladium nanoparticles immobilized on siliceous mesocellular foam. Chem Eur J 2017;23:12886–91.
13. Choi I, Chung H, Park JW, Chung YK. Active and recyclable catalytic synthesis of indoles by reductive cyclization of 2-(2-Nitroaryl)acetonitriles in the presence of Co–Rh heterobimetallic nanoparticles with atmospheric hydrogen under mild conditions. Org Lett 2016;18:5508–11.

14. Safari J, Zarnegar Z. Sulphamic acid-functionalized magnetic Fe_3O_4 nanoparticles as recyclable catalyst for synthesis of imidazoles under microwave irradiation. J Chem Sci 2013;125:835–41.
15. Safari J, Zarnegar Z. Immobilized ionic liquid on superparamagnetic nanoparticles as an effective catalyst for the synthesis of tetrasubstituted imidazoles under solvent-free conditions and microwave irradiation. C R Chimie 2013;16:920–8.
16. Girish YR, Kumar KSS, Manasa HS, Shashikanth S. ZnO: an ecofriendly, green nano-catalyst for the synthesis of pyrazole derivatives under aqueous media. J Chin Chem Soc 2014;61:1175–9.
17. Arde SM, Patil AD, Mane AH, Salokhe PR, Salunkhe RS. Synthesis of quinoxaline, benzimidazole and pyrazole derivatives under the catalytic influence of biosurfactant-stabilized iron nanoparticles in water. Res Chem Intermed 2020;46:5069–86.
18. Chetia M, Ali AA, Bordoloi A, Sarma D. Facile route for the regioselective synthesis of 1,4-disubstituted 1,2,3-triazole using copper nanoparticles supported on nanocellulose as recyclable heterogeneous catalyst. J Chem Sci 2017;129:1211–7.
19. Phukan P, Agarwal S, Deori K, Sarma D. Zinc oxide nanoparticles catalysed one-pot three-component reaction: a facile synthesis of 4-aryl-NH-1,2,3-triazoles. Catal Lett 2020;150:2208–19.
20. Mohammadi B. Microwave assisted one-pot pseudo four-component synthesis of 2,4,6-trisubstituted pyridines using γ-MnO_2 nanoparticles. Monatsh Chem 2016;147:1939–43.
21. Heravi MM, Beheshtiha SYS, Dehghani M, Hosseintash N. Using magnetic nanoparticles Fe_3O_4 as a reusable catalyst for the synthesis of pyran and pyridine derivatives via one-pot multicomponent reaction. J Iran Chem Soc 2015;12:2075–81.
22. Aday B, Yıldız Y, Ulus R, Eris S, Sen F, Kaya M. One-pot, efficient and green synthesis of acridinedione derivatives using highly monodisperse platinum nanoparticles supported with reduced graphene oxide. New J Chem 2016;40:748–54.
23. So MH, Liu Y, Ho CM, Lam KY, Che CM. Silica-supported gold nanoparticles catalyzed one-pot, tandem aerobic oxidative cyclization reaction for nitrogen-containing polyheterocyclic compounds. ChemCatChem 2011;3:386–93.
24. Hemmat K, Nasseri KA, Allahresani A. $CoFe_2O_4$@SiO_2@Co(III) salen complex: a magnetically recyclable heterogeneous catalyst for the synthesis of quinoline derivatives in water. ChemistrySelect 2019;4:4339–46.
25. Sapkal SB, Shelke KF, Shingate BB, Shingare MS. Nickel nanoparticles: an ecofriendly and reusable catalyst for the synthesis of 3,4-dihydropyrimidine-2(1H)-ones via Biginelli reaction. Bull Kor Chem Soc 2010;31:351–4.
26. Ganji P, van Leeuwen PWNM. Phosphine supported ruthenium nanoparticles catalyzed synthesis of substituted pyrazines and imidazoles from α-diketones. J Org Chem 2017;82:1768–74.
27. Hasaninejad A, Shekouhy M, Zare A. Silica nanoparticles efficiently catalyzed synthesis of quinolines and quinoxalines. Catal Sci Technol 2012;2:201–14.
28. Song T, Ren P, Ma Z, Xiao J, Yang Y. Highly dispersed single-phase Ni_2P nanoparticles on N,P-codoped porous carbon for efficient synthesis of N-heterocycles. ACS Sustainable Chem Eng 2020;8:267–77.
29. Kohestani T, Sayyed-Alangi SZ, Hossaini Z, Baei MT. Ionic liquid as an effective green media for the synthesis of (5Z, 8Z)-7H-pyrido[2,3-d]azepine derivatives and recyclable Fe_3O_4/TiO_2/multi-wall cabon nanotubes magnetic nanocomposites as high performance organometallic nanocatalyst. Mol Divers 2021;26:1441–54.
30. Maleki A. One-pot multicomponent synthesis of diazepine derivatives using terminal alkynes in the presence of silica-supported superparamagnetic iron oxide nanoparticles. Tetrahedron Lett 2013;54:2055–9.
31. Naeimi H, Foroughi H. ZnS nanoparticles as an efficient recyclable heterogeneous catalyst for one-pot synthesis of 4-substituted-1,5-benzodiazepines. New J Chem 2015;39:1228–36.

Parvathi Jayasankar* and Rajasree KarthyayaniAmma

7 Surfactants-surface active agents behind sustainable living

Abstract: Surfactants are surface active agents. They are mainly chemicals, when added to water will reduce the surface tension of water and thus increases wettability on the given surface. Surfactants normally carry hydrophilic and hydrophobic ends among which hydrophilic end connects to water layer. The hydrophobic part connects between aqueous phase and the given hydrophobic surface through the hydrophobic end. However, these surfactants act as emulsifying agents or foaming agents. Further the chemistry behind the action of surfactants is introduced for the readers. Surfactants can be classified based on origin, charge on heads, solubility of water etc. and is specified in this paper. Also synthesis of various types of surfactants is carefully incorporated in the chapter. The chapters dwells in detail the various sustainability related applications of surfactants which is relevant for sustainable living in the society.

Keywords: bio surfactants; hydrophilic; hydrophobic; micelle; surfactants.

7.1 Introduction

Surfactants are surface active agents. Surfactants have the tendency to reduce the cohesive force between liquid molecules. It reduces surface tension between two liquids or between a solid and a liquid. The ability to reduce surface tension of water makes it useful as an agent to increase the wettability of surfaces. Surfactants normally carry hydrophilic and hydrophobic ends among which hydrophilic end connects to water layer. The hydrophobic part connects between aqueous phase and the given hydrophobic surface. Surfactants act as foaming agents, dispersants, soaps, medicines and detergents etc. [1].

7.1.1 Surfactants and sustainability

Water repellancy of soil is a major factor which affects irrigation in agriculture. Various factors affect soil water repellency includes microorganisms like fungal pathogens, root exudates, organic matter, organic coatings on soil particles etc. [2]. Water repellancy

***Corresponding author: Parvathi Jayasankar**, JAIN Deemed to be University, Bangalore, India,
E-mail: parvathi.jaysankar@jainuniversity.ac.in
Rajasree KarthyayaniAmma, A M JAIN College, Meenambakkam, Chennai, India

As per De Gruyter's policy this article has previously been published in the journal Physical Sciences Reviews. Please cite as: P. Jayasankar and R. KarthyayaniAmma "Surfactants-surface active agents behind sustainable living" *Physical Sciences Reviews* [Online] 2022. DOI: 10.1515/psr-2022-0130 | https://doi.org/10.1515/9783110913361-007

of soil causes issues like variation in plant growth, soil contamination and environmental property deterioration. Water holding nature of the soil in local areas due to its organic coating in grassy lands results in the formation of turf grass (Figure 7.1) Soil surfactants are more useful for treating these turf grasses which enhances the percolation of water in the soil and reduces its water retention capacity [3]. Surfactants also helps to optimize the formulation of herbicides, pesticides and insecticides.'Leaf Shield'- agrochemical surfactant- helps in enhancing the wettability of water on surface of the leaf and thus improves the distribution of the herbicides (Figure 7.2) [4].

Soil surfactants act as soil conditioners which support the quality of soil like PH, alkalinity, fertility etc. These soil conditioners enhance mechanics of soil and prevent soil erosion [5]. Non-ionic surfactants majorly find application in the area of wetting agents [6]. However, surfactants have its own application in various fields including oil recovery; enhance flow of oil in porous rocks and to produce aerosol [7].

Surfactant enhanced oil recovery is an important area of attention in oil spillage in water bodies. Proper and careful selection of surfactants according to parameters such as temperature, pressure, salinity etc. helps speedy recovery of oil from water bodies. Alteration in the choice of surfactants causes undesirable wettability and rock dissolution which enhances unwanted chemical reactions among rock materials with displacing fluid and blockage of the pore space [8].

Figure 7.1: Turf grass treatment using surfactants.

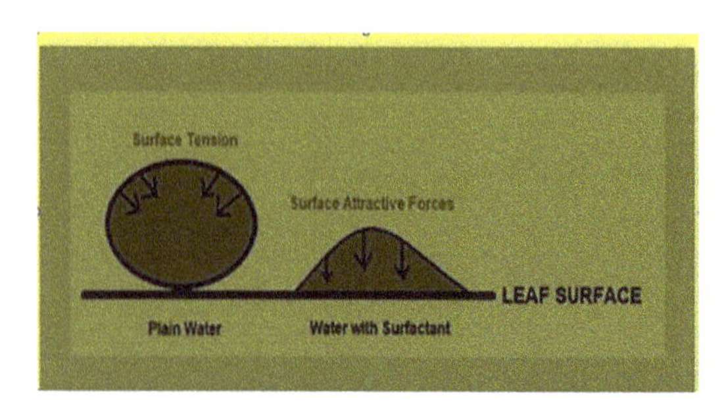

Figure 7.2: Leaf shield.

7.1.2 Aim and objective of the chapter

Major aim of the present chapter is to give awareness about the application of surfactants for sustainable living. Present chapter targets the following objectives
1. Understand the use of surfactants in sustainability
2. Chemistry behind action of Surfactants
3. Classification of Surfactants
4. Synthesis of Biosurfactants
5. Advantages of Biosurfactants
6. Applications of Surfactants

7.2 Chemistry behind surfactant action

Surfactants are amphiphilic molecules. It possesses a hydrophilic head and a hydrophobic tail as shown in Figure 7.3 [9]. Their hydrophilic heads are strongly connected with water, and weaken the hydrogen bond among water molecules (Figure 7.4). Alternatively, their tails adsorb to materials including clay minerals, air molecules in pores, or hydrophobic organic substances in soil [10]. These surfactant molecules form a cluster like structure on interface of surfaces of two liquids or a solid and a liquid. These clusters known as micelle (Figure 7.5) which reduces surface tension in water, water–air and soil–water surfaces.

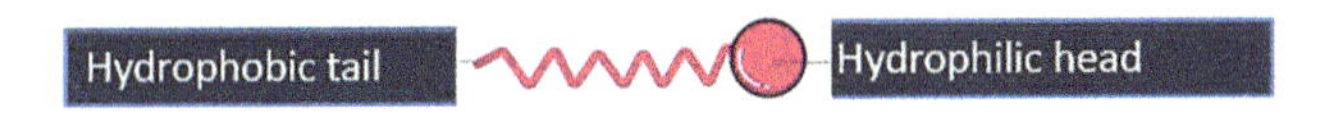

Figure 7.3: Surfactant molecule.

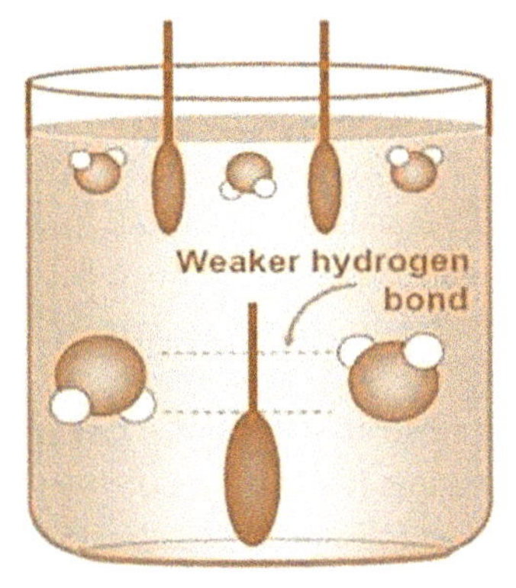

Figure 7.4: Action of surfactant on water molecule.

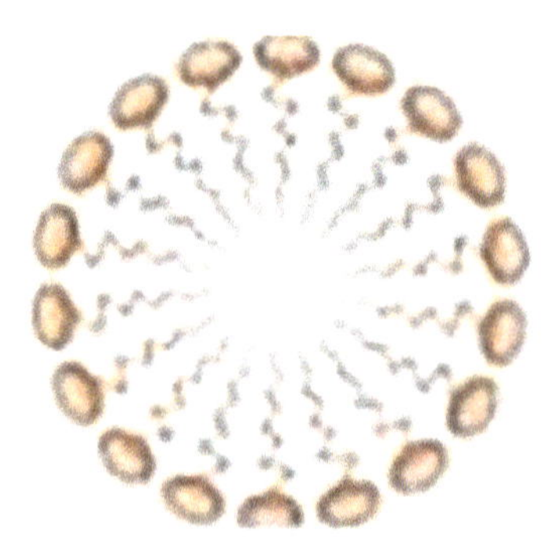

Figure 7.5: Micelle formation.

These surface-active agents (surfactants) help to percolate water through soil, which have hydrophobic, water repellent and organic coatings [11]. Thus, surfactants can connect between two phases and it forms strong interactons with both of the phases and it reduces surface tension.

7.3 Classification of surfactants

There are various types of classification of surfactants.

- Based on the solubility in water surfactants are are classified as hydrophilic and hydrophobic. Surfactants have affinity towards water are considered as hydrophilic. Hydrophobic surfactants are insoluble in water
- Based on the origin and composition of surfactants they are classified into synthetic, bio based and microbial biosurfactants
- Synthetic, chemical surfactants which are widely used have harmful effects to human health and environment [12]. Application of synthetic surfactants above the permitted concentration causes toxicity [13] and the need of availability of natural surfactant arose.
- Biosurfactants are naturally occurring surfactants in plants, animals and microorganisms. These biosurfactants are sustainable and environmental friendly as plants, animals and microorganisms are ecofriendly and less hazardous [14]. Agriculture products like nutmeg, industrial waste or residual items, soya bean waste, sun flower oil refinery waste, bio diesel waste are some examples of biosurfactants.
- Micro biosurfactants are produced by microbial organisms viz bacteria, fungus and yeast extracellularly or linked to plasmatic membrane [15].
- Based on the polarity of the head group, synthetic surfactants are further classified into anionic, cationic, zwitter ionic and non-ionic Surfactants (Refer Table 7.1) [10]. The structure of these surfactants is shown in Figure 7.6.

- Non ionic surfactants do not pocess any charge in their head group. Solubility of non ionic surfactants in water decreases with temperature. Non ionic surfactants do not disintegrate in aqueous solution. Soil surfactants used for soil conditioning are nonionic in nature due to its adhesive nature towards organic matter in the soil [16].
- Anionic surfactants: Anionic surfactants pocess negative charge such as sulphate (SO_4^-), sulphonate (SO_3^-), phosphate, and carboxylates (COO–). The alkyl sulphates include ammonium lauryl sulphate, sodium lauryl sulphate and the related alkyl-ether sulphates
- Cationic surfactants: They have positive charge on their head group. They act on all PH ranges and the action of the surfactant depends on its ability to adsorb on negatively charged surfaces. Benzyl quats, Epichloro hydrin quats etc. are among cationic surfactants
- Zwitterionic surfactants: They are amphoteric surfactants in which both cationic and anionic parts attached to the same molecule [10]. The biological zwitterionic surfactants have a phosphate anion with an amine or ammonium, such as phospholipids phosphatdylserine, phosphatdylethanolamine, phosphatdylcholine and sphigomyelins.Sultaines and cocamido propyl betaine are examples of zwitter ionic surfactants.

Table 7.1: Types of surfactants.

Type of surfactant	Example	Use
Anionic	Alkyl sulfates, texapon, soaps	Laundry detergent, dishwashing liquid, shampoos
Cationic	Quaternary ammonium salts	Softners in textile industry, anti-static additives
Nonionic	Ethoxylated aliphatic alcohol, polyoxy-ethylene surfactants	Food ingradient, wetting agent in coatings
Zwitterionic	Amphoacetates, betaines	Cosmetics

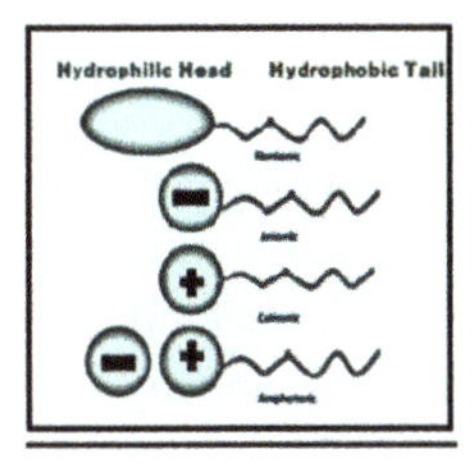

Figure 7.6: Classification of ionic surfactants.

7.4 Preparation of surfactants

Synthetic surfactants can be manufactured using chemicals such as thiols, alkyl halides sulphonate etc., whereas biosurfactants are extracted from various plants and animals.

7.4.1 Synthetic surfactants

Thiols are coming under cationic synthetic surfactants. Thiol surfactants are synthesized by the reaction between halo alkanes and 4-mercapto pyridine in 1:1 ratio in the presence of ethanol. The above mixture has to heat for long hours followed by cooling to room temperature. The product obtained is the cationic surfactant which is recrystallised from diethyl ether and ethanol and then filtrated and dried [17].

7.4.2 Synthesis and chemical constitution of biosurfactants

Aspergillus, penicillium etc. are few of the ascomycetes which are studied for the production of biosurfactants [18]. Biosynthesis of biosurfactants involves methods to form hydrophilic and hydrophobic components and then they are combined [18]. An example of biosynthesis of biosurfactant is the production of rhamnolipids by bacteria in a medium containing glycerol [19]. Asperigillium and penicillium are the most commonly reported microorganisms for the synthesis of biosurfactants with emulsions of high stability which enables effective and excellent reduction of interface and interfacial tension [20]. Plant saponins also have excellent surfactant properties [21]. Saponins are composed of amphiphilic glucosides which contain polar glycone structural units (sugars) separated from non-polar aglycones structure moieties (sapogenesis) [22]. Based on the nature of aglycone components present in saponins they are categorized as steroidal counter parts and triterpenoid saponins [23]. Saponins are extracted by ultrasound assisted extraction and microwave assisted extraction and accelerated solvent extraction [24].

7.4.3 Advantages of biosurfactants

The significant advantages of biosurfactants over the synthetic surfactants are low toxicity, high biodegradability, tolerance to pH variation, the use of renewable substrates which ensures cost effectiveness and less pollution to environment [18]. Rhamnolipids and sophorlipids have been recommended for industrial applications globally [19]. Biosurfactants derived from microorganisms reduces the surface tension and shows excellent emulsifying capacity [20]. Biosurfactants differ from synthetic surfactants in its complex structure since they have extracted from various biomolecules [21]. Biosurfactants are classified as glycolipids, lipopeptides and polymeric biosurfactants based on their chemical constitution [22].

7.5 Application of surfactants

Surfactants have wide range of applications as emulsifying agents or foaming agents, dispersants, soaps and detergents. However, in this chapter focus is given for sustainability related applications of surfactants.

7.5.1 Water and soil treatment

Petroleum and other hydrocarbons are major contaminants for soil and water bodies and ground water as well. Surface active agents-surfactants are chemicals used to remove the petrochemicals from soil and water source through the reduction of interfacial tension [25]. Recently techniques like addition of nanoparticles to surfactants enhance aquifer recovery [26].

7.5.2 Minimize water repellancy

Some of the soil surfactants reduce water repellancy and increase the wettability of the soil. Soil surfactants reduce surface tension of water and increases the percolation power of water through soil. These soil surfactants are mainly used in treating turf grasses in golf courses. Soil surfactants play a big role in various water conservation techniques in grassy area by maximizing irrigation efficiency and minimizing surface runoff of water [27].

7.5.3 Clinical application

Some of the synthetic surfactants are surface active reagents which reduces surface tension at the air water interface of alveolar lining. It helps to reduce the tendency for alveolar collapse. These pulmonary surfactants are a special type of chemicals called lipoproteins [28].

7.5.4 Cosmetic application

Biosurfactants derived from plants and microorganisms shows better performance and more preferred in removing heavy metal from contaminated soil [29]. Plant saponins found wide range of applications in Industry including cosmetic, pharmaceutical and food industry [30]. They are used as antioxidants, antimicrobial, anti-inflammatory and antidiabetic agents [31].

7.5.5 Application as corrosion inhibitor

Extensive use of acid solutions in industries leads to severe corrosion on metal surfaces. Surfactants form micelle at the interphase of metal surface and solution prevents the exposure of metal towards further contact with the liquid surface. The corrosion inhibition character of a surfactant is directly related to CMC (critical Micelle concentration is the concentration of surfactant above which micelle formation is happening) [17].

7.5.6 Surfactants in agriculture

Surfactants also have very interesting applications in agriculture. They help in soil water percolation and uniform application of pesticides and herbicides on the surface of leaves. An overview of the usage of surfactants in agriculture is given in Figure 7.7.

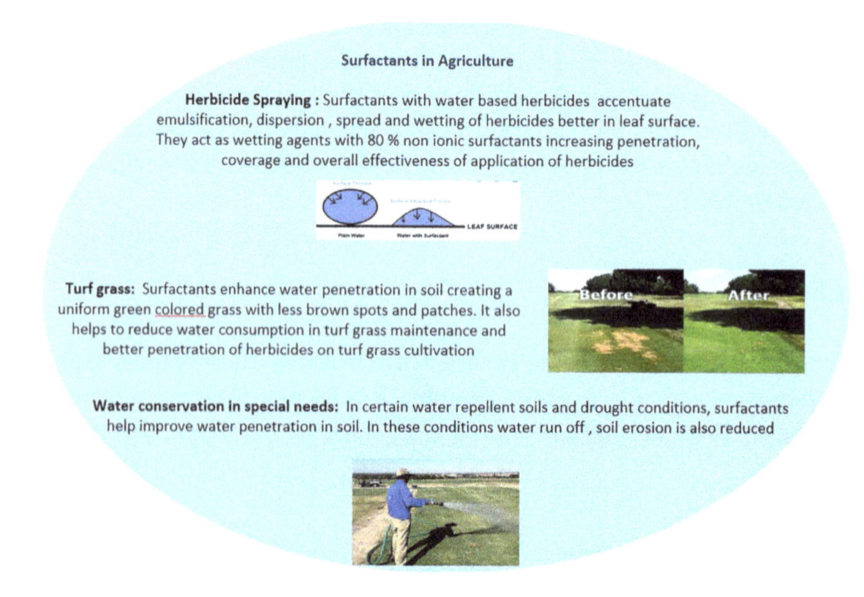

Figure 7.7: Soil surfactant activity.

7.6 Conclusions

Surfactants are chemicals which act as surface active agents. Surfactants take wide range of application in everyday life. They are used as soaps, detergents, dispensers etc. Surfactants find application in sustainable living. It reduces water retention in soil, improve distribution of herbicides and insecticides on leaves and increases penetration power of water through soil.

Surfactant molecules are amphiphilic in nature. It has a hydrophilic head and hydrophobic tail which forms micelle structure at the interface of two liquids or between a solid and liquid phase which deteriorates surface tension.

Surfactants can be classified into various types based on their origin, solubility, ionic nature etc. The major classification includes non-ionic, anionic, cationic and zwitter ionic surfactants. However, soil surfactants are majorly ionic and nonionic.

Synthetic surfactants like thiols can be synthesized by various chemical methods and the surfactant products can be recrystallized and dried to improve its purity. However, biosurfactants like rhamnolipids are synthesized by bacteria in glycerol medium.

Biosurfactants are ecofriendly and less toxic in nature. They are useful in soil purification, drug manufacture and in cosmetic industry. They are well known for their use as antioxidants, anti-inflammatory and antidiabetic agents. Surfactants are widely applied in various fields of life.

In addition to its usage in agriculture, surfactants are used in treatment of soil and recover oil from soil and water bodies. However, as we move into sustainable way of living biosurfactants will gain momentum provided the manufacturing cost is substantially reduced.

References

1. Texter J, editor. Reaction and synthesis in surfactant system. Rochester, Newyork: Strider Research Corporation.
2. Water management: using soil surfactants, SportsTurf; 2011. Available from: www.sportsturfonline.com.
3. Mobbs TL, Peters RT, Davenport J, Evans M, Wu J. Effects of four soil surfactants on four soil water properties in sand and silt loam. J Soil Water Conserv 2012;67:275–83.
4. Ishiguro M, Fujii T. Upward infiltration into porous media as affected by wettability and anionic surfactants. Soil Sci Soc Am J 2008;72:741–9.
5. Moore D, Boerth TJ, Kostka SJ, Franklin M. The effect of soil surfactants on soil hydrological behavior, the plant growth environment, irrigation efficiency and water conservation. J Hydrol Hydromechanics 2010;58:142–8.
6. Song E, Goyne KW, Kremer RJ, Anderson SH, Xiong X. Certain soil surfactants could become a source of soil water repellency after repeated application. Nanomaterials 2021;11:2577.
7. Britannica, The Editors of Encyclopaedia. Surfactant. In: Encyclopedia Britannica; 2020. Available from: https://www.britannica.com/science/surfactant [Accessed 30 May 2022].

8. Ali S, Xie Q, Negin C. Most common surfactants employed in chemical enhanced oil recovery. Petroleum 2017;3:197–211.
9. Salehi M, Johnson SJ, Liang J-T. Enhanced wettability alteration by surfactants with multiple hydrophilic moieties. J Surfactants Deterg 2010;13:243–6.
10. Mouton J, Mercier G, Bleis, Blazier JF. Amphoteric surfactants for PAH and lead polluted-soil treatment using flotation. Wat Air Soil Pollut 2009;197:381–93.
11. Kostka SL, Dekker WL, Ritsema CJ, Cisar JL, Franklin MK. Surfactant as Management tools for ameliorating soil water repellency in turf grass. Columbus, Ohio, USA: International Society for Agrochemical Adjuvants (ISAA); 2007. Available from: https://s3.amazonaws.com/aquatrols/2009111694129.pdf.
12. Rebello S, Asok AK, Mundayoor S, Jisha MS. Surfactants: Chemistry, Toxicity and Remediation. Cham, Switzerland: Springer; 2013.
13. L´emery E, Briançon S, Chevalier Y, Bordes C, Oddos T, Gohier A, et al. Skin toxicity of surfactants: structure/toxicity relationships. Colloids Surf A 2015;469:166–79.
14. Helmy Q, Kardena E, Funamizu N, Wisjnuprapto. Strategies toward commercial scale of biosurfactant pro duction as potential substitute for it's chemically counter parts. Int J Biotechnol 2011;12:66–86.
15. Moldes AD, Lopez LR, Rincon-Fontan M, Lopez-Prieto A, Vecino X, Cruz JM. Synthetic and bio-derived surfactants versus microbial biosurfactants in the cosmetic industry: an overview. Int J Mol Sci 2021;22:2371.
16. Kuhnt G. Behaviour and fate of surfactants in soil. Environ Toxicol Chem 1993;10:1813–20.
17. Azzama EMS, Hegazya MA, Kandil NG, Badawi AM, Samia RM. The performance of hydrophobic and hydrophilic moieties in synthesized thiol cationic surfactants on corrosion inhibition of carbon steel in HCl. Egypt J Pet 2015;24:493–503.
18. Silva ACS, Santos PND, e Silva TAL, Andrade RFS, Campos-Takaki GM. Biosurfactant production byfungi as a sustainable alternative. Agric Microbiol 2018;85:1–12.
19. Laine RA, Griffin PFS, Sweeley CC, Brennan PJ. Monoglucosyloxyoctadecenoic acid, a glycolipid from aspergillus niger. Biochemistry 1972;11:2267.
20. Aghel N, Moghimipour E, Raies A. Formulation of a herbal shampoo using total saponins of acanthophyllum squarrosum. Iran. J Pharm Res 2007;6:167–72.
21. Sparg SG, Light ME, Van Staden J. Biological activities and distribution of plant saponins. J Ethnopharmacol 2004;94:219–43.
22. Belwal T, Chemat F, Venskutonis PR, Cravotto G, Jaiswal DK, Bhatt ID, et al. Recent advances in scaling-up of non-conventional extraction techniques: learning from successes and failures. Trac Trends Anal Chem 2020;127:115895.
23. Belwal T, Ezzat SM, Rastrelli L, Bhatt ID, Daglia M, Baldi A, et al. A criticalanalysis of extraction techniques used for botanicals: trends, priorities, industrial uses and optimization strategies. Trac Trends Anal Chem 2018;100:82–102.7.
24. Ramamurthy AS, Chen Z, Li X, Azmal M. Surfactant assisted removal of engine oil from synthetic soil. Environ Protect Eng 2015;41:67–79.
25. Mao X, Jiang R, Xiao W, Yu J. Use of surfactants for remediation of contaminated soil,a review. J Hazard Mater 2015;285:419–35.
26. Oostindie K, Dekker LW, Wesseling JG, Ritsema CJ. Soil surfactant stops water repellency and preferential flow paths. Soil Use Manag 2008;24:409–15.
27. Poynter SE, LeVine AMMD. Surfactant biology and clinical application. Crit Care Clin 2003;19:459–72.
28. Pioselli B, Salomone F, Mazzola G, Amidani D, Sgarbi E, Amadei F, et al. Pulmonary surfactant: a unique biomaterial with life-saving therapeutic applications. Curr Med Chem 2022;29:526–90.

29. Hu J, Nie S, Huang D, Li C, Xie M. Extraction of saponin from Camellia oleifera cake and evaluation of its antioxidant activity. Int J Food Sci Technol 2012;47:1676–87.
30. Sur P, Chaudhuri T, Vedasiromoni JR, Gomes A, Ganguly DK, Wiley J. Antiinflammatory and antioxidant property ofsaponins of tea [Camellia sinensis (L) O. Kuntze] root extract. Phytother Res 2001;176:174–6.
31. Liu ZF, Li ZG, Zhong H, Zeng GM, Liang YS, Chen M, et al. Recent advances in the environmental applications of biosurfactant saponins: a review. J Chem Environ Eng 2017;15:6030–8.

Supplementary Material: The online version of this article offers supplementary material (https://doi.org/10.1515/psr-2022-0130).

Piyali Bhattacharya and Swati De*

8 Simple naturally occurring β-carboline alkaloids – role in sustainable theranostics

Abstract: This review is a brief treatise on some simple β-carboline alkaloids that are abundantly available in plants, animals and foodstuff. These alkaloids are well known for their pharmacological action as well as their allelopathic behaviour. The focus of this review is on sustainable use of naturally occurring compounds in safeguarding human health and protecting our environment at large i.e. the prospective applications of these molecules for *Sustainable Theranostics*. The review commences with an initial introduction to the β-carboline alkaloids, followed by an outlay of their geographical distribution and natural abundance, then the basic structure and building units of the simplest β-carboline alkaloids have been mentioned. This is followed by a discussion on the important methods of extraction from natural sources both plants and animals. Then the foundation for the use of these alkaloids in *Sustainable Theranostics* has been built by discussing their interesting photophysics, interactions with important biological molecules and an extensive survey of their therapeutic potential and allelopathic behaviour. Finally the review ends with a silver lining mentioning the future prospective applications of these alkaloids with special relevance to sustainability issues.

Keywords: β-carboline alkaloids; allelopathy; sustainable; therapeutics.

8.1 Introduction

β-carbolines more popularly known as harmala alkaloids have a tricyclic ring system comprising a pyridoindole group. This class of compounds is well known for their therapeutic applications [1–5]. The β-carboline alkaloids are well known stimulants, hallucinogens and muscle paralysants [6, 7]. Their anti-carcinogenic activity is enhanced upon UV irradiation [8].

Besides their biological applications, β-carboline alkaloids are useful as fluorophores as their fluorescence is extremely sensitive to solution parameters such as polarity and pH. These features have made these compounds important for research in biology, medicine, photophysics and photochemistry. The tricyclic system containing two categories of nitrogen atoms i.e. in the pyridinic ring and in the pyrrole part

***Corresponding author: Swati De,** Department of Chemistry, University of Kalyani, Kalyani, 741235, India, E-mail: deswati1@gmail.com
Piyali Bhattacharya, Department of Chemistry, University of Kalyani, Kalyani, 741235, India

As per De Gruyter's policy this article has previously been published in the journal Physical Sciences Reviews. Please cite as: P. Bhattacharya and S. De "Simple naturally occurring β-carboline alkaloids – role in sustainable theranostics" *Physical Sciences Reviews* [Online] 2022. DOI: 10.1515/psr-2022-0132 | https://doi.org/10.1515/9783110913361-008

strongly influence their photophysical properties. The pyridinic nitrogen is more basic in nature, moreover its basicity increases when excited. This basicity increase is dependent on the substituents present in the structure [9–11]. β-Carboline alkaloids occur as four ionic species-cationic, neutral, zwitterionic and anionic forms. The equilibria between the various forms is guided by the pH of the solution and solvent [12].

This special class of compounds is found to be present in plants, animal tissues and human urine and often appear as a photo initiated by product of Tryptophan present in the lens of human eye. The focus of this review is on the prospective applications of these molecules for *Sustainable Theranostics*, their interesting photophysical behaviour particularly fluorescence can be useful as a diag*nostic* tool while their well-known pharmacological action can be harnessed for *thera*peutic purpose. More importantly, some of these harmala alkaloids can be easily extracted from the abundant natural sources thus providing a pathway for *Sustainable Theranostics*.

Keeping the central theme of sustainability in mind, the present review focuses on four naturally available β-carboline alkaloids *viz.* norharmane (NHM), harmane, harmine and harmol – their natural occurrence in plants and animals; simple methods of extraction; interesting photophysical behaviour; interactions with biological molecules like proteins, enzymes and DNA; pharmacological applications and allelopathic applications.

8.2 Natural abundance of the β-carboline alkaloids

8.2.1 Plant sources

8.2.1.1 Worldwide abundance

β-Carboline alkaloids were first extracted from the plant *Peganum harmala*. This plant (Figure 8.1) is mainly found in the Mediterranean regions and temperate deserts. This plant has been used as a source of abortifacients and emmenagogue drugs in North Africa and the Middle East for a long time [13]. In Northwest China, the seeds of this plant were used to treat malaria and alimentary tract cancers [14]. In some Middle East countries and in Turkey too, seeds and in fact the entire plant *P. harmala* has been used as medicine [15, 16]. Many medicinal components e.g. flavones, amino acids, polysaccharides and alkaloids have been found in the plants of this group [17–19]. It was later proven that it is actually the β-carboline alkaloids e.g. harmane, NHM, harmol and harmine that were the most important components showing antidepressant, analgesic, vasorelaxant, antitumor and antimicrobial action [15, 20–24].

Figure 8.1: Some natural plant sources of β-carboline alkaloids: (a) *Rauvolfia vomitoria*, (b) *Peganum harmala*, (c) *Geissospermum vellosii*. [Source: Google images].

Other β-carboline alkaloids can be extracted from other plants, one example being *Geissospermum vellosii* which is a native tree of Brazil. Concentrated extract from the bark of the *G. vellosii* (Figure 8.1) induced apoptosis of tumor cells and inhibited proliferation of tumor cells, when applied to a human prostate cancer cell line, LNCaP [25]. Another plant *Rauvolfia vomitoria* found in Senegal, east Sudan, Tanzania and south Angola also contains β-carboline alkaloids. *R. vomitoria* (Figure 8.1) is also found in Bangladesh, China, parts of the Himalayan ranges and in Puerto Rico. The concentrated plant extract from *R. vomitoria* induced apoptosis in prostate cancer cells *viz.* the LNCaP cell line and helped in reducing tumor volumes. The anticancer activity was monitored by the suppression of growth of the cancer cells, progress of the cell cycle and by facilitating the building up of G1 phase cells and up-regulation of genes involved in DNA damage signaling pathway and apoptosis [26]. Two novel β-carboline alkaloids namely Oppositinines A and B were obtained from the plant *Neisosperma oppsitifolia* [27]. The genus *Neisosperma* is found in Western Pacific Islands, Tropical Asia and Polynesia. The specific plant *N. oppsitifolia* is found only in Malaysia. Passion flowers are another abundant natural source of β-Carboline alkaloids [28]. The world-wide distribution of these trees are summarized in Table 8.1 and indicated in the world map shown in (Figure 8.2).

Table 8.1: Location-wise availability of plant sources of β-carboline alkaloids around the world.

Various plant sources of β-carbolines	Available locations
Peganum harmala	Mediterranean regions and temperate deserts
Geissospermum vellosii	Brazil
Rauvolfia vomitoria	Senegal, east Sudan, Tanzania and south Angola, Bangladesh, China, parts of the Himalayan ranges and in Puerto Rico
Neisosperma oppositifolia	Malaysia
Passion flowers	95% from the South American forests and 5% from the Asian continent, Australian sub-continent and parts of North America

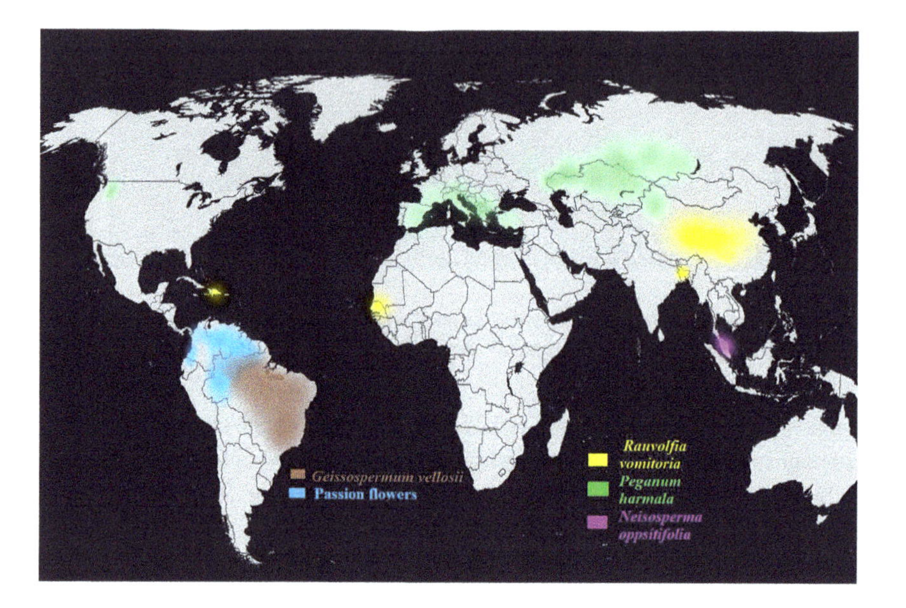

Figure 8.2: World Map indicating the abundance of various natural sources of β-carboline alkaloids.

8.2.1.2 Plant sources in India

In India, *P. harmala* which is a rich source of harmine, harmaline, harmalol and harmane is found in the Kashmir and Ladakh region. *R. vomitoria* is found in parts of the Himalayas (Figure 8.3). The plants of this genus are commonly known as '*Sarpagandha*' (*Rauvolfia serpentina*) in India.

8.2.2 Animal sources

Marine invertebrates are sources of several β-carboline alkaloids (Figure 8.4). These include hydroids [29] (*Aglaophenia*), bryozoans [30–32] (*Cribricellina*, *Catenicella*), soft

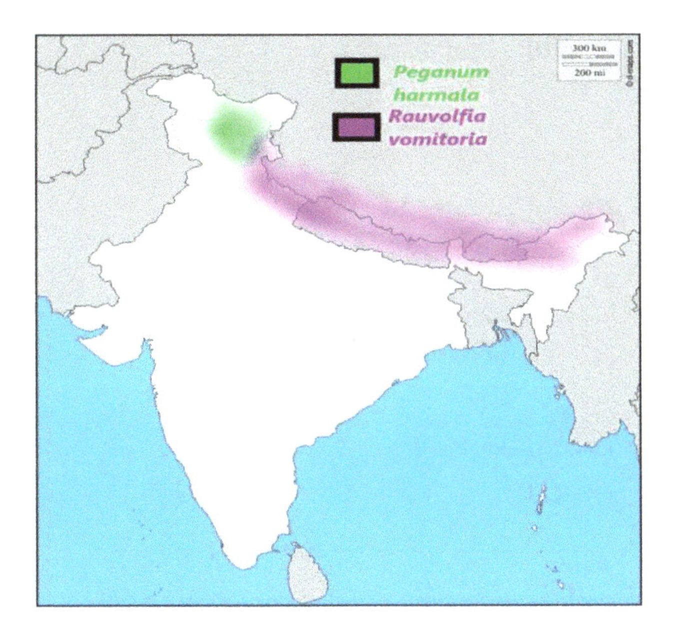

Figure 8.3: The natural abundance of *P. harmala* and *R. vomitoria* in India.

corals [33] (*Lignopsis*), tunicates [34–44] (*Eudistoma*, *Didemnum*, *Lissoclinum*, *Ritterella*, *Pseudodistoma*) and sponges. Sea ascidians of the genus Eudistoma also comprise few biologically active β-carbolines. A number of β-carbolines like eudistomins A-T [45–48], isoeudistomin U [40] and eudistomin U, eudistomidins A-F [49–51], eudistalbins [52] and some trypargine derivatives [53] have been isolated from *Eudistoma* species (Scheme 8.1).

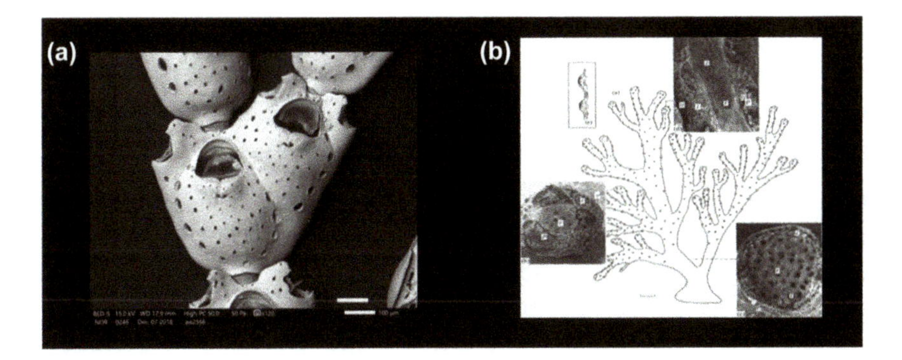

Figure 8.4: Marine invertebrate sources of β-carboline alkaloids: (a) *Cribricellina cribraria* and (b) *Lignopsis spongiosum*. [Source: Google images].

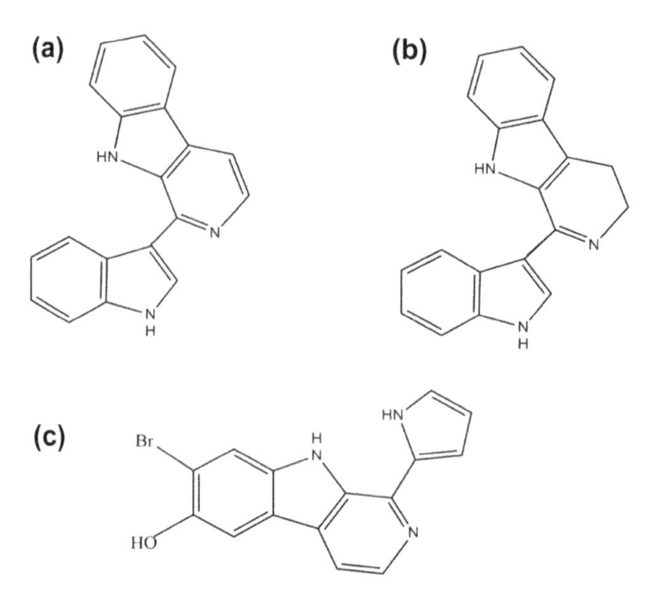

Scheme 8.1: Some β-carboline alkaloids extracted from *Eudistoma sp.* (a) Eudistomin U; (b) Isoeudistomin U and (c) Eudistomin A.

8.3 Structure of some simple β-carboline alkaloids

Norharmane (NHM) has the simplest structure among the β-carboline alkaloids. All others have various substitutions in the norharmane skeleton. Scheme 8.2 depicts a chart where some specific β-carboline alkaloids are listed based on where the substitution is present in the basic NHM skeleton. These are:

A. Substitution in the pyridine ring e.g. harmane, harmalan and β-carboline acid carboxylic acid ethyl ester (βCCE).
B. Substitution in both the indole and the pyridine ring e.g. harmine, harmol.
C. Substitution in the pyrrolic nitrogen e.g. 9-(4′-aminophenyl)-9*H*-pyrido[3,4-*b*] indole (aminophenylnorharmane) (APNH).
D. Substitution of indole moiety e.g. pinoline.

This review is primarily focused on four β-carboline alkaloids – the basic β-carboline alkaloid i.e. NHM and three others with simple substituents in either the pyridine ring e.g. harmane or in both the indole and pyridine rings e.g. harmine and harmol. The reason for focusing on these four molecules being:

(i) these four β-carboline alkaloids have been obtained from natural sources thus making the process of extraction sustainable;
(ii) the photophysics of these compounds is very interesting and they have been widely used as fluorescent probes in the past;
(iii) the biological applications of these β-carbolines are well documented.

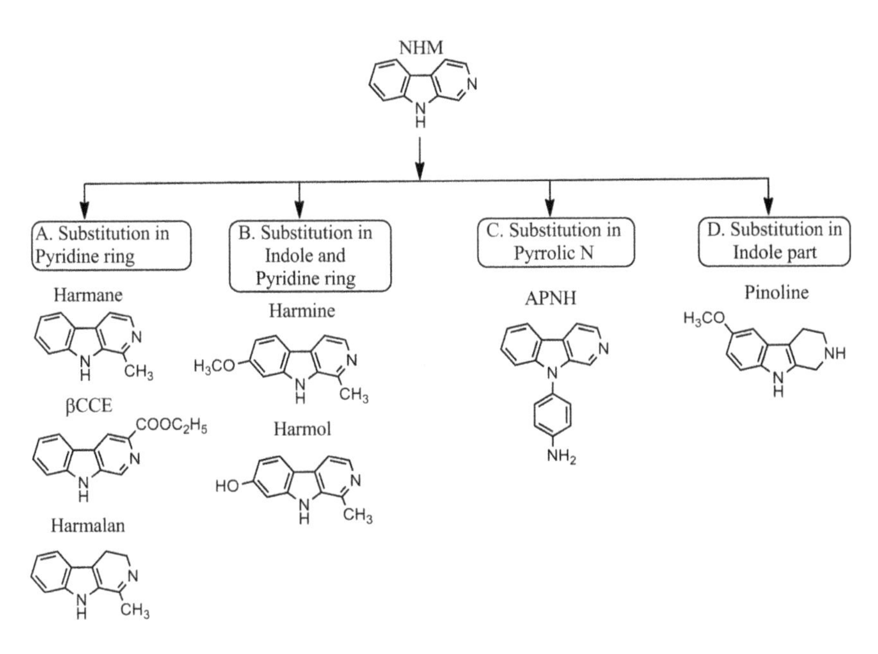

Scheme 8.2: Structures of some simple β-carboline alkaloids.

The focus of this review was mainly on: (i) the extraction/synthesis of these four β-carboline alkaloids; (ii) purification and characterization/identification of these compounds; (iii) their photophysical behaviour in confined systems and (iv) their biological applications with special emphasis on *thera*peutic and diag*nostic* applications (*Theranostics*).

8.4 Extraction from natural sources and structure determination of NHM, Harmane, Harmine and Harmol

Many β-carboline alkaloids are present in food stuff, beverages, fruits and fruit-based products. Past research in this field has documented strategies for the extraction of the four β-carboline alkaloids NHM, harmane, harmine and harmol. In this section, we have summarized some of the extraction techniques used.

One of the commonly used techniques is separation of these β-carboline alkaloids using some resins based on polymers e.g. Amberlite XAD-1180 and Amberlite XAD-16 [54–56]. In the resin-based extraction techniques, selection of the proper resin is critical as it has a direct bearing on the selectivity and adsorption ability [57]. Separation of the β-carboline alkaloids depends on the compound-resin interactions. Methanol extraction can separate the β-carbolines from the resins. Subsequently, the solid alkaloid powder obtained after solvent evaporation is once again dissolved in ethanol. Ethanol evaporation, dissolving in water and then freeze drying gives the crude extract which can be used for qualitative and quantitative analysis [58].

8.4.1 Norharmane (NHM)

The simplest β-carboline alkaloid NHM [9*H*-Pyrido (3,4-*b*) indole] has been extracted from some cyanobacteria [54, 58–60]. Although most metabolites originating from cyanobacteria get accumulated in the biomass, some also find their way into the surroundings [61, 62]. Generally biologically active compounds from cyanobacteria can be found in the medium of growth used as can be found during the extraction of NHM from *Nodularia harveyana* [54].

A typical methodology for extraction of NHM from algae was reported by Erenler et al. [63] They first collected algal samples and incubated the samples for two weeks for 12 h each in light and in dark environment at ambient temperature. The algal samples were then identified and isolated under an inverted microscope using microinjector and micropipette. Total seven samples were thus isolated. Subsequently the cyanobacteria (Figure 8.5) were cultivated in suitable culture medium. Then the extraction procedure as outlined in Scheme 8.3 was performed.

Component (x) in the HPLC chromatogram which was noted at retention time 10 min corresponds to NHM (Figure 8.6). Quantification of NHM in the cyanobacteria was done as shown in the table in Figure 8.5.

8.4.2 Harmane

The β-carboline alkaloid harmane [1-methyl-9*H*-pyrido (3,4-*b*) indole] was isolated from the algicidal bacterium *Pseudomonas sp.* [64, 65]. Harmane was extracted from

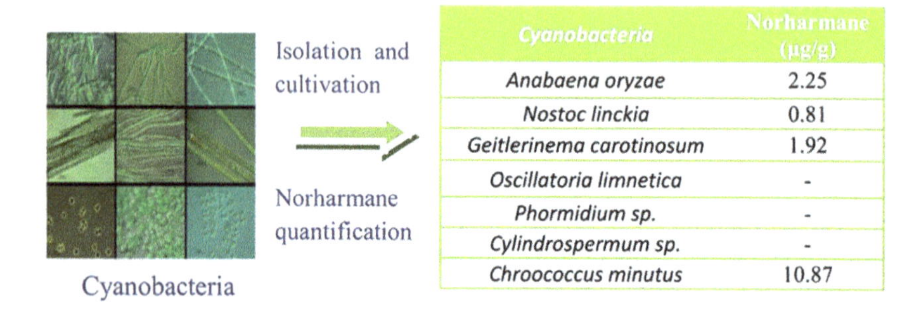

Cyanobacteria	Norharmane (µg/g)
Anabaena oryzae	2.25
Nostoc linckia	0.81
Geitlerinema carotinosum	1.92
Oscillatoria limnetica	-
Phormidium sp.	-
Cylindrospermum sp.	-
Chroococcus minutus	10.87

Figure 8.5: Some cyanobacteria used in NHM quantification. [Taken from the open access source: Ref. 63].

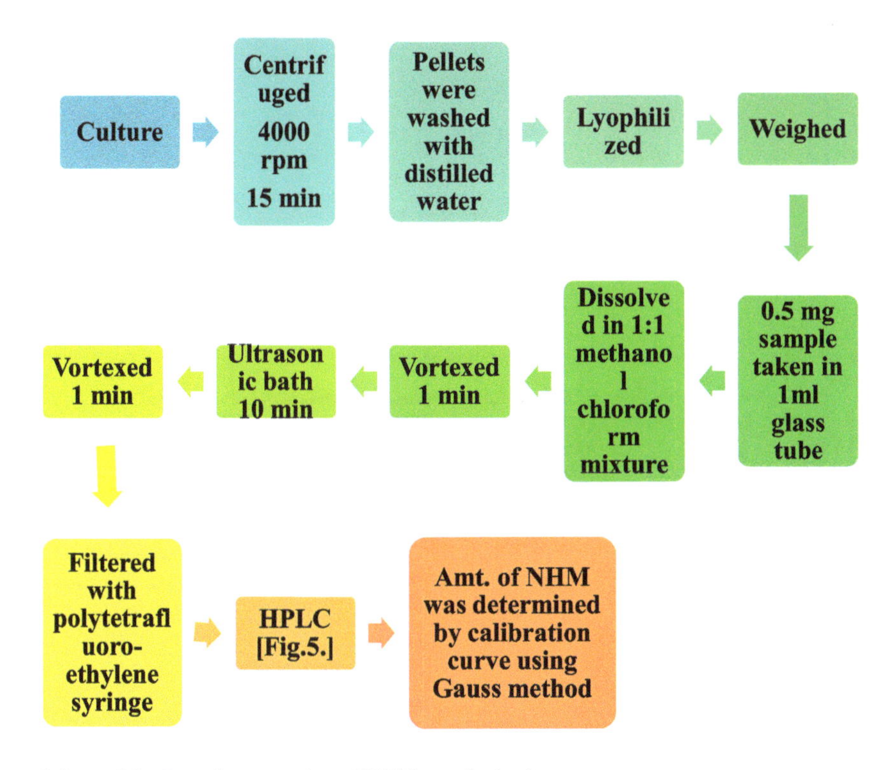

Scheme 8.3: Steps for extraction of NHM from algal culture.

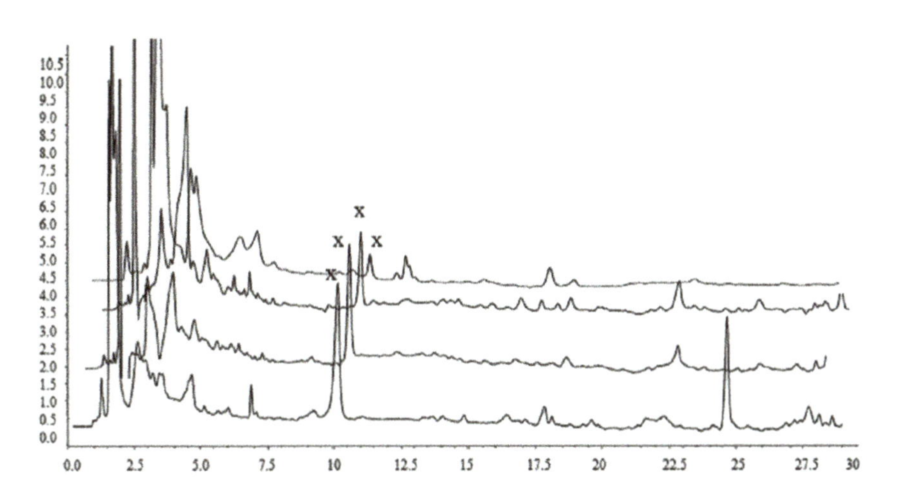

Figure 8.6: HPLC based detection of NHM (x) in extracts of various cyanobacteria. [Taken from the open access source: Ref. 63].

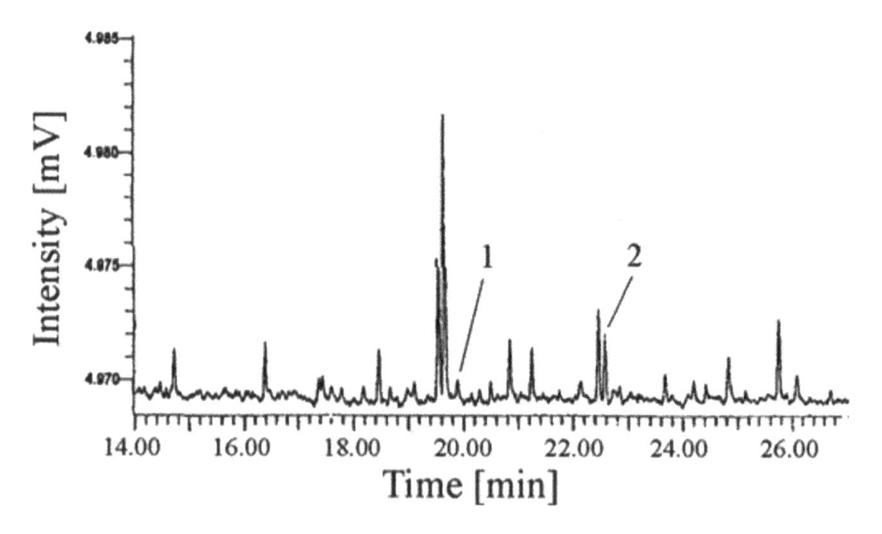

Figure 8.7: The extract of the medium in which *Geitlerinema sp.* was cultured was subsequently silylated and the GC analysis of that is shown above. The peak labeled 1 at retention time 19.89 min corresponds to harmane; the peak labeled 2 corresponds to 4,4′-dihydroxybiphenyl. [Taken from the open access source: Ref. 66].

the marine Cyanobacterium *Geitlerinema sp.* by Kumirska et al. [66] The cyanobacteria were cultivated in three different columns under favorable conditions. Then the supernatant was taken from the cultures after suitable number of days and centrifuged. The bioactive compounds were isolated on the resin Amberlite XAD-1180. The isolated components were then identified by GC-FID techniques upon co-injection with standards. Finally, the intensity peak obtained at 19.89 min in GC-MS analysis confirms the presence of harmane (Figure 8.7).

8.4.3 Harmine

Zhang et al. [67] extracted harmine [7-methoxy-1-methyl-9*H*-pyrido (3,4-*b*) indole] from *P. harmala* seeds. The seeds were first powdered ground, refluxed thrice with 80% ethanol and subsequently this extract was concentrated. The dark red solid so obtained was dissolved in 5% HCl and filtered. The aqueous portion was shaken and then partitioned with chloroform several times. The chloroform extracts obtained after each partitioning were combined and dried under vacuum. The water portion was next taken and its pH adjusted to 9 using NH_4OH. This was then partitioned against chloroform four times. The extract was dried and recrystallized using 95% ethanol and passed through silica gel column in various compositions of solvent mixtures (EtOAc/MeOH). Out of several fractions, only fraction six was selected and passed through Sephadex LH-20 column. Finally harmine was isolated.

Figure 8.8: *Passiflora incarnata*. [Source: Google images].

8.4.4 Harmol

Frye et al. extracted and analyzed harmala alkaloids (harmine and harmol) from three species of passion flower (Figure 8.8) [28]. They noted that *P. incarnate* could produce harmol [1-methyl-2,9-dihydropyrido (3,4-*b*) indol-7-one] in good yield (Table 8.2). The procedure for extraction of Harmol from passion flower is as follows:

Table 8.2: Areas of HPLC peaks with harmol concentration in the extract.

Area of HPLC peak	Harmol concentration (mg/g)
246.8	0.012
334.8	0.049

Stems, leaves and tendrils of the plants were dried in a dessicator. The dried plants were ground to powder in a coffee grinding machine. This powder was mixed with excess acetic acid solution, stirred briefly and filtered with a Buchner funnel. The aqueous portion was washed thrice with ether and ethyl acetate mixture to remove all the organic impurities. The resultant solution was poured into a separating funnel, collecting the lower portion and neutralizing using $NaHCO_3$. Excess water was removed from the top layer using sodium sulfate. The residue obtained after running the rotary evaporator was then dried.

Components could be identified from HPLC retention time. The HPLC chromatograms showed two peaks at retention times 4.641 and 4.686 min corresponding to harmol. The calibration curves were plotted using standard and the amount of harmol was determined to be 0.031 mg/g in *Passiflora incarnata*.

8.5 Photophysics of the β-carboline alkaloids

The photophysics of β-carboline alkaloids has been exploited to study bio-mimicking systems and actual biological systems. Such studies have mainly employed fluorescence to study the interactions of the alkaloids with receptor systems. The photophysics of β-carboline alkaloids is very interesting and thus many research groups have used these molecules to probe restricted systems like micelles, reverse micelles, vesicles and cyclodextrins (Scheme 8.4) [68–71]. The photophysical behaviour of β-carbolines are strongly dependent upon solvent parameters. The interesting photophysical behaviour of β-carboline alkaloids can be assigned to the following structural aspects:

(i) The presence of electron donor sites in the β-carboline skeleton and the overall ring planarity of these molecules favor the formation of π–π stacked complexes by their derivatives;

(ii) Additionally the pyrrolic NH group and the pyridinic nitrogen atom act as potential donor and acceptor sites respectively and thus facilitate hydrogen bond formation by the β-carbolines;

(iii) It was also shown that the complex forming ability of the β-carbolines gets altered in their excited state and this can be easily followed by the fluorescence technique.

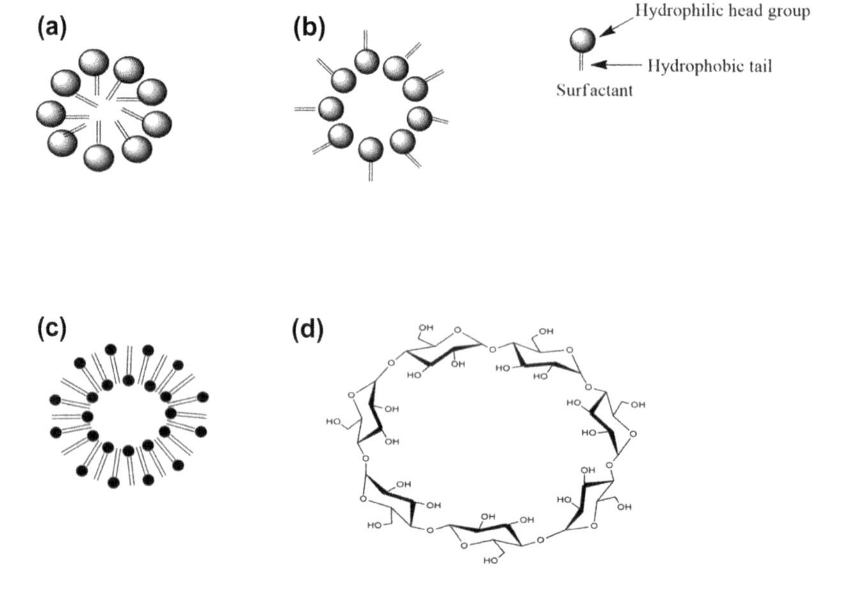

Scheme 8.4: Schematic representation of some restricted media like (a) micelle, (b) reverse micelle, (c) vesicle and (d) β-cyclodextrin.

8.5.1 Photophysics of NHM

The acid–base equilibrium of NHM is pH sensitive as shown in Scheme 8.5.

pH has a marked effect on the photophysics of NHM. Many research groups have done extensive work on the effect of pH on the photophysics of NHM. The collective concepts that emerged from the various studies can be summarized below [72, 73]:

(i) Between pH 1 and 10, the predominant species is NHM cation. The cation shows absorption maximum ($\lambda_{abs}{}^{max}$) at 370 nm in aqueous medium. The maximum of the fluorescence emission ($\lambda_{em}{}^{max}$) of the cation is at 445 nm in aqueous medium.

(ii) The neutral NHM species has $\lambda_{abs}{}^{max}$ at 348 nm while $\lambda_{em}{}^{max}$ for the neutral species was observed at 385 nm.

(iii) The three forms zwitterion, cation and neutral all show emission at pH 12.3. The zwitterion shows emission maximum at 510 nm.

(iv) Above pH 14, NHM anion is the predominant spectroscopic species with $\lambda_{abs}{}^{max}$ at 390 nm. The anion is formed in highly alkaline solution due to deprotonation of the indolic –NH group [74].

One research group studied the detailed photochemistry of NHM in aqueous media, in both acidic and alkaline solutions [75]. The progress of the photoreactions was followed by various methods e.g. chromatography, HPLC, mass spectroscopy and UV-vis absorption spectroscopy. It was found that in inert Argon atmosphere, NHM cation and neutral species were photostable even under long periods of irradiation. However, NHM was unstable when subjected to irradiation in oxygen atmosphere. NHM undergoes photochemical changes in oxygen environment and forms higher aggregates like dimers, trimers and tetramers. One usefulness of studying the photochemistry of NHM is that upon irradiation NHM can sense H_2O_2 and singlet O_2 ($^1O_2$).

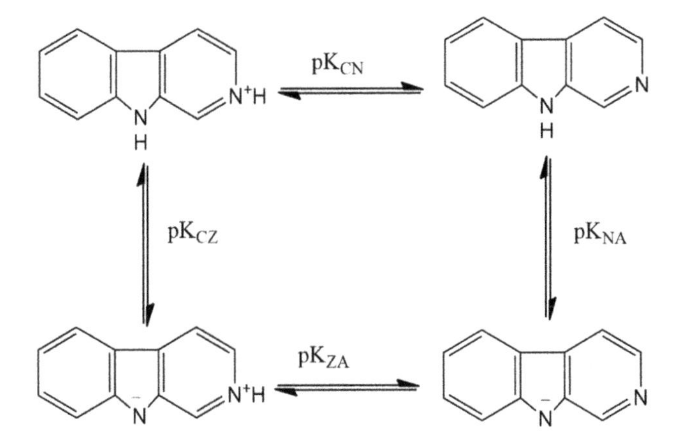

Scheme 8.5: Acid–base equilibria of β-carbolines; CZ, cation-zwitterion; CN, cation-neutral; NA, neutral-anion; ZA, zwitterion-anion.

Ghiggino et al. showed that NHM in 1N H_2SO_4 medium can act as a better fluorescence quantum yield standard than even quinine bisulphate [76]. Now, λ_{em}^{max} of NHM cation (450 nm) and its quantum yield ($\varphi_f = 0.60$) are similar to those of quinine bisulphate. However, an added advantage in using NHM as fluorescence standard is the single-exponential fluorescence decay of NHM across its emission band, in contrast to quinine bisulphate. In another work, Ghiggino et al. studied the acid-base equilibrium of NHM in the excited state using fluorescence [77]. Their findings also supported earlier works showing that the dominant species in acidic medium is the NHM cation whereas in alkaline medium, the neutral species dominates. However, in alkaline medium under irradiation the neutral species forms the cation by proton exchange with solvent.

Vert et al. measured the ground state and excited state acidity constants of NHM [73]. Their conclusions support the work of Ghiggino et al. [77] i.e. the formation of the zwitterion from the cation in presence of OH^-.

Varela et al. studied the fluorescence of NHM in the surfactant Aerosol OT (AOT) based microemulsions [78]. They varied the water pool size of the microemulsions and also the pH within the water pool. The effects of water pool size on the fluorescence decay of the various forms were assigned to "quenching" of the neutral species due to it getting trapped within the microemulsion droplets.

Chattopadhyay et al. [79] used steady state fluorescence to study the photophysics of NHM in aqueous cyclodextrin media. They showed that NHM can form either 1:1 or 1:2 inclusion complexes (Figures 8.9 and 8.10) with β-CD. They also explained that the inner cavity of α-CD is too small to form any inclusion complex with NHM. β-CD has the optimum cavity size for complex formation. Such complex formation has a marked effect on the fluorescence of NHM. Although γ-CD has a cavity large enough for encapsulating NHM, strangely such encapsulation has no effect on the spectra of NHM. The researchers assigned this to trapped water molecules within the larger γ-CD cavity along with NHM molecule.

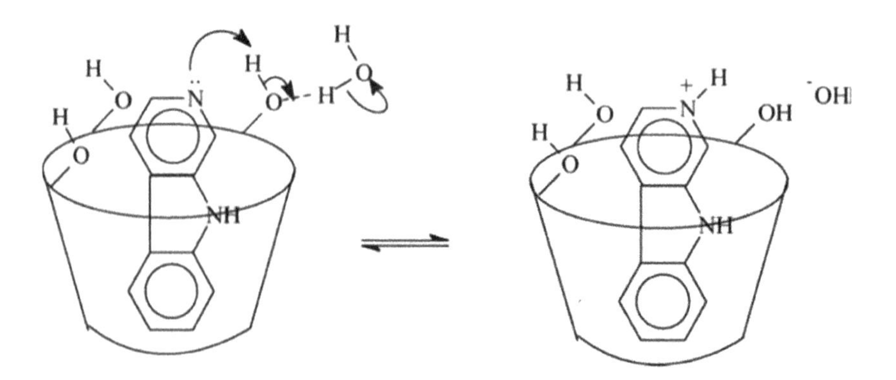

Figure 8.9: Interaction between encapsulated NHM with β-CD. [Reprinted from Ref. 79, with permission from Elsevier].

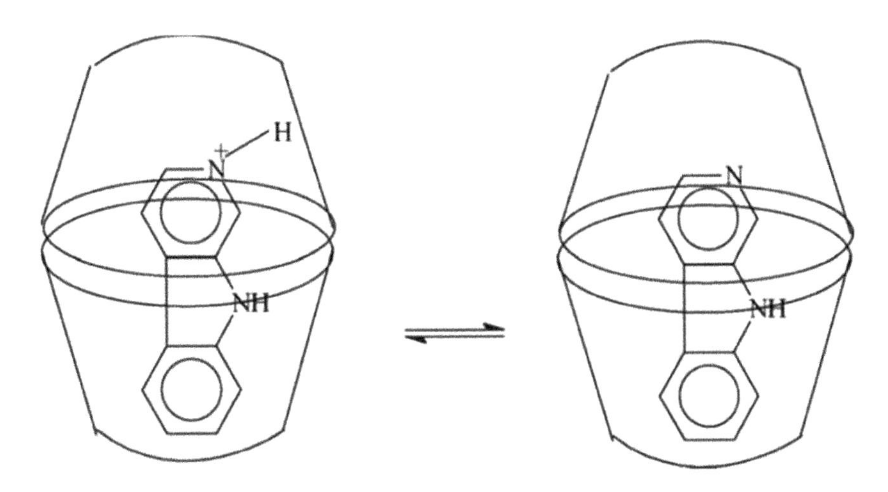

Figure 8.10: Schematic representation of complex formation between NHM and β-CD in 1:2 stoichiometry. [Reprinted from Ref. 79, with permission from Elsevier].

In another work, the same group studied the detailed photophysical behaviour of NHM in cationic, neutral and anionic micelles (Figure 8.11). [80] They found that in the anionic micelles formed by the surfactant sodium dodecyl sulfate (SDS), NHM primarily exists in the cation form. In contrast, in the cationic micelles formed by the

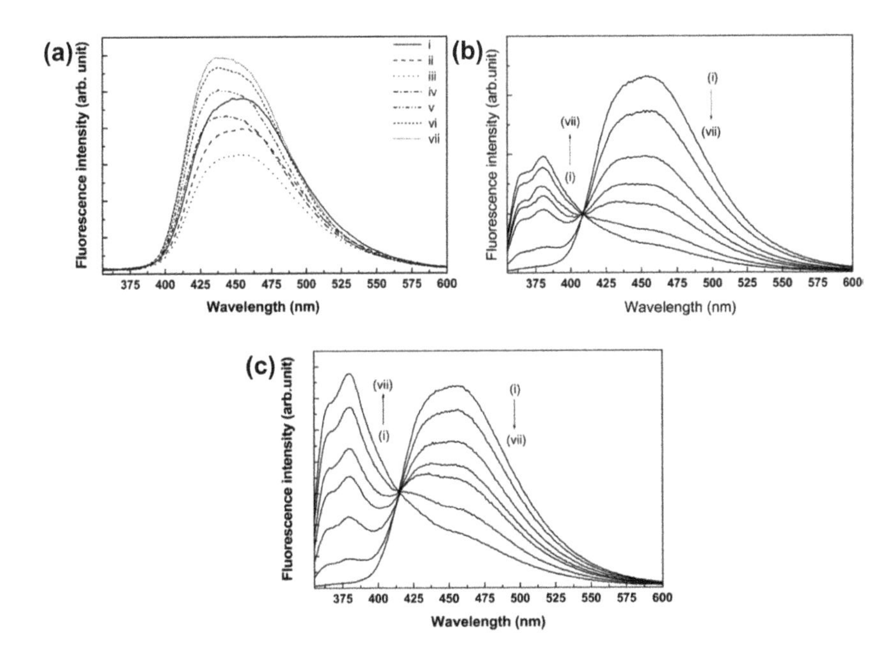

Figure 8.11: Emission spectra of NHM in micelles formed by (a) SDS, (b) CTAB and (c) TX-100. Curves at various concentrations of the respective surfactants. [Reprinted from Ref. 80, with permission from Elsevier].

surfactant cetyltrimethylammonium bromide (CTAB) and in the micelles formed by the neutral surfactant Triton X-100, NHM exists in its neutral form. The same group also concluded that NHM cannot penetrate into the hydrophobic interior but rather resides at the micelle–water interface.

Chattopadhyay et al. showed that surfactant chain length often determines which prototropic form of NHM will predominate in micellar media. They concluded this by studying the fluorescence of NHM in a series of cationic surfactants with varying hydrophobic tail length *viz.* cetyltrimethylammonium bromide (CTAB), dodecyltrimethylammonium bromide (DTAB) and tetradecyltrimethylammonium bromide (TTAB) [81]. The same group showed that a similar effect also arises by varying the chain length of anionic surfactants sodium decyl sulfate ($S_{10}S$), sodium dodecyl sulfate ($S_{12}S$) and sodium tetradecyl sulfate ($S_{14}S$) [82].

Mukherjee et al. showed that the prototropic equillibria of NHM could be modulated within the niosomal environment (Figure 8.12) [83]. They showed that the prototropic equilibrium of NHM was favored toward the neutral species at increasing salt concentrations and assigned the reason for this to increased water uptake by the niosomal bilayer. This group also studied the rotational relaxation behaviour of NHM from fluorescence anisotropy decay (Figure 8.13). They also studied the effect of addition of β-cyclodextrin (β-CD) on niosome-encapsulated NHM.

They also showed that β-CD can be employed for effective release of NHM from the niosomes (Figure 8.13).

The photophysics of NHM in different organic solvents was probed by Dias et al. [9] Their results showed that in hydrocarbon solvents and non protic solvents, only the neutral form of NHM exists. However, in protic solvents upon excitation by light the neutral form gets converted to the zwitterionic and cationic forms in the excited state. A kinetic model was developed to explain this conversion. The experimental results were supported by theoretical calculations based on charge densities of the ground and

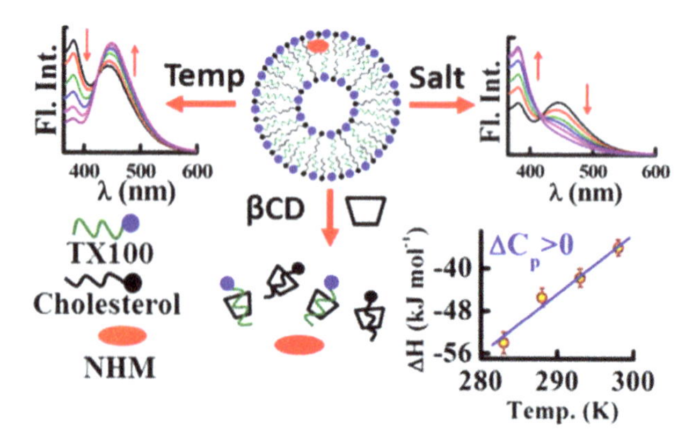

Figure 8.12: NHM in niosomal environment. [Reprinted (adapted) with permission from Ref. 83, American Chemical Society].

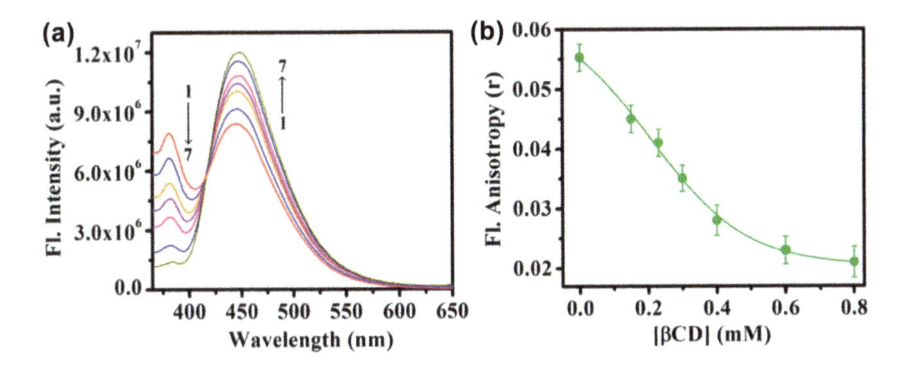

Figure 8.13: Effect of β-CD on the fluorescence behavior of NHM encapsulated in niosomes. (a) Effect of β-CD on the emission spectra of NHM encapsulated in niosomes. (b) Effect on fluorescence anisotropy of NHM encapsulated in niosomes. [Reprinted (adapted) with permission from Ref. 83, American Chemical Society].

excited states. Reyman et al. focused on the proton transfer dynamics of NHM in various organic solvents [72].

Chattopadhyay et al. [80] used mixed solvents of varying polarity to study NHM fluorescence. They observed an interesting correlation between the micropolarity of the medium and the cation: neutral ratio (Figure 8.14).

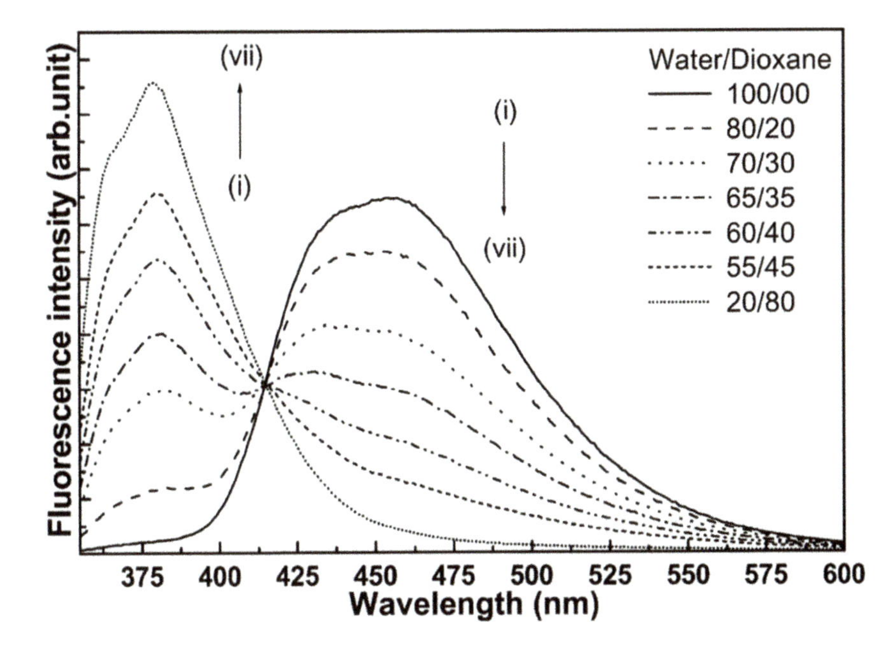

Figure 8.14: Fluorescence of NHM in water–dioxane mixtures, λ_{exc} = 352 nm [Reprinted from Ref. 80, with permission from Elsevier].

Solvents play an important role in the proton transfer process of NHM [84, 85]. In non-polar solvents like benzene or dioxane, the spectral signatures of the neutral, cationic and zwitterionic species were observed. In dichloromethane, the pyridinic nitrogen of NHM participates in H-bonding with the C–H group of the solvent [86, 87]. Thus in acetic acid–dichloromethane mixtures, the spectral signatures of all the four fluorescent species – neutral, cationic, zwitterionic and the hydrogen bonded form are observed. In polar solvents, only the cationic and neutral emission is seen as proton transfer is hindered due to the solvent–cage. Carboxylic acids exert a strong influence on the fluorescence behaviour of NHM. Thus NHM can be used for the recognition of carboxylic acids [72].

Some workers have studied the ability of NHM to sense various inorganic ions. This has important implications particularly for sensing the lethal cyanide ion [88]. Chattopadhyay et al. [88] performed a detailed study on the fluorescence and absorption response of NHM in presence of various inorganic ions like Cl^-, Br^-, CN^-, HSO_4^-, NO_3^- and I^-. The ratiometric response of NHM absorbance i.e. A_{348}/A_{372} (neutral band/cationic band) showed that the highest response was obtained for CN^- (Figure 8.15).

The high sensitivity towards ratiometric sensing of CN^- by NHM has been assigned to the low acid dissociation constant (K_a) of the conjugate acid of CN^-. It was shown that the ratiometric response of NHM spectra can be used to detect as well as quantitatively estimate even micro molar concentrations of CN^-.

Mallick et al. studied the interaction of NHM with different anions using fluorescence, NMR and density functional theory (DFT) [89]. They found that the spectral bands of NHM showed specific sensitivity towards fluoride ions (F^-). They assigned this to the fact that F^- extracts out the acidic proton of NHM, thus forming NHM anion. This gives rise to distinct spectral changes. Support for the acidic proton abstraction by F^- comes from DFT calculations too.

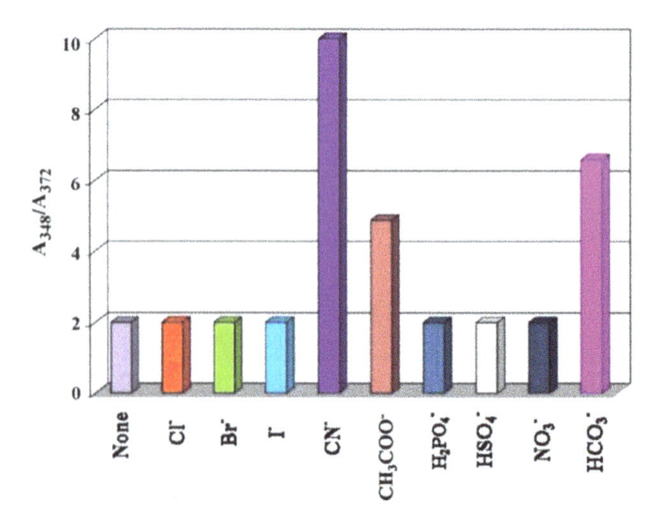

Figure 8.15: Ratiometric sensing of various inorganic anions using NHM. [Reprinted from Ref. 88, with permission from Elsevier].

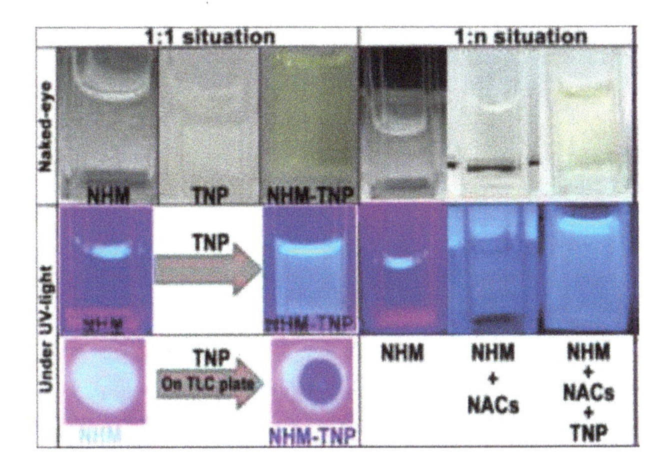

Figure 8.16: Specific molecular recognition of TNP by NHM. [Reprinted from Ref. 90, with permission from Elsevier].

Majumder et al. reported on the sensing of 2,4,6-trinitrophenol (TNP) by NHM. Their procedure involves the reaction of NHM with various nitro compounds. However, they found that only TNP reacts with NHM due to the ground state proton transfer. This result is very important as it is a method of high fidelity molecular recognition. This detection is so sensitive that it can be observed by our bare eyes even on an ordinary TLC plate under UV light (Figure 8.16) [90].

The same workers also showed that proton transfer from HSO_4^- to the pyridinic nitrogen of NHM results in a hydrogen bonded ion pair that provides an optical response in the visible region [91]. Thus their work provides a new reversible "*ON-OFF-ON*" chemo sensor based on NHM.

This chemo sensor behaves as a molecular lock, HSO_4^- ion acts as the "key" for the lock and F^- ion the "hand" to remove the key (HSO_4^-) and thus close the lock.

8.5.2 Photophysics of Harmane

Vert et al. reported that the λ_{abs}^{max} for the harmane cation was at 364 nm [73], while the corresponding values for the neutral and anionic forms of harmane were at 346 and 375 nm respectively. The λ_{em}^{max} of the cationic, neutral and zwitterionic forms of harmane are at 430 nm, 374 and 483 nm, respectively. In highly alkaline solution, the anionic form of harmane exists [74] and this arises due to the deprotonation of the indolic –NH group. Independent studies of harmane by Vert et al. [73] and Ghiggino et al. [77] led to the same conclusion i.e. that harmane zwitterion was formed by OH^- induced quenching of excited harmane cation.

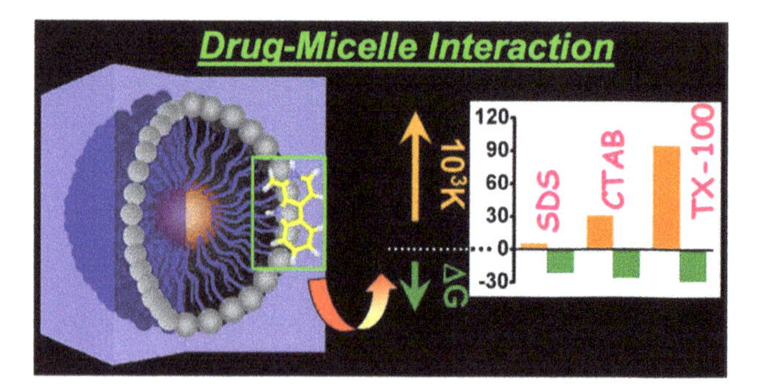

Figure 8.17: Schematic representation of harmane-micelle interaction. [Reprinted (adapted) with permission from Ref. 93, American Chemical Society].

Marques et al. studied the various protolytic forms of harmane in micelles by fluorescence methods and molecular orbital calculations [92]. In another work, Guchhait et al. showed that the spectra of harmane were distinctly different in various types of micelles formed by the neutral surfactant Triton X-100 (TX-100), the cationic surfactant CTAB and the anionic surfactant SDS [93]. They showed that the nature of the micellar environment modulates the extent of harmane binding to the micelle (Figure 8.17).

Varela et al. probed the photophysics of harmane in AOT based microemulsions [78]. The study was performed at different pH and by varying the size of the water pool of the micro emulsions. In these *w/o* micro emulsions, the harmane molecules preferentially reside at the AOT/water interface. However, with pH lowering harmane moves into the water pool region. From the spectra, distinct differences are observed in the tautomeric species of harmane within the microemulsion as compared to neat water.

Ahmed et al. used harmane as a guest in cucurbit[7]uril (CB7) as host and showed that the emission spectrum of harmane can be modulated by CB7. Isothermal titration calorimetry showed that complexation between harmane and CB7 is enthalpically favored (Figure 8.18) [94].

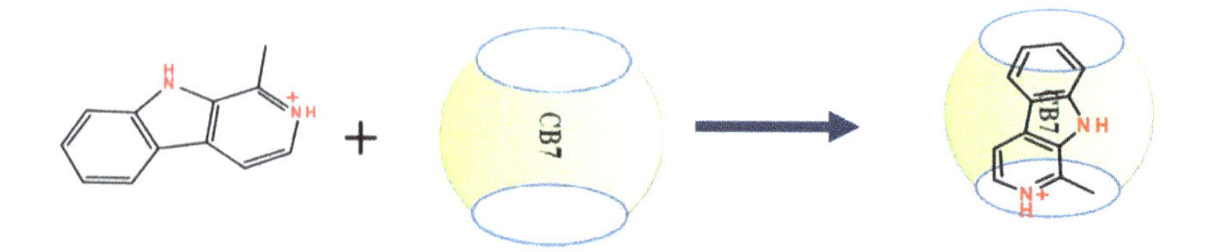

Figure 8.18: Schematic representation of complex formation between harmane with CB7. [Reprinted from Ref. 94, with permission from Elsevier].

Guchhait et al. studied the photophysics of harmane in the liposomes formed by the lipids dimyristoylphosphatidycholine (DMPC) and dimyristoylphosphatidyglycerol (DMPG) [95]. From fluorescence and docking studies, it was shown that harmane associated faster with the membrane bilayer of DMPG vesicles rather than DMPC vesicles, due to favorable electrostatic interactions between harmane and the surface charges on DMPG liposomes. Their work is particularly relevant as it gave valuable insight on the partitioning of harmane within the liposomes.

In another study, the fluorescence of harmane was investigated (Figures 8.19 and 8.20) in mixtures of β-CD and surfactants [96]. From this work, it could be shown that β-CD can disrupt the micellar structures (Figure 8.19).

From a study of the photophysics of harmane in different organic solvents, Dias et al. [9] showed that the neutral form exists in hydrocarbons and non protic solvents. However, in protic solvents, the predominant excited state species are the cations and zwitterions. Other workers investigated the excited state proton transfer between harmane and hexafluoroisopropanol in cyclohexane – toluene mixture [97]. The spectral behaviour of harmane was also investigated by systematically varying the solvent polarity [98]. It was shown that harmane acts as a proton acceptor in both its excited and ground states. Fluorescence emission of harmane showed a red shift with increasing solvent polarity.

Reyman et al. showed that harmane [99] forms a 1:3 cyclic complex with acetic acid whereas NHM forms a 1:2 complex. Their work gave important insight on the H-bond forming ability of harmane and NHM in acidic solution from ^{1}H and ^{13}C NMR studies.

Detailed spectroscopic studies were carried out to study the interactions between harmane and heterocyclics such as pyridine and quinolone [100, 101]. The results indicated that complex forming tendency is higher in the excited singlet state. It was also shown that the stability of the complex between harmane and pyridine decreases as polarity and H-bonding ability of the solvent increases.

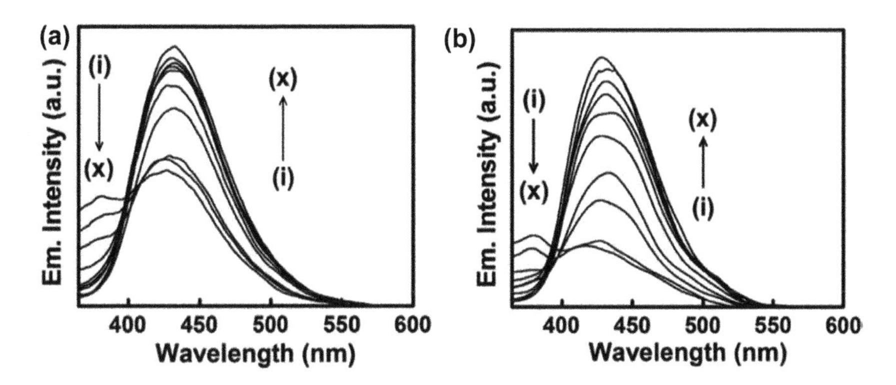

Figure 8.19: Emission spectra of harmane in (a) TX-100 micelles, (b) CTAB micelles with added β-CD. [Reprinted from Ref. 96, with permission from Elsevier].

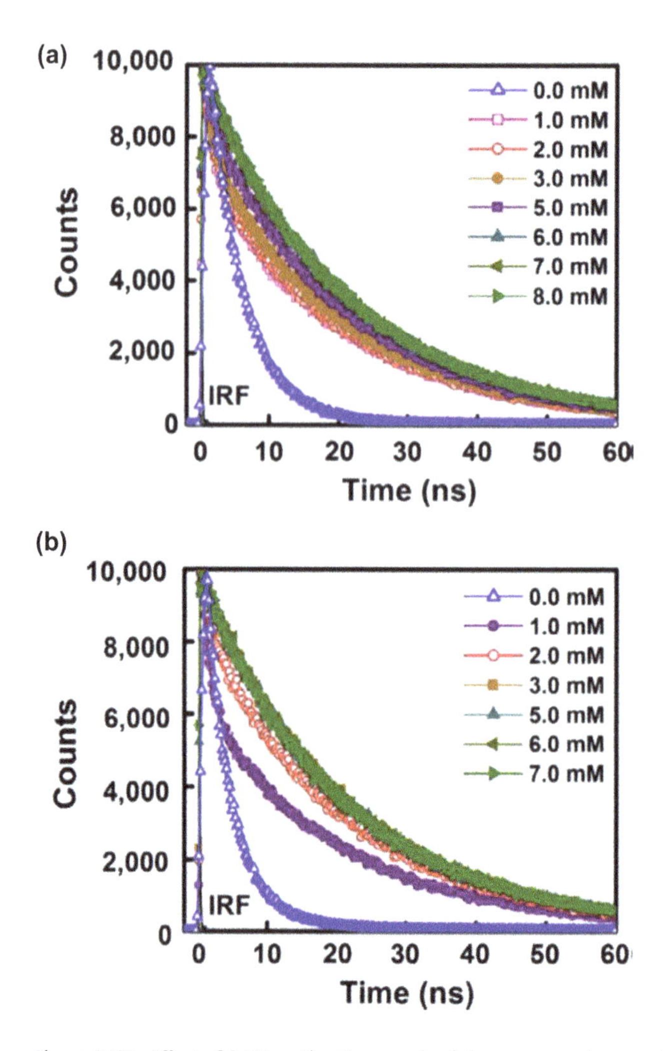

Figure 8.20: Effect of β-CD on the time-resolved fluorescence decays of harmane (λ_{ex} = 375 nm and λ_{em} = 435 nm) in the (a) TX-100 micelles and (b) CTAB micelles. [Reprinted from Ref. 96, with permission from Elsevier].

8.5.3 Photophysics of Harmine

In aqueous medium, $\lambda_{abs}{}^{max}$ for harmine cation, neutral harmine and harmine anion is at 355 nm, 336 and 370 nm respectively [73]. $\lambda_{em}{}^{max}$ for harmine cation is at 418 nm and $\lambda_{em}{}^{max}$ for the neutral and zwitterionic forms are at 373 and 476 nm respectively. In highly alkaline medium, the predominant emissive species is harmine anion which arises due to deprotonation of the indolic –NH group [74]. Ghiggino et al. studied the acid-base equilibria of harmine in excited state [77]. Their results indicated that the cationis are the main emissive species in acidic solution. Vert et al. measured the acidity constants of harmine [73]. Their results support the work of Ghiggino et al. [77].

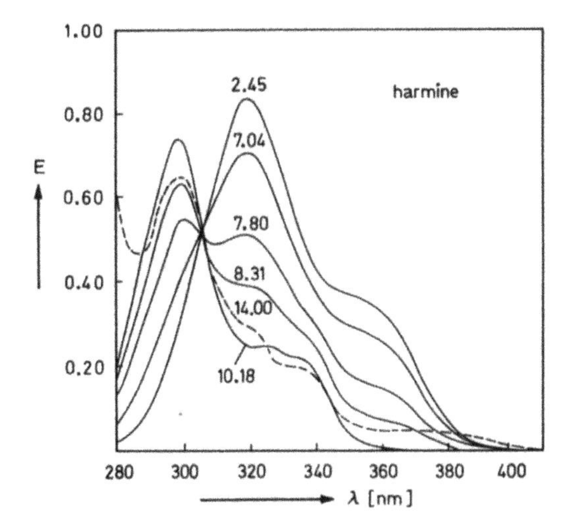

Figure 8.21: Effect of pH on the absorption spectra of harmine. [Taken from the open access source: Ref. 102].

The photo-induced tautomerism of harmine was studied by Wolfbeis et al. in both excited and ground states (Figures 8.21 and 8.22) [102].

They showed that three protolytic forms of harmine could be identified from absorption spectra while spectral signatures of four protolytic forms were seen in the fluorescence spectra, $\lambda_{em}{}^{max}$ at 419 nm corresponds to the cation, $\lambda_{em}{}^{max}$ ~ 370 nm to the neutral form, $\lambda_{em}{}^{max}$ = 430 nm to the anion and $\lambda_{em}{}^{max}$ 480 nm to the zwitterion. The pK_a's of the ground state species were obtained from the absorption spectra while the excited state pK_a's were obtained from fluorescence spectra and then compared with those calculated from Förster equation. It was shown that harmine forms the zwitterionic species by adiabatic proton transfer.

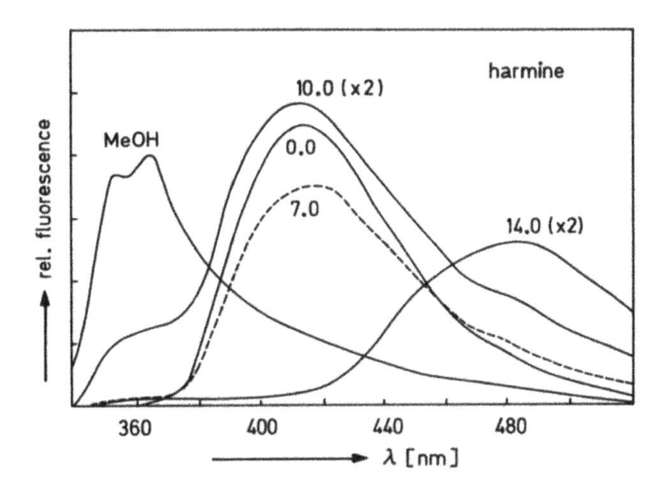

Figure 8.22: Effect of pH on the fluorescence emission spectra of harmine. [Taken from the open access source: Ref. 102].

Becker et al. studied the solvent effect on the photophysics of harmine [103]. In ethanol, both the neutral and cationic forms of harmine showed phosphorescence in addition to fluorescence. However, triplet transients of only the neutral form of harmine were obtained in nonprotic solvents. Summarizing, the results showed that the triplet states of the neutral form of harmine had long lifetime and could also produce significant amounts of singlet oxygen.

It was also shown that the surfactant CTAB and the cyclic sugar β-CD could regulate the reversible photo transformation of harmine between the cationic and neutral forms [104]. This dual-band ratiometric emission responseof harmine (Figure 8.23) was proposed to have prospective applications in designing molecular security devices based on β-carbolines (Figure 8.24).

8.5.4 Photophysics of Harmol

The photophysical studies of harmol [105–108] in aqueous solution show that the absorption maxima of the four protolytic forms are at 357 nm, 338 nm, 367 nm and 332 nm for the cationic, neutral, zwitterionic and anionic forms, respectively. The emission maxima are at 418 nm, 360 nm, 440 nm and 450 nm for the cationic, neutral, zwitterionic and anionic forms, respectively. Wolfbeis et al. carried out extensive studies on the protolytic equilibria of harmol (Figure 8.25) [102]. They reported that harmol exists in the neutral and zwitterionic forms at neutral pH.

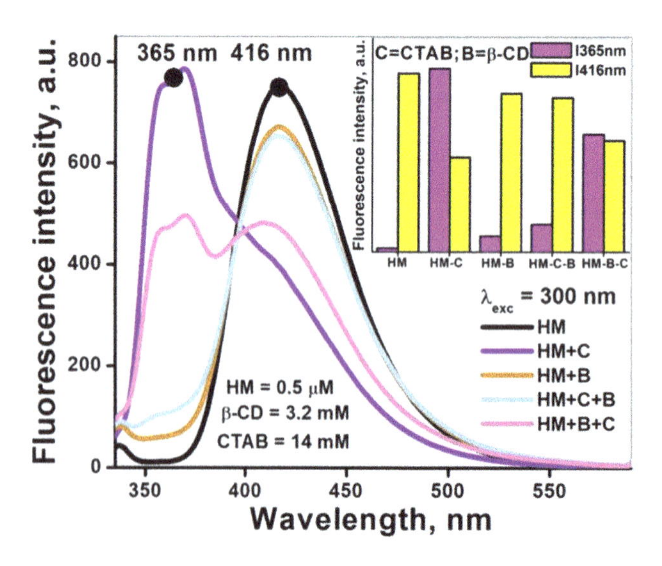

Figure 8.23: Ratiometric emission response of aqueous solution of harmine induced by sequential treatment with CTAB and β-CD. [Reprinted from Ref. 104, with permission from Elsevier].

D3
D4
D3
300 nm
300 nm
300 nm
1
2
3
4
5
Fl@365 nm
Fl@416 nm
Fl@365 nm
D4
D3
D3

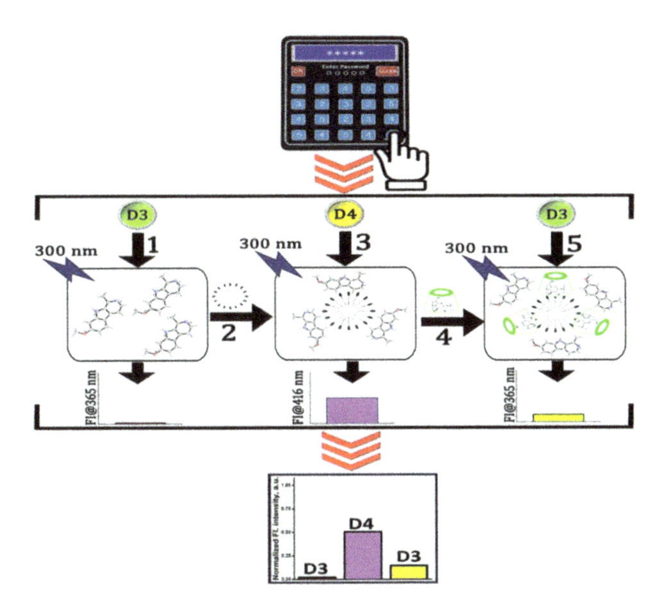

Figure 8.24: Design of the proposed molecular security device based on β-carbolines (23452 is used as password). [Reprinted from Ref. 104, with permission from Elsevier].

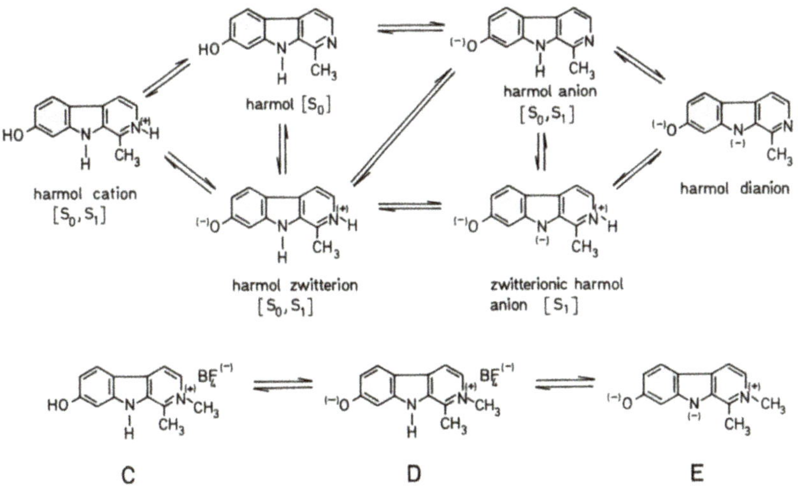

harmol [S0]
harmol anion [S0, S1]
harmol cation [S0, S1]
harmol dianion
harmol zwitterion [S0, S1]
zwitterionic harmol anion [S1]
C
D
E

Figure 8.25: Tautomerie species of harmol. [Taken from the open access source: Ref. 102].

As seen from Figure 8.25 above, harmol has two groups having the potential to exhibit acidic behaviour: (a) the –OH group of the phenol and (ii) the weakly acidic –NH group; while it possesses a single basic moiety i.e. the pyridinic nitrogen.

Another group showed that the hydroxyl group of harmol in excited singlet state is more acidic than in the ground state while the pyridinic nitrogen is less basic in excited state than in the ground state [108].

Olba et al. studied the fluorescence and phosphorescence spectra of harmol at 77 K in solid phase [109]. They found spectral signatures of the three forms of harmol *viz.* cation, anion and zwitterion in both types of spectra.

Summarizing the literature on the photophysics of the four β-carbolines reviewed in this work *viz.* NHM, harmane, harmine and harmol, their spectral data are given in Table 8.3, to help future researchers.

8.6 Interaction of β-carboline alkaloids with biological systems

The interactions of the harmala alkaloids with biological receptors like proteins and nucleic acids are a topic of considerable research interest [110]. Hence, the interactions of β-carbolines with the common biological receptors have been briefly reviewed in this section. The photophysical studies of these compounds gives valuable insight into their binding with the physiologically important proteins bovine serum albumin (BSA) and human serum albumin (HSA).

Chattopadhyay et al. have done extensive work on NHM-BSA interactions [111]. They found that the neutral form of NHM exists in the protein interior while the cationic form was almost exclusively present outside.

Another group of researchers studied the interactions of SDS, an anionic surfactant with the two serum proteins BSA and HSA, using NHM as a proton transfer probe [112]. Their findings indicated that protein-bound NHM was dislodged by SDS and subsequently at excess SDS concentrations; NHM selectively binds to SDS micelles. Circular dichroism (CD) spectra indicate that the secondary structure of the protein gets destroyed when SDS concentration is relatively high (Figure 8.26).

Table 8.3: The spectral data of the four β-carboline alkaloids.

Compound and types of spectra		Cation	Neutral	Zwitterion	Anion
Norharmane (NHM)	Absorption	370 nm	348 nm	–	390 nm
	Fluorescence	445 nm	385 nm	510 nm	–
Harmane	Absorption	364 nm	346 nm	–	375 nm
	Fluorescence	430 nm	374 nm	483 nm	–
Harmine	Absorption	355 nm	336 nm	–	370 nm
	Fluorescence	418 nm	373 nm	476 nm	–
Harmol	Absorption	357 nm	338 nm	367 nm	332 nm
	Fluorescence	418 nm	360 nm	440 nm	450 nm

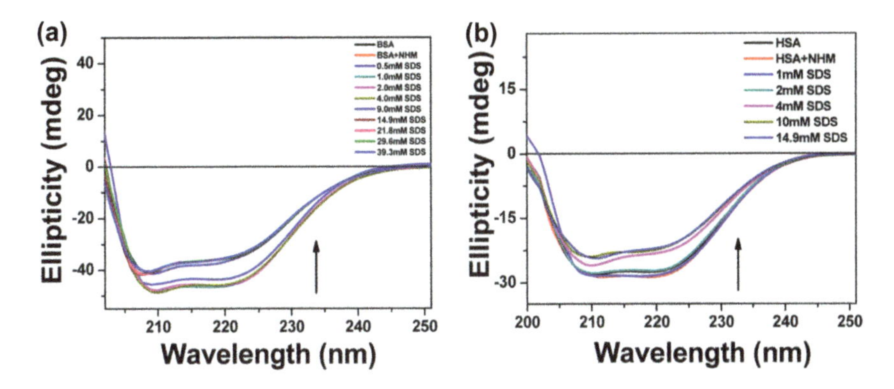

Figure 8.26: Effect of adding SDS on the CD spectra of (a) BSA and (b) HSA. [Reprinted (adapted) with permission from Ref. 112, American Chemical Society].

Another group probed the effect exerted by fatty acids on the binding of NHM and harmane with HSA [113]. Induced circular dichroism (CD), absorption and fluorescence spectra of protein-bound NHM showed anomalous behaviour in presence of fatty acids indicating the formation of complexes via cooperative allosteric mechanism.

Some groups have shown that the intercalation of β-carboline alkaloids into DNA can affect the important DNA replication and DNA repair processes [114–119].

It has been reported that NHM can inhibit the DNA relaxation activity of both topoisomerase I (Topo I) and topoisomerase II (Topo II) at 20–250 µg/mL [120]. NHM is also reported to be a potent inhibitor of the aggregation of human platelets by the vasoconstrictor 5-hydroxytryptamine [121]. It has been reported that NHM shows preferential inhibitory action towards monoamine oxidase-B (MAO-B) [122–125] than towards MAO-A. These have been implicated in Parkinson's disease. Harmane shows preferential inhibitory activity towards MAO-A rather than MAO-B. Harmine can also act as an inhibitor of MAO-A activity [126]. Studies show that harmine and related β-carbolines can act as potent inhibitors of cyclin-dependent kinases (CDKs) [127, 128]. Action of harmine is specific towards CDK1, CDK2 and CDK5.

Harmine extracted from *P. harmala* seeds can inhibit the activity of human DNA Topo I [129]. *In vitro* studies show that *P. harmala* seed extract inhibits human DNA Topo I and based on the results of high performance thin-layer chromatography (HPTLC) analysis, it seems that the biological activity of the extract arises due to its β-carboline content. Harmine is also a potent inhibitor of 5-hydroxytryptamine (5-HT) induced aggregation of human platelets [121] which can cause inhibition in uptake of 5-HT by human platelets.

Earlier studies showed that the pharmacological action of the β-Carboline alkaloids on cells of prokaryotes and eukaryotes is due to fact that these alkaloids can intercalate DNA which in turn affects replication of their DNA and thus the DNA repair process in general.

It was reported by another group that both harmane and NHM intercalate DNA. This was inferred from the fluorescence quenching of the β-carbolines and corresponding red shift in the absorption spectra. It was also shown that interaction of Harmine with calf thymus DNA occurs via the process of intercalation [130]. Harmine is also known to intercalate DNA like NHM and harmane [130, 131].

8.7 Therapeutic potential of the four β-carboline alkaloids

β-Carboline alkaloids have been quite abundantly used for their therapeutic potential. They are particularly relevant in the context of sustainability as most of them can be derived from natural resources. The β-carboline alkaloids are popular in the drug community as they are known to interact with several neurotransmitters and neuromodulators. Besides the central nervous system (CNS), some β-carboline alkaloids can target diseases affecting other human organs too. Different β-carboline alkaloids are known to interact with different targets leading to varied pharmacological effects.

8.7.1 Neurological activity

Research during the past few years has shown that the β-carboline alkaloids interact with specific sites on the CNS, particularly with the receptors *viz.* 5-hydroxytryptamine (5-HT) receptors regulating serotonin and benzodiazepine receptor (BZRs) in the CNS. Binding to these receptor sites can have a direct bearing in regulating mood swings, hallucinations and convulsions.

Harmine has been demonstrated to produce hallucinogenic response in humans [132, 133]. Ample evidence for this came from the plasma of alcoholic people and heroin addicts, which showed high levels of harmane and NHM [134–136]. On the other hand, it was shown that harmane and harmine showed beneficial action on withdrawal syndrome arising out of naloxone-precipitation in rats [137]. Harmine was also very effective in alleviating the symptoms of morphine withdrawal.

Lantz et al. studied the *in vitro* dopaminergic activity of NHM [138]. Nicotine present in tobacco is a drug of abuse. In addition to nicotine, tobacco also contains β-carboline alkaloids such as harmane and NHM which reinforce effects of nicotine. The study by Lantz et al. was directed on the effects of exposure to harmane and NHM alone and in combination with nicotine on neuron cultures extracted from the midbrain. It was found that there is dose-dependent decrease in dopamine intracellular levels upon exposure to NHM (Figure 8.27). Interestingly, this decrease in intracellular dopamine by NHM occurred without causing significant cell death, mitochondrial dysfunction or even without significant formation of reactive oxygen species (ROS). Exposure to harmane on the other hand caused increase in extracellular dopamine levels.

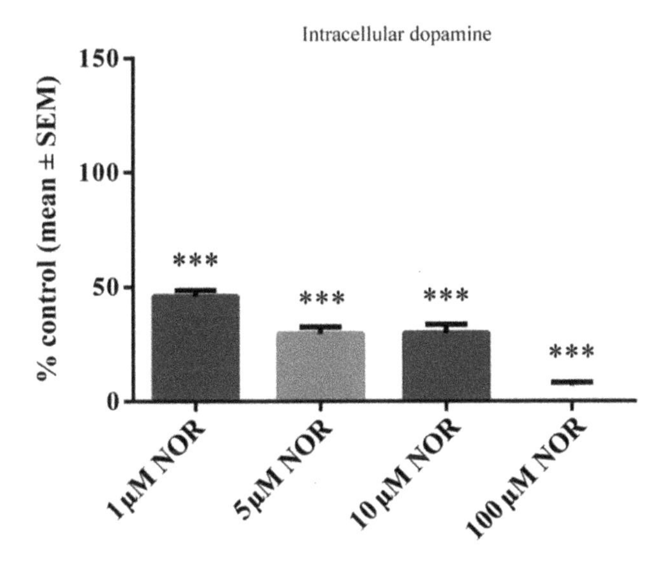

Figure 8.27: Dose dependent decrease in intracellular dopamine levels after one day of exposure to NHM (1–100 µM, $N = 12$). [Taken from the open access source: Ref. 138].

NHM can act as a potent toxin and thus restore the degeneration of nigrostriatal neurons in Parkinson's disease [139]. It can also form a demethylated carbolinium ion *in vivo* that acts as a dopaminergic neurotoxin [140].

Harmane can also modulate dopamine levels in cells [138]. It was shown that there is dose-dependent decrease in dopamine concentrations in the intracellular bodies upon treatment with harmane (Figure 8.28) whereas there was an increase in dopamine

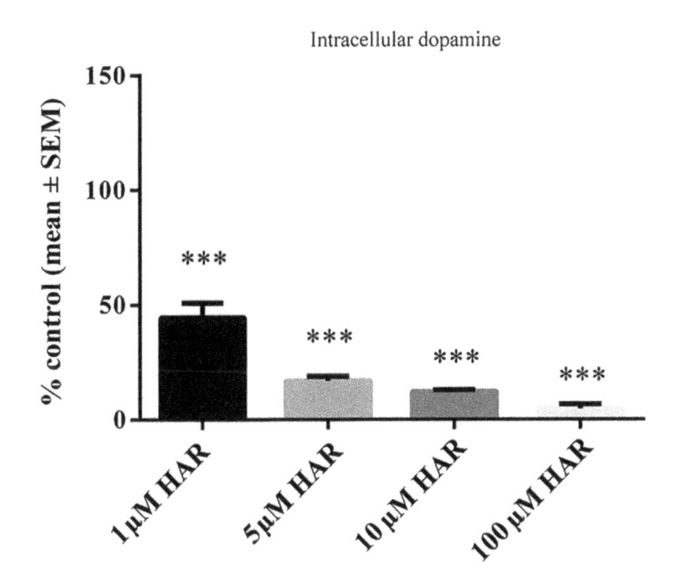

Figure 8.28: Effect of 24 h exposure to harmane on intracellular dopamine concentration. [Taken from the open access source: Ref. 138].

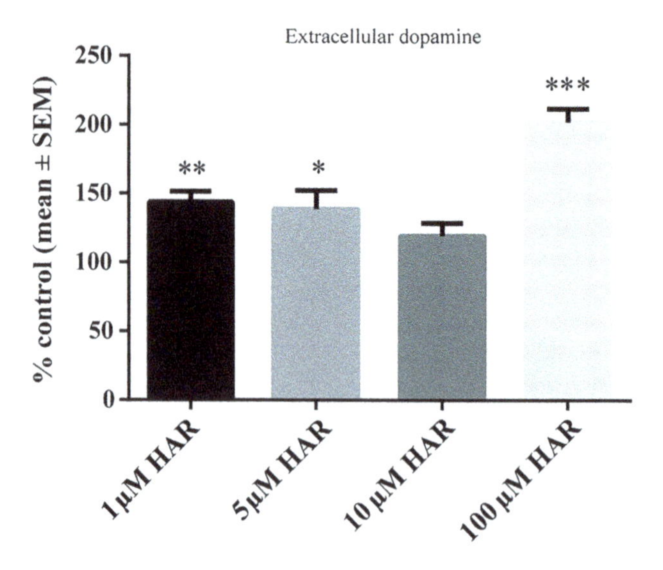

Figure 8.29: Effect of 24 h exposure to harmane on extracellular dopamine concentration. [Taken from the open access source: Ref. 138].

concentration in the extracellular fluid (Figure 8.29). This has important implications in the use of harmane to study dopaminergic activity in neurons.

Originally harmine was administered for Parkinson's syndrome. However, later research also showed that harmine exhibits significant anti-HIV, anti-tumor and vasorelaxant actions [141].

8.7.2 Antimicrobial activity

Every year, the fruit and vegetable sectors incur large economic losses arising out of post-harvest anti-fungal diseases [142]. Among these fungi, *Penicilliumdigitatum*, causing green mold on citrus fruit, is the most common post-harvest pathogen. Synthetic fungicides are used to arrest the proliferation of this pathogen [143]. Fruits like grapes and berries often develop gray mold during storage due to the fungus *Botrytis cinerea* [144, 145]. Benzimidazoles and dicarboximides are used to control this pathogen [146]. However, use of synthetic fungicides is often accompanied by resistance developed by certain fungal strains [147]. To overcome this problem, fungicides extracted from natural resources are attractive. In this context β-carboline alkaloids serve as a sustainable source of antifungal agents.

Olmedo et al. showed that NHM exhibits significant antifungal activity towards *Penicillium digitatum* and *B. cinerea* [148]. At quite low concentrations i.e. between 0.5 and 1.0 mM, NHM, harmane, harmine and harmol could inhibit conidia germination. This finding is extremely significant in preventing post-harvest diseases of fruits and vegetables. Harmol was very effective on these fungal strains inhibiting mycelial growth and arresting sporulation. Thus harmol might be a prospective fungicide to prevent fruit diseases.

NHM was also found to inhibit several fungi [54, 149, 150]. Harmol could inhibit *Trypanosoma cruzi* [151] while NHM, harmane and harmine all three showed significant inhibitory action on *T. gondii* [152].

Olmedo et al. showed that harmol could inhibit growth of fungal mycelia of *P. digitatum* and *B. cinerea* [148] (Figure 8.30a). The most important aspect was that after a week of incubation of these fungi with harmol, sporulation could be totally prevented (Figure 8.30b).

In another study, Olmedo et al. showed that harmol exhibits enhanced antifungal activity when photo activated with UV-A [153]. The role played by some β-carboline alkaloids in ROS generation under UV-A irradiation has been studied [154]. It was proposed that H_2O_2 is generated by electron transfer to O_2 giving rise to the superoxide anion ($O_2\cdot^-$). Thus the UV-A induced enhanced photodynamic effect of harmol can arise due to interaction of the excited state of harmol with structural constituents of the fungi.

8.7.3 Anticancer activity

Several β-carboline alkaloids show significant anticancer activity too. One group of researchers showed that encapsulation of NHM in Cucurbit[7]uril (CB7) increased the cellular uptake of CB7 entrapped NHM (Figure 8.31) [155].

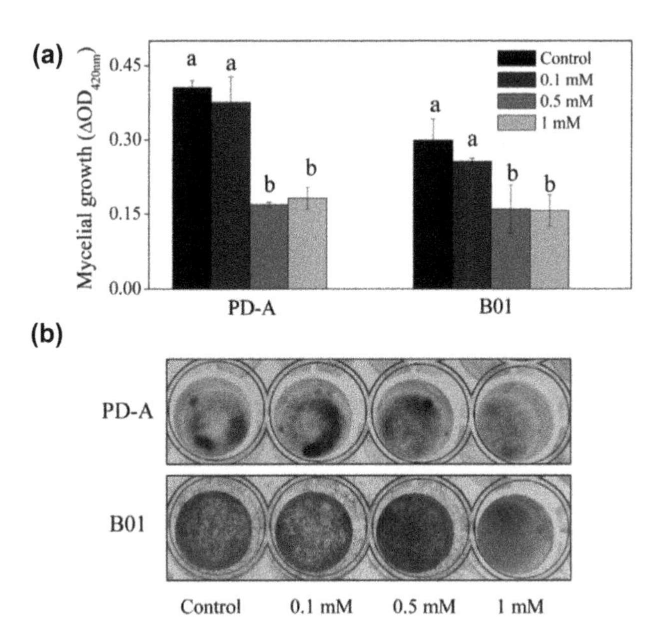

Figure 8.30: Anti-fungal action of harmol on *Penicillium digitatum* (PD-A) and *Botrytis cinerea* (B01). The harmol concentrations in (b) indicate the exposure of pathogens to the alkaloid. [Reprinted from Ref. 148, with permission from Elsevier].

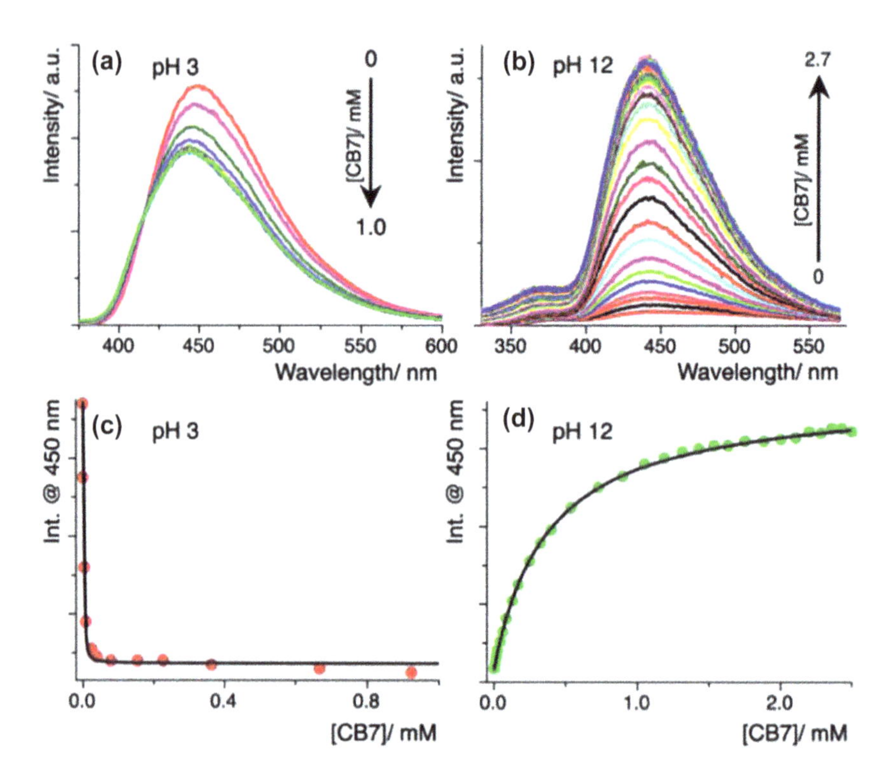

Figure 8.31: Binding of 6.0 μM NHM at increasing CB7 concentration, monitored using fluorescence at (a) pH 3.0 and (b) pH 12.0. Fluorescence intensity at 450 nm was plotted versus CB7 concentration at (c) pH 3.0 and (d) pH 12.0 [1:1 binding model was used]. [Reprinted from Ref. 155, with permission from Elsevier].

CB7 is not too much cytotoxic towards live animals and hence was used for NHM encapsulation and delivery onto human breast cancer cell line (MCF-7). The fluorescence originates from NHM localized in the MCF-7 nuclei, as NHM is a good DNA intercalator (Figure 8.32).

Abe et al. reported anticancer activity of harmol on H596 human lung carcinoma cells [156]. Harmol treatment caused release of cytochrome c from mitochondria to cytosol. Harmol-induced apoptosis was totally inhibited by caspase-8 inhibitor and partially inhibited by caspase-9 inhibitor. Summarizing, these results showed that harmol caused apoptosis in H596 human lung carcinoma cells via caspase 8-dependent pathway.

Another interesting study showed that harmol caused cell death in A549 non-small human lung cancer cells by autophagy, not by apoptosis [157]. Autophagy was seen by electron microscopy in A549 cells treated with 70 mM harmol (Figure 8.33). Apoptosis was not observed. Often the anticancer activity of these alkaloids has been found to be associated with the induction of autophagy.

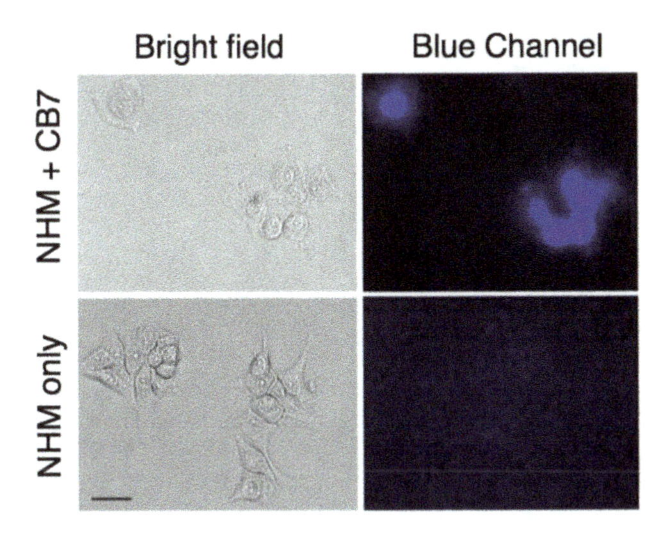

Figure 8.32: Live-cell fluorescence imaging of MCF-7 cells showing increased cellular uptake of CB-7 encapsulated NHM. [Reprinted from Ref. 155, with permission from Elsevier].

In another study, harmol caused apoptosis incells of human giloma [158]. Harmol led to autophagy and also caused suppression of expression of survivin protein. This caused apoptotic cell death in U251MG human giloma cells. The reduction of surviving protein expression was found to be related to autophagy.

A novel trans-palladium (II)-harmine complex showed significant cytotoxicity towards P-388, L1210 and K562 leukaemia cell lines with IC_{50} values of 0.385, 0.385 and 0.364 μM respectively [159]. Also, platinum (II) and palladium (II) complexes of harmine and harmane prevented tumor cell proliferation [160].

The breast cancer resistance protein BCRP has often being implicated in expelling hydrophobic chemotherapy drugs from breast cancer cells. This ultimately culminates in development of drug resistance and hence chemotherapy fails. This problem could be alleviated by the presence of harmine [161].

8.7.4 Harmine based liposomes for drug delivery

Chen et al. synthesized harmine-based liposomes (HM-lip) and then further coated these liposomes with *N*-trimethyl chitosan (TMC-HM-lip) [162]. The liposomes ranged in size from 150 to 200 nm and showed 81% entrapment efficiency (EE). The work demonstrated sustained and pH dependent release of harmine from the liposomes. It was shown that coating by *N*-trimethyl chitosan led to increased retention time of harmine when applied to the gastrointestinal tract (GIT) of mouse model. Such coating could also prevent harmine degradation in presence of enzymes which in turn

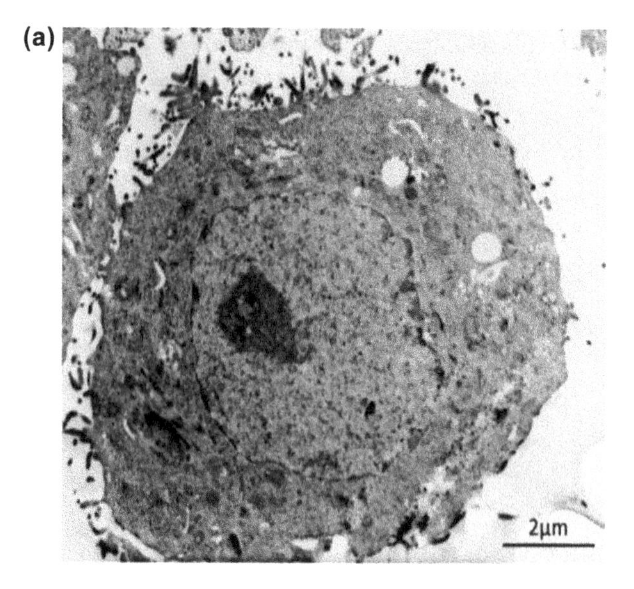

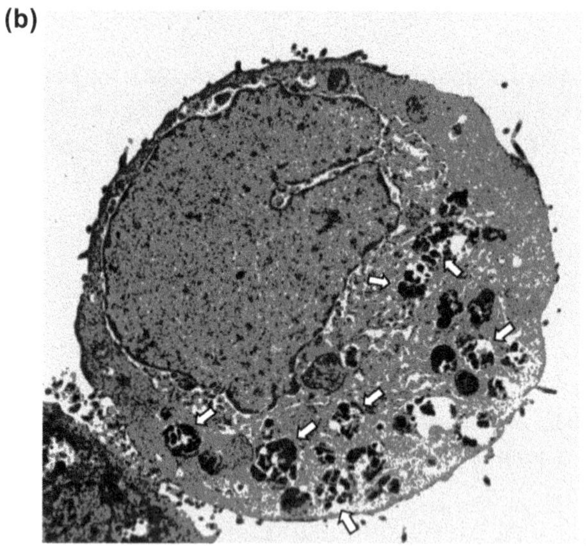

Figure 8.33: Electron microscopy images of Harmol-Treated A549 Cells. [Taken from the open access source: Ref. 157].

promoted the oral bioavailability of the drug. Caco-2 cells, from human colon adenocarcinoma serve as important models to study the effect of oral drugs [163]. The HM-lip and TMC-HM-lip could enhance transport of harmine across Caco-2 cell monolayers.

8.7.5 Antiviral activity of the β-carboline alkaloids

Synthesized platinum (II) and palladium (II) complexes of harmane and harmine demonstrated good antiviral action towards the common influenza virus and also the harmful herpes virus [160].

8.7.6 Antiparasitic activity of the β-carboline alkaloids

In the last two decades, the antiparasitic activities of the β-carboline alkaloids have seen a spurt in research. Both harmine and harmane showed significant inhibition of both the extracellular promastigote and intracellular amastigote forms of *Leishmania* that can cause infections [164]. Harmine and some other β-carboline alkaloids [151] showed trypanosomicidal activities *in vitro* against the disease causing parasite *T. cruzi* epimastigotes.

8.8 Allelopathic applications of β-carboline alkaloids

Some β-carboline alkaloids are known to exhibit allelopathy. Allelopathy is a biological phenomenon whereby an organism produces biochemicals that exert significant influence on the growth, survival, development and reproduction of other organisms. These effects may be beneficial or harmful. Plant allelopathy is a mode of interaction between receptor and donor plants and often results in positive effects e.g. weed control and crop protection. For sustainable agriculture, often advantage is taken of the stimulatory/inhibitory action of plant allelopathy. Some β-carboline alkaloids derived from plants exhibit positive allelopathy.

Another study also showed that β-carboline alkaloids isolated from seeds of *P. harmala* demonstrated strong inhibition towards growth of some plants [67]. Further investigation revealed that harmaline and harmine present in *P. harmala* seeds showed allelopathic behaviour. This result is significant as it demonstrates the prospective use of *P. harmala* for weed management in future. Besides allelopathy, *P. harmala* has certain additional advantages i.e. it is drought resistant, the root of the plant has two to three rings of vascular bundles, which may be an adaptation for dry conditions [165]. Above all, it is unpalatable to animals [166] which is a distinct advantage for use in plant allelopathy.

Summarizing, in the context of development of environment friendly herbicides via a sustainable approach, *P. harmala* offers an excellent source that can be utilized to overcome: (i) the toxicity and environmental hazards caused by synthetic herbicides and (ii) the development of herbicide resistance by some weeds due to overuse of chemical herbicides. A greenhouse experiment showed that [167] plant residues of *P. harmala* could suppress growth of wild weeds. As the natural abundance of

P. harmala is quite high, this plant can be an extremely sustainable source for future weed management minus the use of chemicals.

Harmine has been reported to decelerate the growth and to decrease the food consumption of beet armyworm, *Spodoptera exigua* [168]. Harmine also suppressed the larval weight and also inhibited pupa formation of the Indian meal moth. Harmine showed insecticidal activity towards the mosquitoes, *Culex quinquefasciatus*, and mustard aphids, *Lipaphis erysimi* [169] and significantly hindered the growth and reproduction of grain pests, *Tribolium castaneum* and *Rhyzopertha dominica* [170].

8.9 Harmful effects of some β-carboline alkaloids

However, in addition to the numerous beneficial effects of β-carboline alkaloids, there are some harmful effects too. Some foodstuffs containing β-carboline alkaloids are found to cause tremors in humans. The natural occurrence of these alkaloids in the food chain has ignited genuine concern regarding the potential hazards that may be encountered upon dietary exposure to these alkaloids. Some tremorogenic (tremor causing) β-carboline alkaloids such as harmane, harmine and harmaline have also been traced in plant-based products e.g. cereals, soya beans, fruits such as grapes, vinegar; some beverages and plant-based inhaled products such as tobacco [171, 172]. Such β-carbolines have also been found in animal derived products *viz.* beef and sardines [173].

In fact, β-carboline alkaloids have also been found in human blood in low quantities [174–176]. Some groups have done productive work on the development of methods to quantitatively estimate β-carbolines in body fluids. An example is the fluorescence based method used to quantitatively estimate β-carbolines in rat brains [177]. This method was not specific as the researchers worked with β-carboline mixtures in the tissue samples. Another group overcame this problem by separating the individual β-carbolines by thin-layer gel-plate method, they also used multiple steps of extraction along with HPLC to detect an NHM derivative *viz.*6-OH-tetrahydronorharmane (6-OH-THN) in rat and human samples [176]. Other groups could also estimate harmane, harmine, and harmaline individually in tissue samples by HPLC [178].

8.10 Conclusions

β-Carboline alkaloids also known as harmala alkaloids are fluorescent alkaloids that can be easily extracted from natural resources. β-carboline alkaloids and their derivatives have established themselves as promising drugs showing sedative, anxiolytic, hypnotic, anticonvulsant, antitumor, antiviral, antiparasitic and antimicrobial action. Their wide bio-availability from natural sources and interaction with various biological targets such as DNA, enzymes and proteins have added to their importance in pharmacology.

In this review, the natural occurrence of four important β-carboline alkaloids *viz.* NHM, harmane, harmine and harmol has been outlined in detail. The methods of extraction of these harmala alkaloids from natural sources have also been reviewed here. This has been followed by a detailed overview of the photophysical behaviour of these compounds with particular emphasis on the photophysics in bio mimicking environments. The interactions of these four β-carboline alkaloids with biological systems like proteins and DNA has also been reviewed. Finally the pharmacological activity as well as allelopathic behaviour of these compounds has been reviewed.

The main aim of this review is on sustainable use of naturally occurring compounds in safeguarding human health and protecting our environment at large. To this end, this review provides a bird's eye view on the natural abundance of the four harmala alkaloids, their easy extraction, their prospective use in diagnostics based on their interesting photophysical/fluorescence behaviour and finally their use in safeguarding the environment and providing sustainable solutions to human health issues.

8.11 Future perspective

The β-carboline alkaloids will definitely continue attracting the attention of the scientific community due to their biochemical activities and various pharmacological applications. Thus, the β-carboline alkaloids will lead to further developments in discovery of newer drugs using sustainable methods. However the dark side is that some β-carbolines are harmful too. Due to the wide natural abundance of β-carboline alkaloids, living beings will always be at risk of exposure to endogenous and exogenous β-carboline alkaloids.

Thus it is very important that the scientific community learns to make a *cost-benefit* analysis on utilization of the immense natural abundance of β-carboline alkaloids for sustainable growth and development of the world community at large while minimizing the harmful effects of these alkaloids. One concrete step towards realization of this goal will be designing minimally invasive synthetic strategies to slightly modify the extracted β-carboline alkaloids so as to enhance their positive pharmacological and allelopathic action while simultaneously minimizing their harmful effects. The review ends on this positive note that the community of synthetic chemists will be able to achieve this goal in the near future.

Acknowledgements: S. De thanks University of Kalyani for generous grant of the RUSA Scheme IP/RUSA(C-10)/11/2021. P Bhattacharya thanks DST-INSPIRE, Govt. of India for research fellowship [Ref. No. IF170936]. Above all, the authors thank the Editor, Professor P. Ramasami for the kind invitation to submit a Book Chapter.

References

1. Abrimovitch RA, Spencer ID. The carbolines. Adv Heterocycl Chem 1964;3:79–207.
2. Reyman D, Pardo A, Poyato JML. Phototautomerism of beta-carboline. J Phys Chem 1994;98: 10408–411.
3. Hudson JB, Towers GHN. Antiviral properties of photosensitizers. Photochem Photobiol 1988;48: 289–96.
4. Miguel MG, Burrows HD, Pereira MAE, Varela AP. Probing solute distribution and acid-base behaviour in water-in-oil microemulsions by fluorescence techniques. Col SurfA 2001;176:85–99.
5. Dillon J, Spector A, Nakanishi K. Identification of β carbolines isolated from fluorescent human lens proteins. Nature 1976;259:422–3.
6. Schilitter E, Bein HJ, editors. Medicinal Chemistry. New York: Academic Press; 1967.
7. Sniecks V, Manske RHF, editors. The Alkaloids. New York: Academic Press; 1968.
8. Beljanski M, Beljanski MS. 3 alkaloids as selective destroyers of the proliferative capacity of cancer-cells. IRCS (Int Res Commun Syst) Med Sci 1984;50:587–8.
9. Dias A, Varela AP, Miguel Mda G, Macanita AL, Becker RS. β-carboline photosensitizers. 1. Photophysics, kinetics and excited-state equilibria in organic solvents, and theoretical calculations. J Phys Chem 1992;96:10290–6.
10. Carmona C, Galan M, Angulo G, Munoz MA, Guardado P, Balon M. Ground and singlet excited state hydrogen bonding interactions of betacarbolines. Phys Chem Chem Phys 2000;2:5076–83.
11. Hidalgo J, Balon M, Carmona C, Munoz M, Pappalardo RR, Marcos ES. AM1 study of a β-carboline set: structural properties and potential reactivity. J Chem Soc Perkin Trans 2 1990;2:65–71.
12. Varela AP, Burrows HD, Douglas P, Miguel Mda G. Triplet state studies of β-carbolines. J Photochem Photobiol A 2001;146:29–36.
13. Mahmoudian M, Jalilpour H, Salehian P, Iranian J. Toxicity of peganum harmala: review and a case report. Pharmacol Ther 2002;1:1–4.
14. Chen Q, Chao R, Chen H, Hou X, Yan H, Zhou S, et al. Antitumor and neurotoxic effects of novel harmine derivatives and structure-activity relationship analysis. Int J Cancer 2005;114:675–82.
15. Kartal M, Altun ML, Kurucu S. HPLC method for the analysis of harmol, harmalol, harmin and harmaline in the seeds of Peganum harmala L. J Pharm Biomed Anal 2003;31:263–9.
16. Duan JA, Zhou RH, Zhao SX, Wang MS, Che CT. Studies on the chemical constituents of peganum multisectum Maxim. I. The alkaloids from seeds and antitumour activity. J China Pharm Univ 1998; 29:21–3.
17. Chatterjee A, Ganguly M. Alkaloidal constituents of peganum harmala and synthesis of the minor alkaloid deoxyvascinone. Phytochemistry 1968;7:307–11.
18. Sharaf M, El-Ansari MA, Matlin SA, Saleh NAM. Four flavonoid glycosides from peganum harmala. Phytochemistry 1997;44:533–6.
19. Movafeghi A, Abedini M, Fathiazad F, Aliasgharpour M, Omidi Y. Floral nectar composition of peganum harmala L. Nat Prod Res 2009;23:301–8.
20. Xing M, Shen F, Liu L, Chen Z, Guo N, Wang X, et al. Antimicrobial efficacy of the alkaloid harmaline alone and in combination with chlorhexidine digluconate against clinical isolates of Staphylococcus aureus grown in planktonic and biofilm cultures. Lett Appl Microbiol 2012;54: 475–82.
21. Herraiz T, Gonzaleza D, Ancin-Azpilicuetac C, Aranb VJ, Guillena H. Î^2-Carboline alkaloids in peganum harmala and inhibition of human monoamine oxidase (MAO). Food Chem Toxicol 2010; 48:839–45.

22. Farouk L, Laroubi A, Aboufatima R, Benharref A, Chait A. Evaluation of the analgesic effect of alkaloid extract of peganum harmala L: Possible mechanisms involved. J Ethnopharmacol 2008; 115:449–54.
23. Lamchouri F, Settaf A, Cherra Y, Hassar M, Zemzami M, Atif N, et al. In vitro cell-toxicity of peganum harmala alkaloids on cancerous cell-lines. Fitoterapia 2000;71:50–4.
24. Berrougui H, Cordero M, Khalil A, Hamamouchia M, Ettiab A, Marhuenda E, et al. Vasorelaxant effects of harmine and harmaline extracted from peganum harmala L. seeds in isolated rat aorta. Pharmacol Res 2006;54:150–7.
25. Bemis DL, Capodice JL, Desai M, Katz AE, Buttyan R. β-Carboline alkaloid–enriched extract from the amazonian rain forest tree Pao Pereira suppresses prostate cancer cells. J Soc Integr Oncol 2009;7:59–65.
26. Bemis DL, Capodice JL, Gorroochurn P, Katz AE, Buttyan R. Anti-prostate cancer activity of a β-carboline alkaloid enriched extract from Rauwolfia vomitoria. Int J Oncol 2006;29:1065–73.
27. Ahmad K, Thomas NF, Hadi AH, Mukhtar MR, Mohamad K, Na, et al. New vasorelaxant β-carboline alkaloids from neisosperma oppositifolia. Chem Pharm Bull 2010;58:1085–7.
28. Frye A, Haustein C. Extraction, identification, and quantification of harmala alkaloids in three species of passiflora. Am J Undergrad Res 2007;6:19–26.
29. Aiello A, Fattorusso E, Magno S, Mayol L. Brominaed β-carbolines from the marine hydroid aglaophenia pluma linnaeus. Tetrahedron 1987;43:5929–32.
30. Prinsep MR, Blunt JW, Munro MHG. New cytotoxic β-carboline alkaloids from the marine bryozoan, cribricellina cribraria. J Nat Prod 1991;54:1068–76.
31. Harwood DT, Urban S, Blunt JW, Munro MHG. β-Carboline alkaloids from a New Zealand marine bryozoan, cribricellina cribraria. Nat Prod Res 2003;17:15–9.
32. Beutler JA, Cardellina JH II, Prather T, Shoemaker RH, Boyd MR. A cytotoxic β-carboline from the bryozoan catenicella cribraria. J Nat Prod 1993;56:1825–6.
33. Carbrera GM, Seldes AM. A β-carboline alkaloid from the soft coral lignopsis spongiosum. J Nat Prod 1999;62:759–60.
34. Schuup P, Poehner T, Edrada R, Ebel R, Berg A, Wray V, et al. Two new β-carbolines from the micronesian tunicate Eudistoma sp. J Nat Prod 2003;66:272–5.
35. Rashid MA, Gustafson KR, Boyd MR. New cytotoxic N-methylated β-carboline alkaloids from the marine Ascidian Eudistomagilboverde. J Nat Prod 2001;64:1454–6.
36. Kearns PS, Coll JC, Rideout JA. A β-carboline dimer from an Ascidian, Didemnum sp. J Nat Prod 1995;58:1075–6.
37. Foderaro TA, Barrows LR, Lassota P, Ireland CM. Bengacarboline, a new β-carboline from a marine Ascidian Didemnum sp. J Org Chem 1997;62:6064–5.
38. Oku N, Matsunaga S, Fusetani N. Shishijimicins A–C, novel enediyne antitumor antibiotics from the Ascidian Didemnum proliferum. J Am Chem Soc 2003;125:2044–5.
39. Searle PA, Molinski TF. Five new alkaloids from the tropical ascidian, Lissoclinum sp. lissoclinotoxin A is chiral. J Org Chem 1994;59:6600–5.
40. Badre A, Boulanger A, Abou-Mansour E, Banaigs B, Com-baut G, Francisco C. Eudistomin U and Isoeudistomin U, new alkaloids from the carribean Ascidian Lissoclinum fragile. J Nat Prod 1994; 57:528–33.
41. Lake RJ, Blunt JW, Munro MHG. Eudistomins from the New Zealand ascidian Ritterella sigillinoides. Aust J Chem 1989;42:1201–6.
42. Lake RJ, Brennan MM, Blunt JW, Munro MHG, Pan-nell LK. Eudistomin K sulfoxide – an antiviral sulfoxide from the New Zealand ascidian Ritterella sigillinoides. Tetrohedron Lett 1988;29: 2255–6.
43. Davis RA, Carroll AR, Quinn RJ, Eudistomin V. A new β-carboline from the australian ascidian *Pseudodistoma aureum*. J Nat Prod 1998;61:959–60.

44. Chbani M, Paris M, Delauneux JM, Debitus C. Brominated indole alkaloids from the marine tunicate Pseudodistoma arborescens. J Nat Prod 1993;56:99–104.
45. Rinehart KL, Jr., Kobayashi J, Harbour GC, Hughes RG Jr., Mizask SA, Scahill TA, et al. Potent antiviral compounds containing a novel oxathiazepine ring from the caribbean tunicate Eudistoma olivaceum. J Am Chem Soc 1984;106:1524–6.
46. Kobayashi J, Harbour GC, Gilmore J, Rinehart KL Jr., Eudistomins A, D G, et al. hydroxy, pyrrolyl and iminoazepino.beta.-carbolines from the antiviral caribbean tunicate Eudistoma olivaceum. J Am Chem Soc 1984;106:1526–8.
47. Rinehart L, Jr., Kobayashi J, Harbour GC, Gilmore J, Mascal M, Holt TG, et al. Eudistomins A-Q, beta.-carbolines from the antiviral caribbean tunicate Eudistoma olivaceum. J Am Chem Soc 1987; 109:3378–87.
48. Kinzer KF, Cardellina JH II. Three new β-carbolines from the bermudian tunicate Eudistoma olivaceum. Tetrahedron Lett 1987;28:925–6.
49. Kobayashi J, Nakamura H, Ohizumi Y, Hirata Y. Eudistomidin-A, a novel calmodulin antagonist from the okinawan tunicate eudistoma glaucus. Tetrahedron Lett 1986;27:1191–4.
50. Kobayashi J, Cheng J, Ohta T, Nozoe S, Ohizumi Y, Sasaki T, et al. Novel antileukemic alkaloids from the okinawan marine tunicate Eudistoma glaucus. J Org Chem 1990;55:3666–70.
51. Murata O, Shigemori H, Ishibashi KS, Hayashi K, Kobaya-shi J, Eudistomidins E, et al. New β-carboline alkaloids from the okinawan marine tunicate Eudistoma glaucus. Tetrahedron Lett 1991;32:3539–42.
52. Buckholtz NS. Neurobiology of tetrahydro-β-carbolines. Life Sci 1980;27:893–903.
53. Wagoner RMV, Jompa J, Tahir A, Ireland CM. Trypargine alkaloids from a previously undescribed Eudistoma sp. Ascidian. J Nat Prod 1999;62:794–7.
54. Volk RB, Furkert F. Antialgal, antibacterial and antifungal activity of two metabolites produced and excreted by cyanobacteria during growth. Microbiol Res 2006;161:180–6.
55. Volk RB. Studies on culture age versus exometabolite production in batch cultures of the cyanobacterium Nostoc insulare. J Appl Phycol 2007;19:491–5.
56. Armstrong E, Boyd KG, Pisacane A, Peppiatt CJ, Burgess JG. Marine microbial natural products in antifouling coating. Biofouling 2000;16:221–32.
57. Abdullah M, Chiang L, Nadeem M. Comparative evaluation of adsorption kinetics and isotherms of a natural product removal by Amberlite polymeric adsorbents. Chem Eng J 2008;146:370–6.
58. Volk RB. Screening of microbial culture media for the presence of algicidal compounds and isolation and identification of two bioactive metabolites, excreted by the cyanobacteria Nostoc insulare and Nodularia harveyana. J Appl Phycol 2005;17:339–47.
59. Becher PG, Beuchat J, Gademann K, Juttner F. Nostocarboline: isolation and synthesis of a new cholinesterase inhibitor from Nostoc 78-12A. J Nat Prod 2005;68:1793–5.
60. Volk RB. Screening of microalgae for species excreting norharmane, a manifold biologically active indole alkaloid. Microbiol Res 2008;163:307–13.
61. Kreitlow S, Mundt S, Lindequist U. Cyanobacteria—a potential source of new biologically active substances. J Biotechnol 1999;70:61–3.
62. Tan LT, Goh BP, Tripathi A, Lim MG, Dickinson GH, Lee SS, et al. Natural antifoulants from the marine cyanobacterium Lyngbya majuscula. Biofouling 2010;26:685–95.
63. Karan T, Erenler R. Screening of norharmane from seven cyanobacteria by high-performance liquid chromatography. Phcog Mag 2017;13:S723–5.
64. Totsuka Y, Ushiyama H, Ishihara J, Sinha R, Goto S, Sugimura T, et al. Quantification of the co-mutagenic betacarbolines, norharman and harman, in cigarette smoke condensates and cooked foods. Cancer Lett 1999;143:139–43.
65. Totsuka Y, Takamura-Enya T, Nishigaki R, Sugimura T, Wakabayashi K. Mutagens formed from beta-carbolines with aromatic amines. J Chromatogr B 2004;802:135–41.

66. Caicedo NH, Kumirska J, Neumann J, Stolte S, Thöming J. Detection of bioactive exometabolites Produced by the filamentous marine cyanobacterium Geitlerinema sp. Mar Biotechnol 2012;14: 436–45.
67. Shao H, Huang X, Zhang Y, Zhang C. Main alkaloids of Peganum harmala L. And their different effects on Dicot and Monocot crops. Molecules 2013;18:2623–34.
68. Das P, Chakrabarty A, Mallick A, Chattopadhyay N. Photophysics of a cationic biological photosensitizer in anionic micellar environments: combined effect of polarity and rigidity. J Phys Chem B 2007;111:11169–176.
69. Sengupta B, Sengupta PK. Binding of quercetin with human serum albumin: a critical spectroscopic study. Biopolymers 2003;72:427–34.
70. Guharay J, Sengupta B, Sengupta PK. Protein–flavonol interaction: fluorescence spectroscopic study. Proteins Struct Funct Genet 2001;43:75–81.
71. Mallick A, Purkayastha P, Chattopadhyay N. Photoprocesses of excited molecules in confined liquid environments: an overview. J Photochem Photobiol C Photochem Rev 2007;8:109–27.
72. Reyman D, Vinas MH, Poyato JML, Pardo A. Proton transfer dynamics of norharman in organic solvents. J Phys Chem A 1997;101:768–75.
73. Vert FT, Sanchez IZ, Torrent AO. Acidity constants of β-carbolines in the ground and excited singlet states. J Photochem 1983;23:355–68.
74. Balon M, Muroz MA, Hidalgo J, Carmona MC, Sanchez M. Fluorescence characteristics of β-carboline alkaloids in highly concentrated hydroxide solutions. J Photochem 1987;36:193–204.
75. Gonzalez MM, Salum ML, Gholipour Y, Cabrerizo FM, Erra-Balsells R. Photochemistry of norharmane in aqueous solution. Photochem Photobiol Sci 2009;8:1139–49.
76. Ghiggino KP, Skilton PF, Thistlethwaite PJ. β-Carboline as a fluorescence standard. J Photochem 1985;31:113–21.
77. Sakurovs R, Ghiggino KP. Excited state proton transfer in β-carboline. J Photochem 1982;18:1–8.
78. Varela AP, Miguel Mda G, Maçanita AL, Burrows HD, Becker RS. Beta-carboline photosensitizers. 3. Studies on ground and excited state partitioning in AOT/water/cyclohexane microemulsions. J Phys Chem 1995;99:16093–100.
79. Mallick A, Haldar B, Chattopadhyay N. Encapsulation of norharmane in cyclodextrin: formation of 1:1 and 1:2 inclusion complexes. J Photochem Photobiol B 2005;78:215–21.
80. Mallick A, Chattopadhyay N. Photophysics of norharmane in micellar environments: a fluorometric study. Biophys Chem 2004;109:261–70.
81. Chakrabarty A, Das P, Mallick A, Chattopadhyay N. Effect of surfactant chain length on the binding interaction of a biological photosensitizer with cationic micelles. J Phys Chem B 2008;112: 3684–92.
82. Chakrabarty A, Mallick A, Haldar B, Purkayastha P, Das P, Chattopadhyay N. Surfactant chain-length-dependent modulation of the prototropic transformation of a biological photosensitizer: norharmane in anionic micelles. Langmuir 2007;23:4842–48.
83. Paul BK, Ghosh N, Mondal R, Mukherjee S. Contrasting effects of salt and temperature on niosome-bound norharmane: direct evidence for positive heat capacity change in the niosome: β-cyclodextrin interaction. J Phys Chem B 2016;120:4091–101.
84. Kasha M. Proton-transfer spectroscopy. Perturbation of the tautomerization potential. J Chem Soc, Faraday Trans 2 1986;82:2379–92.
85. Kasha M, Sytnik A, Dellinger B. Solvent cage spectroscopy. Pure Appl Chem 1993;65:1641–46.
86. Pimentel GC, McClellan AL. The Hydrogen Bond. San Francisco: W. H. Freeman; 1960.
87. Green RD. Hydrogen bonding by C-H groups. New York: MacMillan Press; 1974.
88. Sarkar D, Mallick A, Haldar B, Chattopadhyay N. Ratiometric spectroscopic response of pH sensitive probes: an alternative strategy for multidimensional sensing. Chem Phys Lett 2010;484: 168–72.

89. Mallick A, Roy UK, Majumdar T, Haldar B, Pratihare S. Photophysical, NMR and density functional study on the ion interaction of norharmane: proton transfer vs. hydrogen bonding. RSC Adv 2014; 4:16274–80.
90. Paul S, Karar M, Paul P, Mallick A, Majumdar T. Dual mode nitro explosive detection under crowded condition: conceptual development of a sensing device. J Photochem Photobiol A 2019; 379:123–9.
91. Paul S, Karar M, Mitra S, Sher Shah SA, Majumdar T, Mallick A. A molecular lock with hydrogen sulfate as "Key" and fluoride as "hand": computing based insights on the functioning mechanism. ChemistrySelect 2016;1:5547–53.
92. Marques ADS, Souza HF, Costa IC, Azevedo WMde. Spectroscopic study of harmane in micelles at 77 K using fluorescent probes. J Mol Struct 2000;520:179–90.
93. Paul BK, Ray D, Guchhait N. Binding interaction and rotational-relaxation dynamics of a cancer cell photosensitizer with various micellar assemblies. J Phys Chem B 2012;116:9704–17.
94. Ahmed SA, Chatterjee A, Maity B, Seth D. Supramolecular interaction of a cancer cell photosensitizer in the nanocavity of cucurbit[7]uril: a spectroscopic and calorimetric study. Int J Pharm 2015;492:103–08.
95. Paul BK, Guchhait N. Differential interactions of a biological photosensitizer with liposome membranes having varying surface charges. Photochem Photobiol Sci 2012;11:661.
96. Paul BK, Ray D, Ganguly A, Guchhait N. Modulation in prototropism of the photosensitizer Harmane by host:guest interactions between β-cyclodextrin and surfactants. J Colloid Interface Sci 2013;411:230–39.
97. Carmona C, Balon M, Galan M, Guardado P, Munoz MA. Dynamic study of excited state hydrogen-bonded complexes of harmane in cyclohexane–toluene mixtures. Photochem Photobiol 2002;76: 239–46.
98. Krishnamurthy M, Dogra SK. Electronic spectra of harmane: study of solvent dependence. Int J Chem A 1986;25:178–80.
99. Reyman D, Hallwass F, Goncalvesda Cruz SM, Camacho JJ. Coupled hydrogen-bonding interactions between β-carboline derivatives and acetic acid. Magn Reson Chem 2007;45: 830–34.
100. Balon M, Guardado P, Munoz MA, Carmona C. A spectroscopic study of the hydrogen bonding and π–π stacking interactions of harmane with quinoline. Biospectroscopy 1998;4:185–95.
101. Balon M, Munoz MA, Guardado P, Carmona C. Hydrogen-bonding interactions between harmane and pyridine in the ground and lowest excited singlet states. Photochem Photobiol 1996;64: 531–36.
102. Wolfbeis OS, Füriinger E. The pH-dependence of the absorption and fluorescence spectra of harmine and harmol: drastic differences in the tautomeric equilibria of ground and first excited singlet state. Z Phys Chem 1982;129:171–83.
103. Becker RS, Ferreira LFV, Elisei F, Machado I, Latterini L. Comprehensive photochemistry and photophysics of land- and marine-based β-carbolines employing time-resolved emission and flash transient spectroscopy. Photochem Photobiol 2005;81:1195–204.
104. Karar M, Paul P, Mistri R, Majumdar T, Mallick A. Dual macrocyclic chemical input based highly protective molecular keypad lock using fluorescence in solution phase: a new type approach. J Mol Liq 2021;331:115679.
105. Windholz M, editor. The Merck Index, 9th ed. Rahway, NJ: Merck; 1976. 4471 p.
106. Robinson T. The Biochemistry of Alkaloids. Berlin: Springer; 1968. 132 p.
107. Mitra C, Guha SR. Inhibition patterns of monoamine oxidase in sub-fractions of rat brain mitochondria in presence of some selective inhibitors. Biochem Pharmacol 1979;28:1135–7.
108. Tomas F, Zabala I, Olba A. Acid-base and tautomeric equilibria of harmol in the ground and first excited singlet states. J Photochem 1985;31:253–63.

109. Olba A, Medina P, Codoñer A, Monsó S. Fluorescence and phosphorescence of harmol and harmalol at 77 K. J Photochem 1987;39:273–83.
110. Airaksinen MM, Kari I. β-Carbolines, psychoactive compounds in the mammalian body. Med Biol 1981;59:21–34.
111. Mallick A, Chattopadhyay N. Photophysics in motionally constrained bioenvironment: interactions of norharmane with bovine serum albumin. Photochem Photobiol 2005;81:419–24.
112. Ghosh S, Chakrabarty S, Bhowmik D, Kumar GS, Chattopadhyay N. Stepwise unfolding of bovine and human serum albumin by an anionic surfactant: an investigation using the proton transfer probe norharmane. J Phys Chem B 2015;119:2090–102.
113. Domonkos C, Fitos I, Visy J, Zsila F. Fatty acid modulated human serum albumin binding of the β-carboline alkaloids norharmane and harmane. Mol Pharm 2013;10:4706–16.
114. Meester C. Genotoxic potential of β-carbolines: a review. Mutat Res 1995;339:139–53.
115. Funayama Y, Nishio K, Wakabayashi K, Nagao M, Shimoi K, Ohira T, et al. Effects of β- and γ-carboline derivatives on DNA topoisomerase activities. Mutat Res 1996;349:183–91.
116. Duportail G. Linear and circular dichroism of harmine and harmaline interacting with DNA. Int J Biol Macromol 1981;3:188–92.
117. Taira Z, Kanzawass S, Dohara C, Ishida S, Matsumoto M, Sakiya Y. Intercalation of six β-carboline derivatives into DNA. Jpn J Toxicol Environ Health 1997;43:83–91.
118. Balon M, Munoz MA, Carmona C, Guardado P, Galan M. A fluorescence study of the molecular interactions of harmane with the nucleobases, their nucleosides and mononucleotides. Biophys Chem 1999;80:41–52.
119. Remsen JF, Cerutti PA. Inhibition of DNA-repair and DNA-synthesis by harman in human alveolar tumor cells. Biochem Biophys Res Commun 1979;86:124–29.
120. Funayama Y, Nishio K, Wakabayashi K, Nagao M, Shimoi K, Ohira T, et al. Effects of β- and γ-carboline derivatives on DNA topoisomerase activities. Mutat Res 1996;349:183–91.
121. Manabe S, Kanai Y, Ishikawa S, Wada O. Carcinogenic tryptophan Pyrolysis Products Potent inhibitors of type A monoamine oxidase and the Platelet response to 5-hydroxytryptamine. J Clin Chem Clin Biochem 1988;26:265–70.
122. Herraiz T, Chaparro C. Human monoamine oxidase enzyme inhibition by coffee and β-carbolines norharman and harman isolated from coffee. Life Sci 2006;78:795–802.
123. May T, Rommelspacher H, Pawlik M. [3H]Harman binding experiments. I: a reversible and selective radioligand for monoamine oxidase subtype A in the CNS of the rat. J Neurochem 1991; 56:490–9.
124. Rommelspacher H, May T, Salewski B. Harman (1-methyl-β-carboline) is a natural inhibitor of monoamine oxidase type A in rats. Eur J Pharmacol 1994;252:51–9.
125. Rommelspacher H, Meier-Henco M, Smolka M, Kloft C. The levels of norharman are high enough after smoking to affect monoamineoxidase B in platelets. Eur J Pharmacol 2002;441:115–25.
126. Kim H, Sablin SO, Ramsay RR. Inhibition of monoamine oxidase A by β-carboline derivatives. Arch Biochem Biophys 1997;337:137–42.
127. Song Y, Wang J, Teng SF, Kesuma D, Deng Y, Duan J, et al. β-Carbolines as specific inhibitors of cyclin-Dependent kinases. Bioorg Med Chem Lett 2002;12:1129–32.
128. Song Y, Kesuma D, Wang J, Deng Y, Duan J, Wang JH, et al. Specific inhibition of cyclin-dependent kinases and cell proliferation by harmine. Biochem Biophys Res Commun 2004;317:128–32.
129. Sobhani AM, Ebrahimi SA, Mahmoudian M. An in vitro evaluation of human DNA topoisomerase| inhibition by peganum harmala L. seeds extract and its β-carboline alkaloids. J Pharm Pharmaceut Sci 2002;5:19–23.
130. Hayashi K, Nagao M, Sugimura T. Interactions of norharman and harman with DNA. Nucleic Acids Res 1977;4:3679–86.

131. Cao R, Peng W, Chen H, Ma Y, Liu X, Hou X, et al. DNA binding properties of 9-substituted harmine derivatives. Biochem Biophys Res Commun 2005;338:1557–63.
132. Slotkin TA, Distefano VI, Au WYW. Blood levels and urinary excretion of harmine and its metabolites in man and rats. J Pharmacol Exp Therapeut 1970;173:26–30.
133. Naranjo C. In Ethnopharmacologic Search for Psychoactive Drugs. Efron DK, Holmstedt B, Kline NS, editors. Washington, DC: US Govement Printing Office; 1967. p. 385–91.
134. Rommelspacher H, Schmidt LG, May T. Plasma norharman (β-carboline) levels are elevated in chronic alcoholics. Alcohol Clin Exp Res 1991;15:553–9.
135. Stohler R, Rommelspacher H, Ladewig D, Dammann G. Beta-carbolines (harman/norharman) are increased in heroin dependent patients. Ther Umsch Rev ther 1993;50:178–81.
136. Stohler R, Rommelspacher H, Ladewig D. The role of beta-carbolines (harman/norharman) in heroin addicts. Eur Psychiatr 1995;10:56–8.
137. Aricioglu-Kartal F, Kayır H, Uzbay IT. Effects of harman and harmine on naloxone-precipitated withdrawal syndrome in morphine-dependent rats. Life Sci 2003;73:2363–71.
138. Lantz SM, Cuevas E, Robinson BL, Paule MG, Ali SF, Imam SZ. The role of harmane and norharmane in in Vitro dopaminergic function. J Drug Alcohol Res 2015;4:1–8.
139. Matsubara K, Gonda T, Sawada H, Uezono T, Kobayashi Y, Kawamura T, et al. Endogenously occurring β-carboline induces parkinsonism in nonprimate animals: a possible causative protoxin in idiopathic Parkinson's disease. J Neurochem 1998;70:727–35.
140. Matsubara K, Neafsey EJ, Collins MA. Novel s-adenosylmethionine-dependent indole-N-methylation of β-carbolines in brain particulate fractions. J Neurochem 1992;59:511–18.
141. Sanchaita L, Swapan P, Sibabrata M, Santu B, Mukul KB. Harmine: evaluation of its antileishmanial properties in various vesicular delivery systems. J Drug Target 2004;12:165–75.
142. Marquenie D, Geeraerd AH, Lammertyn J, Soontjens C, Van Impe JF, Michiels CW, et al. Combinations of pulsed white light and UV-C or mild heat treatment to inactivate conidia of Botrytis cinerea and Monilia fructigena. Int J Food Microbiol 2003;85:185–96.
143. Palou L, Usall J, Munoz J, Smilanick J, Viñas I. Hot water, sodium bicarbonate and sodium carbonate for the control of green and blue mold of Clemetine mandarins. Postharvest Biol Technol 2002;24:93–6.
144. Cantu D, Blanco-Ulate B, Yang L, Labavitch JM, Bennett AB, Powell AL. Ripening-regulated susceptibility of tomato fruit to Botrytis cinerea requires NOR but not RIN or ethylene. Plant Physiol 2009;150:1434–49.
145. Elad Y, Evensen K. Physiological aspects of resistance to Botrytis cinerea. Phytopathology 1995; 85:637–43.
146. Garber MP, Hudson WG, Norcini JG, Thomas WA, Jones RK, Bondari K. Biologic and Economic Assessment of Pest Management in the United States Greenhouse and Nursery Industry. Athens: Coop Ext Serv Univ. GA; 1997.
147. Latorre B, Flores V, Sara A, Roco A. Dicarboximide-resistant strains of Botrytis cinerea from table grapes in Chile: survey and characterization. Plant Dis 1994;78:990–4.
148. Olmedo GM, Cerioni L, Gonzalez MM, Cabrerizo FM, Rapisarda VA, Volentini SI. Antifungal activity of β-carbolines on penicillium digitatum and botrytis cinerea. Food Microbiol 2017;62:9–14.
149. Chouvenc T, Su NY, Elliott MI. Antifungal activity of the termite alkaloid norharmane against the mycelial growth of Metarhizium anisopliae and Aspergillus nomius. J Invertebr Pathol 2008;99: 345–7.
150. Xing M, Shen F, Liu L, Chen Z, Guo N, Wang X, et al. Antimicrobial efficacy of the alkaloid harmaline alone and in combination with chlorhexidine digluconate against clinical isolates of Staphylococcus aureus grown in planktonic and biofilm cultures. Lett Appl Microbiol 2012;54: 475–82.

151. Rivas P, Cassels BK, Morello A, Repetto Y. Effects of some beta-carboline alkaloids on intact Trypanosoma cruzi epimastigotes. Comp Biochem Physiol C Pharmacol Toxicol Endocrinol 1999; 122:27–31.
152. Alomar ML, Rasse-Suriani FA, Ganuza A, Coceres VM, Cabrerizo FM, Angel SO. In vitro evaluation of beta-carboline alkaloids as potential anti-Toxoplasma agents. BMC Res Notes 2013;6:193.
153. Olmedo GM, Cerioni L, Gonzalez MM, Cabrerizo FM, Volentini SI, Rapisarda VA. UVA photoactivation of harmol enhances its antifungal activity against the phytopathogens Penicillium digitatum and Botrytis cinerea. Front Microbiol 2017;8:1–9.
154. Gonzalez MM, Salum ML, Gholipour Y, Cabrerizo FM, Erra- Balsells R. Photochemistry of norharmane in aqueous solution. Photochem Photobiol Sci 2009;8:1139–49.
155. Chandra F, Kumar P, Koner AL. Encapsulation and modulation of Protolytic equilibrium of β-carboline-based norharmane drug by cucurbit[7]uril and micellar environments for enhanced cellular uptake. Colloids Surf B Biointerfaces 2018;171:530–7.
156. Abe A, Yamada H. Harmol induces apoptosis by caspase-8 activation independently on Fas/Fas ligand interaction in human lung carcinoma H596 cells. Anti Cancer Drugs 2009;20:373–81.
157. Abe A, Yamada H, Moriya S, Miyazawa K. The β-carboline alkaloid harmol induces cell death via autophagy but not apoptosis in human non-small cell lung cancer A549 cells. Biol Pharm Bull 2011;34:1264–72.
158. Abe A, Kokuba H. Harmol induces autophagy and subsequent apoptosis in U251MG human glioma cells through the downregulation of survivin. Oncol Rep 2013;29:1333–42.
159. Al-Allaf TA, Rashan LJ. Synthesis and cytotoxic evaluation of the first trans-palladium (II) complex with naturally occurring alkaloid harmine. Eur J Med Chem 1998;33:817–20.
160. Al-Allaf TA, Ayoub MT, Rashan LJ. Synthesis and characterization of novel biologically active platinum (II) and palladium (II) complexes of some β-carboline alkaloids. J Inorg Biochem 1990; 38:47–56.
161. Ma Y, Wink M. The beta-carboline alkaloid harmine inhibits BCRP and can reverse resistance to the anticancer drugs mitoxantrone and camptothecin in breast cancer cells. Phytother Res 2010; 24:146–9.
162. Chen W, Yuan Z-Q, Liu Y, Yang S, Zhang C, Li J, et al. Liposomes coated with N-trimethyl chitosan to improve the absorption of harmine in vivo and in vitro. Int J Nanomed 2016;11:325–36.
163. Artursson P, Palm K, Luthman K. Caco-2 monolayers in experimental and theoretical predictions of drug transport. Adv Drug Deliv Rev 2012;64:280–9.
164. Di Giorgio C, Delmas F, Ollivier E, Elias R, Balansard G, Timon-David P. In vitro activity of the β-carboline alkaloids harmane, harmine, and harmaline toward parasites of the species Leishmaniainfantum. Exp Parasitol 2004;106:67–74.
165. Zhang H, Chen L, Hu Z. Xeromorphic characters in the vegetative organs of Peganumharmala. Acta Phytoeco Geobot Sin 1992;16:243–8.
166. Asgarpanah J, Ramezanloo F. Chemistry, pharmacology and medicinal properties of Peganumharmala L. Afr J Pharm Pharmacol 2012;6:1573–80.
167. Sodaeizadeh H, Rafieiolhossaini M, van Damme P. Herbicidal activity of a medicinal plant, Peganum harmala L, and decomposition dynamics of its phytotoxins in the soil. Ind Crop Prod 2010;31:385–94.
168. Cavin JC, Rodriguez E. The influence of dietary β-carboline alkaloids on growth rate, food consumption, and food utilization of larvae of Spodoptera exigua (Hubner). J Chem Ecol 1988;14: 475–84.
169. Zeng Y, Zhang Y, Weng Q, Hu M, Zhong G. Cytotoxic and insecticidal activities of derivatives of harmine, a natural insecticidal component isolated from Peganum harmala. Molecules 2010;15: 7775–91.

170. Nenaah G. Toxicity and growth inhibitory activities of methanol extract and the β-carboline alkaloids of Peganum harmalaL.against two coleopteran stored-grain pests. J Stored Prod Res 2011;47:255–61.
171. Adachi J, Mizoi Y, Naito T, Yamamoto K, Fujiwara S, Ninomiya I. Determination of b-carbolines in foodstuffs by high performance liquid chromatography and high performance liquid chromatograph–mass spectrometry. J Chromatogr 1991;538:331–9.
172. Poindexter EH, Carpenter RD. The isolation of harmane and norharmane from tobacco and cigarette smoke. Phytochemistry 1962;1:215–21.
173. Rommelspacher H, Barbey M, Strauss S, Greiner B, Fahndrich E. Beta-Carbolines and Tetrahydroisoquinolines Bloom F, Barchas J, Sandler M, Usdin E, editors. 41–55. New York: A. R. Liss; 1982.
174. Allen RF, Beck O, Borg S, Skroder R. Analysis of 1-methyl-1,2,3,4-tetrahydro-b-carboline in human urine and cerebrospinal fluid by gas chromatography–mass spectrometry. Eur J Mass Spectrom 1980;1:171–7.
175. Bidder TA, Shoemaker DW, Boettger HG, Evans M, Cummins JT. Harmane in human platelets. Life Sci 1979;25:157–64.
176. Rommelspacher H, Strauss S, Lindemann J. Excretion of tetrahydroharmane and harmane into the urine of man and rat after load with ethanol. FEBS Lett 1980;109:209–12.
177. Zetler G, Singbart G, Schlosser L. Cerebral pharmacokinetics of tremor-producing harmala and iboga alkaloids. Pharmacology 1972;7:237–48.
178. Moncrieff J. Determination of pharmacological levels of harmane, harmine and harmaline in mammalian brain tissue, cerebrospinal fluid and plasma by high-performance liquid chromatography with fluorimetric detection. J Chromatogr 1989;496:269–78.

Taskeen F. Docrat, Naeem Sheik Abdul and Jeanine L. Marnewick*

9 The phytotherapeutic potential of commercial South African medicinal plants: current knowledge and future prospects

Abstract: South Africa, a country considered affluent in nature, ranks third in global biodiversity and encompasses approximately 9% of higher plants on planet Earth. Many indigenous plants have been utilised as herbal medicine, proving successful in treating numerous ailments. From the common cold to pandemic maladies such as COVID-19 in the 21st century and the treatment of incurable diseases, South African inhabitants have found great promise in the healing properties of these plants. Phytomedicine is a rapidly evolving topic, with in-depth bioactive composition analysis, identifying therapeutic action mechanisms, and disease prevention. While we are now poised to take advantage of nature's medicine cabinet with greater scientific vigour, it remains critical that these practises are done with caution. Overharvesting significantly impacts biodiversity and cultivation practices amidst the beautiful nature of these nutraceuticals. This book chapter focuses on the therapeutic potential of commonly used South African medicinal plants, their ethnopharmacological properties, and how we can conserve this treasure cove we call home for future generations.

Keywords: bioactives; medicinal plants; South Africa; sustainability.

9.1 Introduction: phytomedicine

Traditional medicine is an essential category for alternative and complementary medicine and has been used in developing and developed countries for centuries [1]. Before accepting herbal medicine as an orthodox treatment route, many individuals in the Healthcare system would raise an eyebrow at the efficacy of such a route, deeming it "pseudo-medicine." Scientific and technological advances have provided a platform for extensive research into the practical rationale for using plant-based remedies and mechanistic insight based on individual plant constituents.

Phytomedicine is an umbrella term to describe medicinal plants with a range of pharmacologically active metabolites with potential remedial action [2]. Traditionally, naturopathic users try to preserve the integrity of a plant to obtain maximum benefits

***Corresponding author: Jeanine L. Marnewick,** Applied Microbial and Health Biotechnology Institute, Cape Peninsula University of Technology, Bellville, South Africa, E-mail: marnewickj@cput.ac.za
Taskeen F. Docrat and Naeem Sheik Abdul, Applied Microbial and Health Biotechnology Institute, Cape Peninsula University of Technology, Bellville, South Africa

As per De Gruyter's policy this article has previously been published in the journal Physical Sciences Reviews. Please cite as: T. F. Docrat, N. S. Abdul and J. L. Marnewick "The phytotherapeutic potential of commercial South African medicinal plants: current knowledge and future prospects" *Physical Sciences Reviews* [Online] 2022. DOI: 10.1515/psr-2022-0136 | https://doi.org/10.1515/9783110913361-009

in treatment [3]. The widespread practice of plant-based medicine extends from allopathic to Ayurvedic and Traditional Chinese medicine [4]. Plant-based therapies are primarily used for health promotion and chronic treatment and have prospered into a commercial enterprise globally [5]. Many pharmaceutical drugs are scientifically developed by isolating the bioactive compounds found in plant material. This process involves large-scale manufacturing that ensures the final product contains the appropriate concentration of the target compound for the treatment to deliver the desired dose of the active component. It focuses on the importance of natural products in drug development, with over 300 essential drugs mentioned in the basic medicinal product list by the World Health Organisation (WHO) [6]. It has become a common consensus that many prescription drugs are directly or indirectly derived from herbal sources. The probability of this pattern extending to the future is likely, as new plant entities and natural chemicals are discovered [7]. This alludes to confirming that biogenic products are more than just remnants of historical natural medicine; they have essential roles in modern medicine and pharmaceutical developments.

Holistic medical treatments are more common in the 21st century as patients and practitioners gear toward phytomedicine after unsuccessful responses to general drug-based therapies. A paradigm shift occurred in medical practice with phytotherapy promoting multitargeted natural approaches directing the body to defend, protect, and repair instead of isolated targets, e.g., tumor cell death or pathogenic microbes [8]. This strategy reduces side effects if present and rigorous collaborative studies accomplish this by standardising phytotherapy by molecular biologists, pharmacologists, and clinicians. Extensive efforts are put into screening and profiling plant extracts with potential benefits to characterise synergism and add to the absence of information on their molecular machinery.

9.2 Prevention is better than cure

Disease prevention to promote health and well-being has been a primary global goal in South Africa for several years. This is focused on building knowledge-based systems to maintain good health and longevity. Dietary factors provide great supplemental value, especially in herbal medicine and health promotion. The interest in herbal dietary supplementation has increased in recent years [9]. Proactive health approaches allow for well-managed lifestyles, ultimately preventing disease development altogether. Pharmacological advances using natural products can lead to enhancements in discovering novel medicinal metabolites. The use of natural products has been considered safe. However, there is a need to identify herb-drug interactions with possible contraindications. Various safety assessments and quality control protocols are implemented to determine mechanisms of action to enable an optimal holistic experience. Currently, primary prevention strategies focus on socio-economic strategic plans to impede disease manifestation. In this approach, the scientific focus of herbal medicine should include minimising short-term and long-term ailments such as diabetes, cancer, cardiovascular disorders, asthma, arthritis, and associated risk factors.

The western medical practice relies heavily on active ingredients found in natural products for drug development. At the same time, advances in organic chemistry promote the development of synthetic compounds derived from these natural sources [10]. The diversity offered by natural products dated to the 17th century when plant-based malarial treatment was generated [11]. This was successfully followed by morphine, digitalis, and aspirin, frequently used today. Ethnopharmacology provides a basis for investigating folklore and traditional resources, contributing to the scientific backbone of research in herbal medicine. Furthermore, this application aids the development of innovative herbal drug transport systems with enhanced solubility, absorption, and distribution properties.

9.3 South African indigenous plants

The pluralistic nature of the South African healthcare system allows the current practices of modern medicine to coexist with unconventional options such as the use of traditional medicine. A significant proportion of the country's population relies on the latter to fulfil their primary medical requirements due to ease of access and culturally accepted affordability. The diverse number of medicinal plants on South African soil is extensive; here, we review and discuss some of the most popular plants that play essential roles in biodiversity and highlight their therapeutic value.

9.3.1 Aspalathus linearis [12]

Rooibos (pronounced "Roy-boss"– the Afrikaans word for "red bush") herbal tea, produced by the *A. linearis* (Burm.*f.*) R.Dahlgren plant has received the spotlight for its superior health benefits and low caffeine profile [13]. Endemic to the Western Cape's floristic region, the people of Khoi-descend 14 traditionally enjoyed this beverage. In 1904, Clanwilliam, located in the Western Cape Province of South Africa, was established as the home of rooibos, and one can still indulge in the ultimate rooibos experience here today. In 1968, the modern history of rooibos birthed its wellness journey when Mrs Annetjie Theron claimed it soothed her baby's cholic. This led to the development of a booming industry with a range of commercial rooibos health and skincare products still recognised internationally [14]. The beneficial effects of rooibos extend beyond its anti-allergic properties, with enhanced antioxidant anti-ageing, antispasmodic, and anti-inflammatory effects [15, 16]. Women in Africa use herbal tea to alleviate nausea and heartburn during pregnancy [13]. Scientific clinical trials have indicated the positive impact of rooibos on heart health in adults at risk for developing cardiovascular disease [17].

Due to its commercial significance, rooibos' biological activity and phytochemical properties have been extensively studied. Two primary polyphenols attributing to the unique composition and antioxidant effects of rooibos are a dihydrochalcone C-glucoside, aspalathin, and a cyclic dihydrochalcone, aspalalinin [15, 18]. It has been shown to restore stress-related metabolites, endogenous antioxidant enzymes such as superoxide dismutase and catalase, and prevent lipid peroxidation in oxidative stress-induced models [19]. The flavonoid bioactive components quercetin, nothofagin, and isoquercetin have strong antimutagenic properties [20]. Antibody stimulatory effects of rooibos suggest immunomodulatory benefits which have been extensively reviewed [21, 22]. Further investigations demonstrate the antibacterial and antimicrobial activities of rooibos [23, 24]. Recently researchers have proposed using rooibos as dietary support during COVID-19 infections. This ideology stems from this herbal tea's potent anti-inflammatory and antioxidant effects, which may modulate viral inhibition through the ACE-2 receptor, the critical entry point for SARS-CoV-2 into the cells [25]. Indeed, the potential of rooibos to dampen the cytokine storm experienced during COVID-19 infections requires further investigation. These findings collectively emphasise the favorable properties of rooibos and are the rationale behind its global demand.

Of the various indigenous plants in South Africa, rooibos has long been considered an essential commercial plant [26]. Cultivation is primarily in the Cederberg Mountain region, where seedlings are used for plant propagation, leading to variations in rooibos plant genetics. A recent study promoting rooibos plant gene discovery indicates that transcriptomes are unique and dependent on individual plant morphology, growth type, and biological profiles [27]. This information allows for improved cultivation practices and our understanding of various biosynthetic pathways that facilitate and optimise therapeutic plant development. Over the last two decades, the South African rooibos Council indicated that there had been a steady growth in rooibos export from approximately 5000 to 8000 tonnes in 2020. The top 10 largest importers (in chronological order) of rooibos are Germany, Japan, the Netherlands, UK, USA, Botswana, China, Sri Lanka, Poland, and Zimbabwe. A milestone in the history of rooibos is its recognition by the European Union when rooibos and the other famous South African herbal tea, honeybush, received geographical indication status by the EU [28]. This is the first South African product to receive such status from the EU, other than wine and spirits, and will promote economic growth and competitiveness, especially considering the global market. Another success for the industry is that rooibos will receive its own harmonised systems (HS) code, allowing its tracking around the globe [12]. The growing knowledge of the health properties of rooibos contributes to overall health and wellness by promoting rooibos-enriched products.

9.3.2 Agathosma Betulina (Buchu)

The ancient natural herb Buchu is commonly known as a miracle plant. It originates from the Rutaceae flower family and was widely consumed by the Khoi people of South Africa to promote anti-ageing, wound healing, and alleviating gastrointestinal issues [29]. The concise history of Buchu dated to the 1800s and was rendered archaic in the 1900s due to a lack of scientific evidence to support its claims. Recent efforts have been made to validate these anecdotal health claims by determining the efficacy of Buchu *in vivo*. Treatment with the plant significantly improved metabolic disorders such as diabetes, hypercholesterolemia, obesity, and high blood pressure [30]. Today, Buchu is used to prepare essential oils and is found in many herbal over-the-counter concoctions. Buchu is traditionally used to treat urinary tract/kidney infections, flu-like symptoms, spasmolytic, antiseptic, rheumatoid arthritis, and many others [31, 32]. Further anecdotal skin-enhancing properties were documented by the Khoi-San's use of the plant as a lubricant that prevented fungal and bacterial infections and added moisture during the warmer seasons [26]. This theory supports the widespread use of this medicinal plant in the cosmeceutical industry.

The phytochemical properties of Buchu have been well established through fractionation studies. These fractions are comprised of diosphenol and isomenthone, which give rise to distinct flavour properties and other compounds, including *l*-pulegone and limonene [33, 34]. Hybrid plant formation is documented when cultivation is practised outside the natural habitat [35]. More recent phytochemical studies confirmed previous findings when extensively investigated [36]. Other plants can produce monoterpenes and are not distinctive to Buchu. However, the specific blackcurrant flavor produced by diosphenol in Buchu is well sought after by the food industry [37, 38].

Buchu's popular uses and medicinal properties make it an essential contributor to the South African export and trade industry. The only significant threats to the biodiversity of this plant include natural disasters like wildfires, illegal harvesting, deforestation, and poaching. Western Cape government regulations are currently in place for sustainable harvesting to prevent future exploitation [39]. The socioeconomic impact on harvesting methods is commonly neglected and requires attention.

9.3.3 Sutherlandia frutescens (Fabaceae)

The Fabaceae or legume family produces S. frutescens (SF) (L) R.Br. and is indigenous to Southern Africa [26]. This plant is distinguished by its large red flowers and is called the "cancer bush," following claims since 1985 of its ability to counteract cancers internally [40, 41]. However, this claim lacks scientific evidence and requires further investigation. The consumption of SF is predominantly enjoyed as a water infusion and a general tonic where the leaves are used; however, different cultural practices utilise most plant materials above the soil [26]. General ailments such as wound healing, backaches, fever, headaches, stress, and diabetes have been treated with SF [41, 42].

Efforts have been made to understand the molecular machinery responsible for the health-promoting effects of SF. The major bioactive constituents of SF include γ-aminobutyric acid (GABA), D-pinitol, and structural analogues of L-arginine: L-canavanine, triterpenoids, flavonoids, and saponins [41]. D-pinitol possesses antidiabetic effects by mimicking insulin, thus promoting glucose uptake through GLUT4-receptors [43, 44]. Furthermore, the hepatoprotective effects of D-pinitol are evident by preventing toxicity to the liver [45]. L-canavanine may drive the anticancer effects of SF as this compound is known to inhibit cancer growth [46].

Moreover, the documented antiviral effects of L-canavanine [47] may implicate the use of SF in COVID-19-related infections. This concept has been investigated through bioinformatic molecular docking methods. It has revealed the potential of SF, which contains L-canavanine, to bind to the main protease of SARS-CoV-2 3C, thereby disrupting its enzymatic activity [48]. These novel therapeutic findings promote the spotlighted use of SF as an ethnomedicine and could amount to the synergistic effects of the plant constituents.

Another highlighted use of SF is an alternative/complementary medicine for human immunodeficiency virus (HIV)/AIDS, affecting approximately 7.7 million people globally [49]. However, there have been reports of potential herb–drug interaction with current antiretroviral regimens through cytochrome 3A4 inhibition. Thus, the use of SF should be exerted with caution [50]. The medicinal properties of SF differ qualitatively and quantitatively, with the geographical locations of the plant affecting its overall biodiversity. An investigation into various origins of SF revealed that plant material sourced locally had greater antioxidant capacities than that sourced inland [51].

Furthermore, the systemic effects of SF have been demonstrated in the brain and CNS through modulation of oxidative stress in neuronal cells and anti-inflammatory responses in microglial cells [52]. Substantial amounts of the inhibitory neurotransmitter GABA are found in SF, thereby justifying its use in anxiety and stress relief [41]. These findings emphasise the potential of SF to formulate novel therapeutics for neuroinflammatory diseases. Great importance should be placed on improved cultivation methods to prevent threats to our ethnic biodiversity.

9.3.4 Cyclopia species (honeybush)

The medicinal health benefits of honeybush (*Cyclopia intermedia*, *Cyclopia subternata*, *Cyclopia genistoides*) are attributed to its consumption as an herbal tea. Although not a proper fermentation, fermentation of the flowers, leaves, and stems promotes the oxidation of the polyphenolic compounds [53]. This healthy beverage has a pleasant honey-like taste and boasts antioxidant capacities due to its flavonoid content [54]. The adverse effects of consuming nutritional foods with high tannin content have been recognised [55]. However, honeybush has a superior low-tannin range and minimal to no caffeine compared with the *Camellia sinensis* teas, thus promoting its health benefits. Typical uses include alleviating respiratory-related inflammatory disorders, including tuberculosis [56] and pneumonia [57].

Spectrophotometric investigations into the bioactive constituents of honeybush extracts revealed various contents of flavonol, flavanones, isoflavones, flavones, methoxy analogues, and tyrosol that promote its use as a medicinal concoction [58]. The number of polyphenolics found in the extracts varies between geographical locations from which the plant is sourced. Prolonged fermentation periods (72 h) have comparably lower polyphenolic, flavonoid, and tannin concentrations, than fermentation over a 24 h period [59]. Interestingly, unfermented aqueous extracts (4%) had notably increased amounts of total polyphenols compared to fermented aqueous extracts; this promoted the antimutagenic activity of honeybush extracts *in vitro* [60]. Marnewick and colleagues (2005) further investigated the antioxidant potential of honeybush extracts *in vivo*. However, the results revealed that rooibos prevented lipid peroxidation more significantly than honeybush [61]. Isomangiferin and mangiferin, subgroups of xanthones identified in the plant material of honeybush [54], are known to have immunomodulatory and antidiabetic actions [62, 63]. Understanding the composition of an herbal infusion contributes to developing enhanced drug formulations for disease treatment and prevention.

Honeybush plays a vital role in the South African economy, with a steady rise in the amount exported in the last 20 years. Popular export destinations include the United Kingdom, Germany, and the Netherlands. Roughly 15% of the annual harvest is utilised within South Africa [64]. Although emerging markets may add value to the honeybush industry, sustainable trade plans should be implemented to ensure international demands can be met. Unlike rooibos, honeybush is mainly wild-harvested but does have a growing cultivation component. Many agricultural industries promote commercial cultivation with reliable, sustainable harvesting practices [65]. Honeybush has also received geographical indication (GI) status in the European Union [66]. The GI status refers to products containing unique characteristics related to their geographical origin.

9.3.5 Hypoxis hemerocallidea (African potato)

The African potato genus spans five South African provinces [67] and popularsed in folk medicine. For this reason, it is highly sought after and traded in large quantities. Tonics are traditionally made using the plant's corm and used to treat everyday ailments such as flu-like symptoms, TB, diabetes, HIV/AIDS, and some cancers [26, 56]. In the 1990s, researchers used methanolic extracts of African potato to treat HIV patients over prolonged periods and discovered a positive outcome on their inflammatory status [68]. This medicinal plant has various pharmacological benefits; hence, a more profound knowledge base will heighten its therapeutic value.

Of the various phytoglycosides produced by the African potato, hypoxoside is predominantly isolated [69]. When ingested orally, the ß-glucosidase enzyme is responsible for hypoxoside conversion to rooperol in the gut. Rooperol is popularly believed to contribute to the bioactive medicinal properties of the African potato [70]. *Hypoxis* is rich in phytosterols that play vital roles in preventing dietary gut-cholesterol absorption [71] and conform to prophylaxis for atherosclerosis and cardiovascular disease [72]. Additionally, *in vivo* research suggests using phytosterols against prostate, breast, and colon cancers [73]. Further analysis demonstrates that these cancers are inhibited by apoptotic machinery through the sphingomyelin and protein phosphatase A2 pathways [74]. Despite the extensive research into the primary metabolites of the African potato, limited information exists on the secondary metabolites. Recent studies indicate the presence of saponins, tannins, cardiac glycosides, and terpenoids [75]. Terpenoids and saponins have potent antioxidant and antimicrobial effects [76], with the latter having additional antimicrobial, anti-analgesic, and anticonvulsant effects [77]. The phytotherapeutic potential of *Hypoxis* is thus elevated as the multiple constituents produce complementary results.

The list of commercial products containing African potato plant extracts seem to be expanding in horticultural practices [78]. This, together with increased market value and global trade, has negative long-term impacts on plant preservation [79]. Concerns have been raised regarding the sustainability of the African potato as continued unregulated harvesting poses a threat to plant species [80]. Efforts to create ecological balance by plant corm substitution were unsuccessful as the phytochemical components differed between species [81]. However, bioactive compounds' concentration varies seasonally, which can be optimised to diminish the current tension on wild plant species [82].

9.3.6 Asphodelaceae (Aloe ferox/bitter aloe)

Aloe ferox is broadly distributed across South Africa and is recognised for its ethnomedicinal properties [83]. This extract has propagated worldwide for its enhanced supplemental value to Ayurvedic medicine. The historical wisdom and plant heritage correspond to medicinal practices in South Africa, with over 500 Aloe species concentrated in this region [84]. The ethnopharmacology of *Aloe ferox* surpasses that of *Aloe vera* with a superior medicinal and nutritional profile. Locals source the plant from their direct surroundings for medicinal use. Aloe Ferox is commonly used as a laxative to treat skin-related disorders such as psoriasis and eczema, diabetes, cardiovascular disease, and arthritis [85–87].

Over 200 phytotherapeutic bioactive constituents have been recognised in *Aloe ferox* extracts [88, 89]. However, it is crucial to note that interspecies phytochemical variation occurs between *Aloe ferox* compounds, depending on soil and climate exposure [90]. A study by Wintola et al. (2011) revealed the free radical scavenging activity in leaf extracts and promoted its use to treat diseases governed by oxidative stress [87]. Typically, antioxidant capacities are directly proportional to the phytochemical quantity. Although the phytochemicals in *Aloe ferox* are found in lower concentrations, their effects are synergistically superior. The outer plant material is aloe bitter and primarily comprises aloin, aloesin, and aloeresin A [91]. An extensive in-depth review into other phytochemicals and their health-promoting effects in *Aloe ferox* has been outlined [92].

Aloe bitters are widely exported from South Africa, with increased job creation in the rural community through commercialisation. Although this may be a route to eliminating poverty and has been considered by the local government, sustainable harvesting strategies must be employed to promote sustainable financial platforms [93]. The efficacy and superior safety profile has created an international demand and is worth millions to local harvesters. Although *Aloe ferox* remains abundantly available and there is no concern for its extinction, overharvesting will threaten future generations.

9.4 The double-edged sword

Plants and natural systems play an essential role in biodiversity. Evolution has promoted exponential increases in the number of humans, animals, and plant species that coexist and directly impact the survival of the other. Human interference with biodiversity has long been a concern for many a scientist. To prevent the negative impacts on biodiversity, we first need to understand how human activity plays a role. If there are no regulations, many medicinal plants may become extinct. Our footprint on the environment is directly proportional to the health of fragile ecosystems—significant changes affecting the ecosystem result in plant loss and alterations in their medicinal properties. These changes include the rise in population, where natural habitats are removed to accommodate more individuals and go hand in hand with destroying plant-enriched tropical forests, which adds to carbon emissions and global warming [94].

The market value and strategy of using biological resources as remedies incorporate cultural, social, and economic aspects that affect conservation, especially for the most heavily exploited species. Activities involving local ecosystems that promote economic growth impede phytotherapeutic plant accessibility and availability. Indigenous plant purchase by medical personnel involves lengthy travel. Further, storage processing to preserve potency may contribute to the high levels of harvesting wild plant species and over-exploitation. Various medicinal plant species are faced with extinction globally [95]. Thus, posing a risk to prospective local indigenous medicinal practice, given that many impoverished communities rely on natural treatments more than Western medicines.

The destruction of biodiversity contributes to the disappearance of native peoples with vast amounts of indigenous medicinal plant knowledge owing to their years of relying on the land to meet their needs. The native communities with greater insight into local environments can advise on improved control and preservation strategies collaboratively. The omission of local communities from the discussion/decision processes could compromise policies available to the public with minimal social importance and historic influence [95]. A comprehensive assembly is required for public health policies to enable the progression of sustainability, social and cultural values, feasibility, encompassing poltical, and research environments are needed.

There is a growing demand for medicinal plants owing to the upward trend in natural remedies over western practices and commercialising. It is now critical to set up and maintain a sustainable harvest limit for an effective management system of the most utilised medicinal plants based on sound policy backed by conservational science practices [96].

The discovery of plant medicine and interest in its use comes with a misconception that nature provides an endless supply for health improvement. Many developing countries recognise medicinal plants as their primary healthcare option; however, zero efforts are concerted to ensure sustainability alongside increasing demands [97]. Increased trepidation around medicinal plant genetic loss, especially for endangered plant species, is faced as few conservation efforts are implemented. We desire improved therapeutic options, thus emphasising creating 'miracle' drug treatments from natural sources. The narrative should change to focus on plants indigenous to South Africa and how to preserve them.

Joint efforts should be implemented to relieve the pressure of demands in the plant-trade industry by creating alternatives to the supply of therapeutic plant material. The grave dangers of overexploitation on plant biodiversity have a ripple effect on the economy and jobs. The treasures held within the South African plant kingdom have an estimated economic value worth millions, with exorbitant amounts of plant matter utilised annually [98]. It is up to the consumer to create awareness and realise the global impacts of our consumption, support sustainable methods, and overcome complex conservation challenges.

9.5 A greener tomorrow

Although global transformations sharpen the double-edged sword effect of human activity on plant biodiversity, we can take the necessary steps to ensure we maintain our ecosystems for years to come. Sustainable agricultural methods that protect vital resources such as air, water, and soil lead to increased profitability and productivity. Biodiversity can be defined as the sum of all living systems on Earth [99]. If humans adopt good conservation practices, it will support biodiversity conservation. Producers must formulate coherence between socio-economic and environmental systems to ensure the sustainability of plants with superior bioactive constituents. Endangered habitats can be protected by developing national parks, recycling, and initiatives to counteract greenhouse gas emissions.

The rich biodiversity of South Africa is matched by the ethnopharmacology of its people and advances in the field [100, 101]. Traditional communities are the guardians of medicinally valuable knowledge systems often passed through generations. As a signatory to the Nagoya Protocol, South Africa, requires local industries to trade in indigenous biological resources to share benefits with traditional knowledge holders fairly and equitably [12]. In a first-of-its-kind, the rooibos industry, National Khoi-San Council, and the South African San Council signed an access and benefit-sharing agreement allowing these communities to also benefit from the sale of rooibos [12]. South Africa, like many other countries, has shrinking biodiversity. Hence, forward-thinking steps must be made to integrate traditional knowledge while protecting biodiversity. The idea that biodiversity is a priceless nonrenewable, self-sustaining system must always be considered before promoting the direct economic benefits of bioprospecting. It must be highlighted that there is a need to conserve and sustainably utilise natural medicines for present and future generations [95, 100].

Overexploitation is a problem for many species of interest to ethnopharmacologists. It remains a significant threat to species survival. It may come to a point where policy-makers may treat every plant as a limited resource to conserve its natural presence in biodiversity. To combat overexploitation, it is necessary to combat the shortage of knowledge about the biological characteristics of less popular traditional South African medicinal plants; by scientifically exploring and validating anecdotal health claims of these lesser-used plants, we would remove the pressure placed on commonly used plant species. It must be acknowledged that the increased focus on underutilised plant species may not be enough to prevent over-exploitation. Through sustainable farming methods, it is possible to increase the production of popular medicinal plants to meet the needs of the health care system while still maintaining wild types to improve and protect biodiversity [102–104]. For more in-depth knowledge, the recent review by Van Wyk and Prinsloo focuses specifically on South Africa regarding medicinal plant harvesting, sustainability and cultivation, and the current relevant legislation [105].

Synthetic development of plant-derived phytochemicals, like those found in herbal therapeutics, play a role. Using herbal medicine directs the body to an internal healing process. The ability to synthetically replicate active plant compounds poses ecological awareness for the future of drug development. Primary pharmacotherapy approaches advocate the development of site-specific targeted therapies to ensure optimal pharmacodynamic and pharmacokinetic study into their beneficial effects [106]. However, the "herbal shotgun" approach is preferred as the concoction of compounds in plant extracts tends to act synergistically for a multi-targeted treatment option [107]. Multifactorial disease aetiology poses more challenges to treat and thus would benefit from the herbal shotgun method. If scientists can mimic natural plant medicines through synthetic ways that continue to maintain the synergistic effects with limited side-effects seen naturally, it would revolutionise modern medical approaches.

Using a genomic approach is an added benefit to increasing the indigenous plant knowledge base. Compiling plant databases with genome sequencing confers protection to these natural products and regulates their markets and pharmaceutical applications. There has been resurgence in natural products' genomic and proteomic applications [108, 109]. Plant DNA barcoding provides a platform for species identification [110], while single locus combined with chloroplast barcoding offers a more specific optimal method to identify different plant species locally and globally [111]. The threat to biodiversity subsequently endangers plant species and the opportunity to salvage genomic data. With the assistance of omics data, bacterial engineering methodologies allow for improved plant transcriptome sequencing and synthesis of essential phytoconstituents [112]. To generate superior germplasm omics data, good quality resources are required. To enrich and effectively maintain indigenous plant resources, genomic data provides a gateway to breeding techniques using molecular-marker-based methods [113]. This could transform medicinal plant knowledge and data systems and preserve them, leading to molecular refined elite cultivars. There is a need to overcome challenges faced with existing medicinal plant genomic data collection, including formatting inconsistencies and collating various researchers' data. Hence, the development of proper plant database systems requires further attention.

9.6 Conclusions

Over the past 25 years, more than half of all medicines brought to market were derived from or modelled after phytochemicals [114]. Our reliance on nature for health benefits is further supported by mounting scientific evidence that continues to validate once anecdotal plant-based health claims [115–117], suggesting that we are now ready to take advantage of nature's medicine cabinet. The WHO and many governments worldwide have welcomed traditional medicine innovations with valid benefits against diseases [118, 119]; this became especially necessary as part of a reliable strategy to combat the recent COVID-19 pandemic where countries struggled with socio-economic strains and low access to western health care. This has spurred a shift in policy and thinking among health care providers [120–122]. As efforts are underway to validate and improve the way traditional medicines are incorporated into health care systems, it is necessary to assess the risks of human impacts on ecosystems and biodiversity [95]. While the promise of conventional plant-based medicine is alluring, it must be done sustainably to respect and protect biodiversity. Translation of biodiversity and the associated traditional plant knowledge systems into globally used healthcare initiatives with appropriate safeguards must be an essential strategy for ensuring the sustainability of resources for future generations.

Research into phytomedicine provides essential leads for developing new therapies. Creating better drugs through synthetic phytochemical constituent production could promote improved sustainability methods. It is necessary to ensure that significant efforts are made to calibrate Phyto-preparation methods paralleled with bioavailability and pharmacokinetic studies and verify medicinal health claims through regulated double-blind, placebo-controlled clinical trials. In addition, developing genomic databases will allow scientists easy access to genetic data that facilitates medicinal plant research in South Africa and worldwide, powering a systematic renaissance in plant identification. Furthermore, improving DNA barcoding for phytotherapeutics promotes monitoring resources, species identification, and quality control. These methods will facilitate and rationalise regulations in plant industries where the trade of only genuine herbal materials will continue, granting consumers safety for herbal consumption. A contemporary outlook on bioactive utilisation could include the development of recombinant proteins from plants to diminish exploitation. The financial sector of plant commercialisation should not impair the core of the subsistence and sustainability of medicinal plants.

The wealth of South African plants is accompanied by unique blends of phytochemicals that can be further promoted through pragmatic approaches. Despite a general attitude that the fate of humans is separate from that of the natural world, we are deeply dependent on the biosphere that we are intimately part of. Conservation should not be the exclusive field of ecologists and policymakers but rather a collective responsibility. Perhaps when considering all the aspects and role players in this environment, other South African and African medicinal herbals could take their cue from rooibos, as it has made great strides towards being successful in protecting the traditional knowledge product. Physicians and other health care professionals have a critical role in advocating for the sustainable use of medicinal plants and their efficacy in treatment to ensure that the extraordinary biodiversity of our planet is available for the benefit of future generations.

References

1. Zhang Q, Yang H, An J, Zhang R, Chen B, Hao D-J. Therapeutic effects of traditional Chinese medicine on spinal cord injury: a promising supplementary treatment in future. Evid base Compl Alternative Med 2016;2016:8958721.
2. Rashid S, Majeed LR, Nisar B, Nisar H, Bhat AA, Ganai BA. Chapter 1 - phytomedicines: diversity, extraction, and conservation strategies. In: Bhat RA, Hakeem KR, Dervash MA, editors. Phytomedicine. Cambridge, Massachusetts, USA: Academic Press; 2021:1–33 pp.
3. Falzon CC, Balabanova A. Phytotherapy: an introduction to herbal medicine. Prim Care 2017;44: 217–27.
4. Liu J, Feng W, Peng C. A song of ice and fire: cold and hot properties of traditional Chinese medicines. Front Pharmacol 2021;11. https://doi.org/10.3389/fphar.2020.598744.
5. Ekor M. The growing use of herbal medicines: issues relating to adverse reactions and challenges in monitoring safety. Front Pharmacol 2014;4. https://doi.org/10.3389/fphar.2013.00177.
6. Cahlíková L, Šafratová M, Hošťálková A, Jakub C, Hulcová D, Breiterová K, et al. Pharmacognosy and its role in the system of profile disciplines in pharmacy. Nat Prod Commun 2020;15: 1934578X20945450.
7. Jones WP, Chin YW, Kinghorn AD. The role of pharmacognosy in modern medicine and pharmacy. Curr Drug Targets 2006;7:247–64.
8. Wagner H. Natural products chemistry and phytomedicine in the 21st century: new developments and challenges. Pure Appl Chem 2005;77:1–6.

9. Schaffer SD, Yoon S-J, Curry K. Herbal supplements for health promotion and disease prevention. Nurse Pract Am J Prim Health Care 2016;41. https://doi.org/10.1097/01.NPR.0000482381.59982.41.
10. Mukherjee PK, Venkatesh P, Ponnusankar S. Ethnopharmacology and integrative medicine - let the history tell the future. J Ayurveda Integr Med 2010;1:100–9.
11. Permin H, Norn S, Kruse E, Kruse PR. On the history of Cinchona bark in the treatment of Malaria. Dan Medicinhist Arbog 2016;44:9–30.
12. Waters R. Rooibos continues to enjoy gi protection in uk post brexit. In: South African Rooibos Council (SARC); 2020. Available from: https://sarooibos.co.za/rooibos-continues-to-enjoy-gi-protection-in-uk-post-brexit/[Accessed 31 May 2022].
13. Gruenwald J. Novel botanical ingredients for beverages. Clin Dermatol 2009;27:210–6.
14. Joubert E, de Beer D. Rooibos (Aspalathus linearis) beyond the farm gate: from herbal tea to potential phytopharmaceutical. South African J Botany 2011;77:869–86.
15. Joubert E, Gelderblom W, Louw A, de Beer D. South African herbal teas: aspalathus linearis, Cyclopia spp. and Athrixia phylicoides—a review. J Ethnopharmacol 2008;119:376–412.
16. Van Wyk B-E, Gericke N. People's plants: A guide to useful plants of Southern Africa. Pretoria, South Africa: Briza publications; 2000.
17. Marnewick JL, Rautenbach F, Venter I, Neethling H, Blackhurst D, Wolmarans P, et al. Effects of rooibos (Aspalathus linearis) on oxidative stress and biochemical parameters in adults at risk for cardiovascular disease. J Ethnopharmacol 2011;133:46–52.
18. Koeppen B, Roux D. Aspalathin: a novel C-glycosylflavonoid from aspalathus linearis. Tetrahedron Lett 1965;6:3497–503.
19. Hong I-S, Lee H-Y, Kim H-P. Anti-oxidative effects of Rooibos tea (Aspalathus linearis) on immobilization-induced oxidative stress in rat brain. PLoS One 2014;9:e87061–e.
20. Snijman PW, Swanevelder S, Joubert E, Green IR, Gelderblom WC. The antimutagenic activity of the major flavonoids of rooibos (Aspalathus linearis): some dose–response effects on mutagen activation–flavonoid interactions. Mut Res Genet Toxicol Environ Mutagen 2007;631:111–23.
21. Ajuwon OR, Marnewick JL, Davids LM. Rooibos (Aspalathus linearis) and its major flavonoids—potential against oxidative stress-induced conditions. In: Basic principles and clinical significance of oxidative stress. London, UK: IntechOpen; 2015:171 p.
22. Kunishiro K, Tai A, Yamamoto I. Effects of rooibos tea extract on antigen-specific antibody production and cytokine generation in vitro and in vivo. Biosci Biotechnol Biochem 2001;65: 2137–45.
23. Schepers S. Anti-microbial activity of rooibos tea (Aspalathus linearis) on food spoilage organisms and potenial pathogens: Stellenbosch. Western Cape, South Africa: Stellenbosch University; 2001.
24. Simpson MJ, Hjelmqvist D, López-Alarcón C, Karamehmedovic N, Minehan T, Yepremyan A, et al. Anti-peroxyl radical quality and antibacterial properties of rooibos infusions and their pure glycosylated polyphenolic constituents. Molecules 2013;18:11264–80.
25. Sheik Abdul N, Marnewick JL. Rooibos, a supportive role to play during the COVID-19 pandemic? J Funct Foods 2021;86:104684.
26. Van Wyk B-E, Bv O, Gericke N. Medicinal plants of South Africa. Briza; 1997.
27. Stander EA, Williams W, Mgwatyu Y, Heusden P, Rautenbach F, Marnewick J, et al. Transcriptomics of the rooibos (Aspalathus linearis) species complex. BioTech 2020;9:19.
28. Małyjurek Z, Zawisza B, de Beer D, Joubert E, Walczak B. Authentication of honeybush and rooibos herbal teas based on their elemental composition. Food Control 2021;123:107757.
29. Smith CA. Common names of South African plants. South Africa: Government Printer, Pretoria; 1966.
30. Huisamen B, Bouic PJ, Pheiffer C, van Vuuren M. Medicinal effects of Agathosma (Buchu) Extracts. AOSIS; 2019.

31. Simpson D. Buchu—South Africa's amazing herbal remedy. Scot Med J 1998;43:189–91.
32. Watt JM, Breyer-Brandwijk MG. The medicinal and poisonous plants of southern and eastern africa being an account of their medicinal and other uses, chemical composition, pharmacological effects and toxicology in man and animal. the medicinal and poisonous plants of southern and eastern africa being an account of their medicinal and other uses, chemical composition, pharmacological effects and toxicology in man and animal. London, UK: E. & S. Livingstone; 1962.
33. Fluck A, Mitchell W, Perry H. Composition of buchu leaf oil. J Sci Food Agric 1961;12:290–2.
34. Posthumus MA, van Beek TA, Collins NF, Graven EH. Chemical composition of the essential oils of Agathosma betulina, A. crenulata and an A. betulina x crenulata hybrid (Buchu). J Essent Oil Res 1996;8:223–8.
35. Collins NF, Graven EH, van Beek TA, Lelyveld GP. Chemotaxonomy of commercial buchu species (Agathosma betulina and A. crenulata). J Essent Oil Res 1996;8:229–35.
36. Viljoen AM, Moolia A, Van Vuuren SF, van Zyl R, Başer K, Demirci B, et al. The biological activity and essential oil composition of 17 Agathosma (Rutaceae) species. J Essent Oil Res 2006;18:2–16.
37. Moolla A, Viljoen AM. 'Buchu'–agathosma betulina and agathosma crenulata (Rutaceae): a review. J Ethnopharmacol 2008;119:413–9.
38. Viljoen A, Chen W, Mulaudzi N, Kamatou G, Sandasi M. Phytochemical profiling of commercially important South African plants. Cambridge, Massachusetts, USA: Academic Press; 2021.
39. Williams S, Kepe T. Discordant harvest: debating the Harvesting and commercialization of Wild <span class="genus-species">Buchu</span> (<span class= "genus-species">Agathosma betulina</span>) in Elandskloof, South Africa. Mt Res Dev 2008;28:58–64.
40. Drewes SE. Natural products research in South Africa: 1890-2010. South African Journal of Science 2012;108:1–8.
41. Van Wyk B, Albrecht C. A review of the taxonomy, ethnobotany, chemistry and pharmacology of Sutherlandia frutescens (Fabaceae). J Ethnopharmacol 2008;119:620–9.
42. Drewes S, Horn M, Khan F. The chemistry and pharmacology of medicinal plants. African Sun Media, Stellenbosch: Commercialising medicinal plants A Southern African guide Sun Press; 2006:89–95 pp.
43. Bates SH, Jones RB, Bailey CJ. Insulin-like effect of pinitol. Br J Pharmacol 2000;130:1944–8.
44. Dang NT, Mukai R, Yoshida K-I, Ashida H. D-pinitol and myo-inositol stimulate translocation of glucose transporter 4 in skeletal muscle of C57BL/6 mice. Biosci Biotechnol Biochem 2010;74: 1062–7.
45. Zhou Y, Park C-M, Cho C-W, Song Y-S. Protective effect of pinitol against D-galactosamine-induced hepatotoxicity in rats fed on a high-fat diet. Biosci Biotechnol Biochem 2008;72:1657–66.
46. Swaffar DS, Ang CY, Desai PB, Rosenthal GA. Inhibition of the growth of human pancreatic cancer cells by the arginine antimetabolite L-canavanine. Cancer Res 1994;54:6045–8.
47. Green MH. Method of treating viral infections with amino acid analogs. Google Patents; 1992.
48. Dwarka D, Agoni C, Mellem JJ, Soliman ME, Baijnath H. Identification of potential SARS-CoV-2 inhibitors from South African medicinal plant extracts using molecular modelling approaches. South Afr J Bot 2020;133:273–84.
49. Mahy M, Marsh K, Sabin K, Wanyeki I, Daher J, Ghys PD. HIV estimates through 2018: data for decision-making. AIDS 2019;33:S203.
50. Fasinu PS, Gutmann H, Schiller H, James A-D, Bouic PJ, Rosenkranz B. The potential of Sutherlandia frutescens for herb-drug interaction. Drug Metabol Dispos 2013;41:488–97.
51. Zonyane S, Fawole OA, La Grange C, Stander MA, Opara UL, Makunga NP. The implication of chemotypic variation on the anti-oxidant and anti-cancer activities of Sutherlandia frutescens (L.) R. Br.(Fabaceae) from different geographic locations. Antioxidants 2020;9:152.

52. Jiang J, Chuang DY, Zong Y, Patel J, Brownstein K, Lei W, et al. Sutherlandia frutescens ethanol extracts inhibit oxidative stress and inflammatory responses in neurons and microglial cells. PLoS One 2014;9:e89748.
53. McKay DL, Blumberg JB. A review of the bioactivity of South African herbal teas: rooibos (Aspalathus linearis) and honeybush (Cyclopia intermedia). Phytother Res 2007;21:1–16.
54. Ferreira D, Kamara BI, Brandt EV, Joubert E. Phenolic compounds from Cyclopia intermedia (honeybush tea). 1. J Agric Food Chem 1998;46:3406–10.
55. Chung K-T, Wei C-I, Johnson MG. Are tannins a double-edged sword in biology and health? Trends Food Sci Technol 1998;9:168–75.
56. Drewes S, Elliot E, Khan F, Dhlamini J, Gcumisa M. Hypoxis hemerocallidea—not merely a cure for benign prostate hyperplasia. J Ethnopharmacol 2008;119:593–8.
57. Cock IE, Van Vuuren SF. The traditional use of southern African medicinal plants for the treatment of bacterial respiratory diseases: a review of the ethnobotany and scientific evaluations. J Ethnopharmacol 2020;263:113204.
58. Kamara BI, Brandt EV, Ferreira D, Joubert E. Polyphenols from honeybush tea (cyclopia intermedia). J Agric Food Chem 2003;51:3874–9.
59. Du Toit J, Joubert E, Britz T. Honeybush tea–a rediscovered indigenous South African herbal tea. J Sustain Agric 1998;12:67–84.
60. Marnewick JL, Gelderblom WC, Joubert E. An investigation on the antimutagenic properties of South African herbal teas. Mut Res Genet Toxicol Environ Mutagen 2000;471:157–66.
61. Marnewick J, Joubert E, Joseph S, Swanevelder S, Swart P, Gelderblom W. Inhibition of tumour promotion in mouse skin by extracts of rooibos (Aspalathus linearis) and honeybush (Cyclopia intermedia), unique South African herbal teas. Cancer Lett 2005;224:193–202.
62. Garcıa D, Delgado R, Ubeira F, Leiro J. Modulation of rat macrophage function by the Mangifera indica L. extracts Vimang and mangiferin. Int Immunopharm 2002;2:797–806.
63. Muruganandan S, Gupta S, Kataria M, Lal J, Gupta P. Mangiferin protects the streptozotocin-induced oxidative damage to cardiac and renal tissues in rats. Toxicology 2002;176:165–73.
64. Mcgregor G. An overview of the honeybush industry, Cape Town: Department of Environmental Affairs and Development Planning; 2017;2:3 pp.
65. Hobson S, Joubert M. Eastern Cape honeybush tea project: Industry overview, assessment and proposed interventions. Western Cape, South Africa: Unpublished report to the Coega development Corporation; 2011.
66. Biénabe E, Marie-Vivien D. Institutionalizing geographical indications in southern countries: lessons learned from Basmati and Rooibos. World Dev 2017;98:58–67.
67. Singh Y. Hypoxis (Hypoxidaceae) in southern Africa: taxonomic notes. South African J Botany 2007;73:360–5.
68. Albrecht C. Hypoxoside: a putative, non-toxic prodrug for the possible treatment of certain malignancies, HIV-infection and inflammatory conditions. Harare, Zimbabwe: University of Zimbabwe (UZ) Publications; 1996.
69. Boukes GJ, Daniels BB, Albrecht CF, van de Venter M. Cell survival or apoptosis: rooperol's role as anticancer agent. Oncol Res 2009;18:365–76.
70. Mills E, Cooper C, Seely D, Kanfer I. African herbal medicines in the treatment of HIV: Hypoxis and Sutherlandia. An overview of evidence and pharmacology. Nutr J 2005;4:1–6.
71. Bouic PJ. The role of phytosterols and phytosterolins in immune modulation: a review of the past 10 years. Curr Opin Clin Nutr Metab Care 2001;4:471–5.
72. Ling W, Jones P. Dietary phytosterols: a review of metabolism, benefits and side effects. Life Sci 1995;57:195–206.
73. Albrecht C, Kruger P, Smit B, Freestone M, Gouws L, Miller R, et al. The pharmacokinetic behaviour of hypoxoside taken orally by patients with lung cancer in a phase I trial. South African Med J 1995; 85.

74. Awad AB, Gan Y, Fink CS. Effect of β-sitosterol, a plant sterol, on growth, protein phosphatase 2A, and phospholipase D in LNCaP cells. Nutr Cancer 2000;36:74–8.
75. Zimudzi C. African potato (Hypoxis Spp): diversity and comparison of the phytochemical profiles and cytotoxicity evaluation of four Zimbabwean species. India: Creative Pharma Assent (CPA); 2014.
76. Sermakkani M, Thangapandian V. Phytochemical screening for active compounds in pedalium murex L. Recent Res Sci Technol 2010;2.
77. Ali N, Shah SWA, Shah I, Ahmed G, Ghias M, Khan I. Cytotoxic and anthelmintic potential of crude saponins isolated from Achillea Wilhelmsii C. Koch and Teucrium Stocksianum boiss. BMC Compl Alternative Med 2011;11:1–7.
78. Wadley L, Backwell L, d'Errico F, Sievers C. Cooked starchy rhizomes in Africa 170 thousand years ago. Science 2020;367:87–91.
79. Mulholland DA, Drewes SE. Global phytochemistry: indigenous medicinal chemistry on track in southern Africa. Phytochemistry 2004;7:769–82.
80. Moyo M, Amoo SO, Aremu AO, Gruz J, Subrtová M, Doležal K, et al. Plant regeneration and biochemical accumulation of hydroxybenzoic and hydroxycinnamic acid derivatives in Hypoxis hemerocallidea organ and callus cultures. Plant Sci 2014;227:157–64.
81. Katerere D, Eloff J. Anti-bacterial and anti-oxidant activity of Hypoxis hemerocallidea (Hypoxidaceae): can leaves be substituted for corms as a conservation strategy? South African J Botany 2008;74:613–6.
82. Du Toit ES, Von Maltzahn I, Soundy P. Chemical analysis of cultivated Hypoxis hemerocallidea using thin layer chromatography. Hortscience 2005;40:1119B.
83. Bhaludra CSS, Yadla H, Cyprian FS, Bethapudi RR, Basha S, Anupalli RR. Genetic diversity analysis in the genus Aloe vera (L.) using RAPD and ISSR markers. Pakistan: Asian Network for Scientific Information; 2014.
84. McGough HN, Groves M, Mustard M, Brodie C. CITES and plants: a user's guide: Board of Trustees. Surrey, UK: Royal Botanic Gardens; 2004.
85. Loots DT, van der Westhuizen FH, Botes L. Aloe ferox leaf gel phytochemical content, antioxidant capacity, and possible health benefits. J Agric Food Chem 2007;55:6891–6.
86. Loots DT, Pieters M, Islam MS, Botes L. Antidiabetic effects of Aloe ferox and Aloe greatheadii var. davyana leaf gel extracts in a low-dose streptozotocin diabetes rat model. South African Journal of Science 2011;107:1–6.
87. Wintola OA, Afolayan AJ. Phytochemical constituents and antioxidant activities of the whole leaf extract of Aloe ferox Mill. Phcog Mag 2011;7:325.
88. Kambizi L, Sultana N, Afolayan A. Bioactive compounds isolated from Aloe ferox: a plant traditionally used for the treatment of sexually transmitted infections in the eastern cape, South Africa. Pharmaceut Biol 2005;42:636–9.
89. Mabusela WT, Stephen AM, Botha MC. Carbohydrate polymers from Aloe ferox leaves. Phytochemistry 1990;29:3555–8.
90. Radha MH, Laxmipriya NP. Evaluation of biological properties and clinical effectiveness of Aloe vera: a systematic review. J Tradit complement Med 2014;5:21–6.
91. Kokwaro JO. Medicinal plants of east Africa. Nairobi, Kenya: University of Nairobi press; 2009.
92. Chen W, Van Wyk B-E, Vermaak I, Viljoen AM. Cape aloes—a review of the phytochemistry, pharmacology and commercialisation of Aloe ferox. Phytochem Lett 2012;5:1–12.
93. Melin A. A bitter pill to swallow: a case study of the trade and harvest of Aloe ferox in the Eastern Cape, South Africa. Silwood Park: Department of Life SciencesImperial College London; 2009.
94. Tilman D, Lehman C. Human-caused environmental change: impacts on plant diversity and evolution. Proc Natl Acad Sci USA 2001;98:5433–40.

95. Alves R, Rosa IM. Biodiversity, traditional medicine and public health: where do they meet? J Ethnobiol Ethnomed 2007;3:1–9.
96. Schippmann U, Leaman DJ, Cunningham A. Impact of cultivation and gathering of medicinal plants on biodiversity: global trends and issues. Biodiversity and the ecosystem approach in agriculture, forestry and fisheries. New York: FAO Document Repository of United Nations; 2002.
97. Street R, Prinsloo G. Commercially important medicinal plants of South Africa: a review. J Chem 2013;2013. https://doi.org/10.1155/2013/205048.
98. Wiersum KF, Dold AP, Husselman M, Cocks M. Cultivation of medicinal plants as a tool for biodiversity conservation and poverty alleviation in the Amatola region, South Africa. Frontis 2006:43–57.
99. Wilson EO. Biodiversity. Washington, DC: The National Academies Press; 1988.
100. Mulholland DA. The future of ethnopharmacology: a southern African perspective. J Ethnopharmacol 2005;100:124–6.
101. Light M, Sparg S, Stafford G, Van Staden J. Riding the wave: South Africa's contribution to ethnopharmacological research over the last 25 years. J Ethnopharmacol 2005;100:127–30.
102. Garibaldi LA, Gemmill-Herren B, D'Annolfo R, Graeub BE, Cunningham SA, Breeze TD. Farming approaches for greater biodiversity, livelihoods, and food security. Trends Ecol Evol 2017;32: 68–80.
103. Hird V. Farming systems and techniques that promote biodiversity. Biodiversity 2017;18:71–4.
104. Heywood VH. Ethnopharmacology, food production, nutrition and biodiversity conservation: towards a sustainable future for indigenous peoples. J Ethnopharmacol 2011;137:1–15.
105. van Wyk AS, Prinsloo G. Medicinal plant harvesting, sustainability and cultivation in South Africa. Biol Conserv 2018;227:335–42.
106. Carmona F, Pereira AMS. Herbal medicines: old and new concepts, truths and misunderstandings. Revista Brasileira de Farmacognosia 2013;23:379–85.
107. Williamson EM. Synergy and other interactions in phytomedicines. Phytomedicine 2001;8:401–9.
108. Harvey AL, Edrada-Ebel R, Quinn RJ. The re-emergence of natural products for drug discovery in the genomics era. Nat Rev Drug Discov 2015;14:111–29.
109. Sheridan C. Recasting natural product research: can the commercial sector capitalize on the merger of high-throughput technology and natural products? Cormac Sheridan investigates. Nat Biotechnol 2012;30:385–8.
110. Hebert PD, Cywinska A, Ball SL, DeWaard JR. Biological identifications through DNA barcodes. Proc R Soc London, B 2003;270:313–21.
111. Li X, Yang Y, Henry RJ, Rossetto M, Wang Y, Chen S. Plant DNA barcoding: from gene to genome. Biol Rev 2015;90:157–66.
112. Wang X, Zhang J, He S, Gao Y, Ma X, Gao Y, et al. HMOD: an omics database for herbal medicine plants. Mol Plant 2018;11:757–9.
113. Bevan MW, Uauy C, Wulff BB, Zhou J, Krasileva K, Clark MD. Genomic innovation for crop improvement. Nature 2017;543:346–54.
114. Herndon CN, Butler RA. Significance of biodiversity to health. Biotropica 2010;42:558–60.
115. Ovadje P, Roma A, Steckle M, Nicoletti L, Arnason JT, Pandey S. Advances in the research and development of natural health products as main stream cancer therapeutics, 2015;2015. https://doi.org/10.1155/2015/751348.Evid base Compl Alternative Med.
116. Sudhakar K, Mishra V, Hemani V, Verma A, Jain A, Jain S, et al. Reverse pharmacology of phytoconstituents of food and plant in the management of diabetes: current status and perspectives. Trends Food Sci Technol 2021;110:594–610.
117. Reid K, Maes J, Maes A, Tuyiringire N, Pan C, Komba E. Evaluation of the mutagenic and antimutagenic effects of South African plants. J Ethnopharmacol 2006;106:44–50.
118. Benzi G, Ceci A. Herbal medicines in European regulation. Pharmacol Res 1997;35:355–62.

119. Burton A, Smith M, Falkenberg T. Building WHO's global strategy for traditional medicine. European Journal of Integrative Medicine 2015;7:13–5.
120. Xiong Y, Gao M, van Duijn B, Choi H, van Horssen F, Wang M. International policies and challenges on the legalization of traditional medicine/herbal medicines in the fight against COVID-19. Pharmacol Res 2021;166:105472.
121. Ang L, Lee HW, Kim A, Lee JA, Zhang J, Lee MS. Herbal medicine for treatment of children diagnosed with COVID-19: a review of guidelines. Compl Ther Clin Pract 2020;39:101174.
122. Paudyal V, Sun S, Hussain R, Abutaleb MH, Hedima EW. Complementary and alternative medicines use in COVID-19: a global perspective on practice, policy and research. Res Soc Adm Pharm 2022;18:2524–8.

Füreya Elif Öztürkkan* and Hacali Necefoğlu

10 Evaluation of the crystal structures of metal(II) 2-fluorobenzoate complexes

Abstract: When the structures of the complexes of the same primary ligand with different metals and different co-ligands are evaluated in detail, it is thought that it will be easier to reach the target molecules by determining the factors affecting the structure. The study of the structures of metal(II) 2-fluorobenzoates with various N-donor ligands is aimed to contribute to future research. In this study, the crystal structures of transition metal(II) 2-fluorobenzoates with N-donor ligand complexes such as 2,2′-bipyridine, 1,10′-phenanthroline, nicotinamide, isonicotinamide, 4-pyridylmethanol, and 2-aminopyridine were investigated. According to the literature review, it was determined that Cu(II), Zn(II), Cd(II), Pb(II), Co(II), and Ni(II) 2-fluorobenzoates with N-donor ligand exhibit monomeric, dimeric, ionic, and polymeric structures. Although the fluorobenzoate anion has the most electronegative element (fluorine) at the 2-position in the benzene ring, only a few of its complexes have O–H···F, N–H···F, and C–H···F hydrogen bonds. A Pb(II) 2-fluorobenzoate complex without an N- donor ligand is the only complex in which the fluorobenzoate anion coordinates to the metal via the fluorine atom. The structures of the 19 complexes studied included 11 monomers, 5 dimers, 2 ionic, and 1 coordination polymer. In addition, 8 of these complexes are copper(II), 5 of them are Pb(II), 2 of them are Cd(II) and Zn(II), and one of them is Co(II) and Ni(II) metal centered. Many transition metal(II) 2-fluorobenzoate complexes can still be contributed to the literature, according to the findings of this study. Furthermore, because metal carboxylate complexes have various physical and biological potential applications today, their importance in materials science should be emphasized.

Keywords: 2-florobenzoic acid; carboxylate; crystal structure; N-donor ligand; transition metal complexes.

Abbreviations

FBA	2-fluorobenzoic acid
FBz	2-fluorobenzoate

***Corresponding author: Füreya Elif Öztürkkan,** Department of Chemical Engineering, Kafkas University, Kars, 36100, Turkey, E-mail: fozturkkan36@gmail.com. http://orcid.org/0000-0001-6376-4161

Hacali Necefoğlu, Department of Chemistry, Kafkas University, Kars, 36100, Turkey; and International Scientific Research Centre, Baku State University, Baku, 1148, Azerbaijan,
E-mail: alinecef@hotmail.com. http://orcid.org/0000-0003-2901-3748

As per De Gruyter's policy this article has previously been published in the journal Physical Sciences Reviews. Please cite as: F. E. Öztürkkan and H. Necefoğlu "Evaluation of the crystal structures of metal(II) 2-fluorobenzoate complexes" *Physical Sciences Reviews* [Online] 2022. DOI: 10.1515/psr-2022-0152 | https://doi.org/10.1515/9783110913361-010

NA	nicotinamide
INA	isonicotinamide
2,2′-BPy	2,2′-bipyridine
PHEN	1,10′-phenanthroline
4-PM	4-pyridilmethanol
2-APy	2-aminopyridine

10.1 Introduction

Metal-organic coordination compounds have an important place in crystal engineering due to their structural diversity. The contributions of compounds to the science of materials, which offer new contributions to crystal engineering with their unique structures, continue to be investigated [1, 2]. Many factors, including the transition metal's oxidation state, N-, O-, and S-donor ligands, ring-bound functional groups of arylcarboxylates, solvent environment, pH, pressure, temperature, and synthesis process, enrich the structural architecture of these complexes [3]. In addition, noncovalent interactions such as C–H···O, O–H···O, and N–H···O hydrogen bonds and C–O···π, C–H···π, and π···π stacking are also effective in the formation of supramolecular networks in structures [4–6]. Transition metal complexes of benzene carboxylic acids play an essential role in coordination chemistry literature. With their magnetic, catalytic, optical, and pharmacological potentials, as well as their crystal shapes, these complexes are attractive materials. These complexes have been the subject of numerous investigations in computational chemistry in recent years, both in terms of spectroscopic data comparison and structure-activity studies to evaluate pharmacological characteristics [7–9]. In this context, we studied the structures of complexes of Cu(II), Zn(II), Cd(II), Pb(II), Co(II), and Ni(II) 2-fluorobenzoates with N- donor ligands such as 2,2′-bipyridine, 1,10′-phenanthroline, nicotinamide, isonicotinamide, 4-pyridylmethanol, and 2-aminopyridine in this study (Figures 10.1 and 10.2). According to the data obtained, even though there are numerous complexes with interesting and diverse structures, it is believed that there are still many materials that can be synthesized.

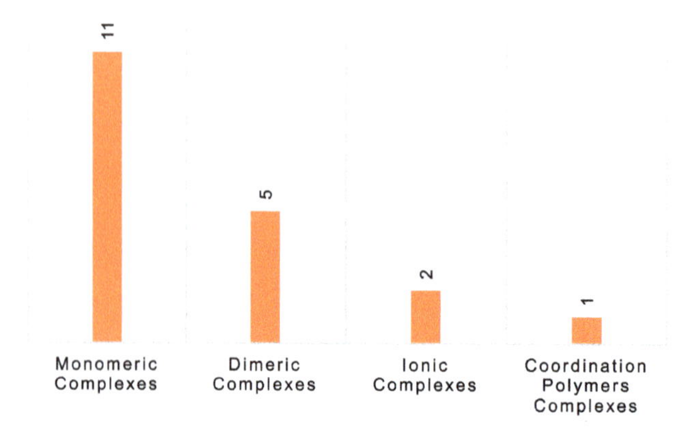

Figure 10.1: According to the structural properties of metal(II) 2-FBz with N- donor ligand complexes.

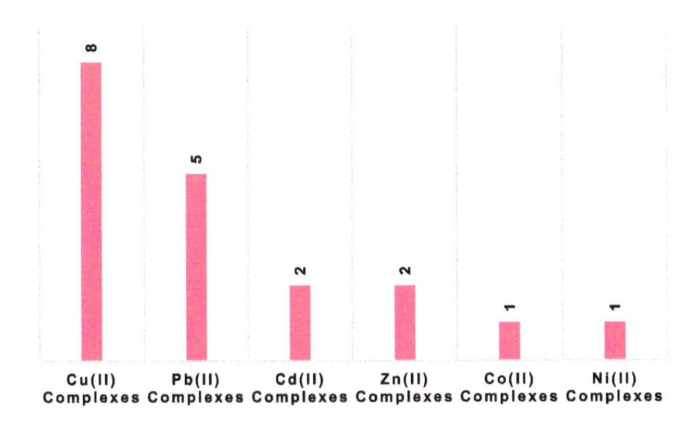

Figure 10.2: According to the central metal atom of metal(II) 2-FBz with N- donor ligand complexes.

10.2 Metal(II) 2-fluorobenzoate complexes

10.2.1 Monomeric complexes

Monomeric complexes of transition metal(II) 2-fluorobenzoates with N- donor ligands such as 2,2′-bipyridine, 1,10′-phenanthroline, 4-pyridylmethanol, nicotinamide isonicotinamide and 2-aminopyridine have been reported. Monomeric complexes outnumber those with dimeric, ionic, and polymeric structures. It was determined that there are Cd^{+2}, Cu^{+2}, Pb^{+2}, and Zn^{+2} cations at the center of the complexes (Figures 10.1 and 10.2).

When the structure of the [Cd(2-FBz)(2,2′-BPy)(H_2O)] [10] monomeric complex was examined, it was determined that cadmium had a coordination number of seven and that the 2-fluorobenzoate anions and 2,2′-bipyridine molecules were bidentate chelate coordinated. In the structure with pentagonal bipyramid geometry, the lengths of the Cd–O and Cd–N bonds were reported to be in the 2.31–2.44 and 2.35–2.36 ranges, respectively. It has also been stated that hydrogen bonds (O–H···O bonds) in the structure contribute significantly to the stability of the structure (Figure 10.3).

When the complexes of metal(II) 2-fluorobenzoates with N-donor ligands were investigated, copper complexes were found to be the most numerous monomeric complexes. The coordination number of copper(II) in these complexes is 4, 5, or 6. In complexes containing monomeric *trans* structured nicotinamide [11] and isonicotinamide [12] ligands (Figure 10.4a and b), the Cu–O bond length and Cu–N bond length were determined as 1.98 Å, 1.93 Å, and 1.99 Å, 2.04 Å, respectively, and it is seen that the bond lengths are close to each other. In the [Cu(2-FBz)(NA)(H_2O)] [11] complex (Figure 10.4c), the coordination number of copper is 5. In this complex, the Cu–O and Cu–N bond lengths were determined as 1.96 Å and 2.01 Å, respectively. In another complex where copper has a coordination number of 5, it is seen that both the 2-FBz anion and the 2-FBA molecule are coordinated (Figure 10.4d) [12]. In this structure with a square pyramidal coordination

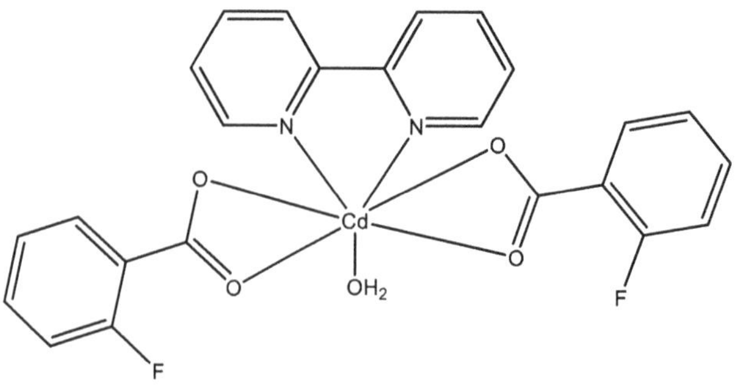

Figure 10.3: The molecular structure of [Cd(2-FBA)(2,2′-BPy)(H_2O)] complex [10].

geometry, the average bond length between the oxygen atoms of the 2-FBz anion and the copper atoms was 1.96 Å, while the length of the bond between the oxygen atom of 2-fluorobenzoic acid and copper was recorded as 2.23 Å. d(Cu–N) = 2.00 Å. All ligands in the aforementioned compounds were monodentate coordinated. It has been found that there are hydrogen bonds such as N–H···O and O–H···O.

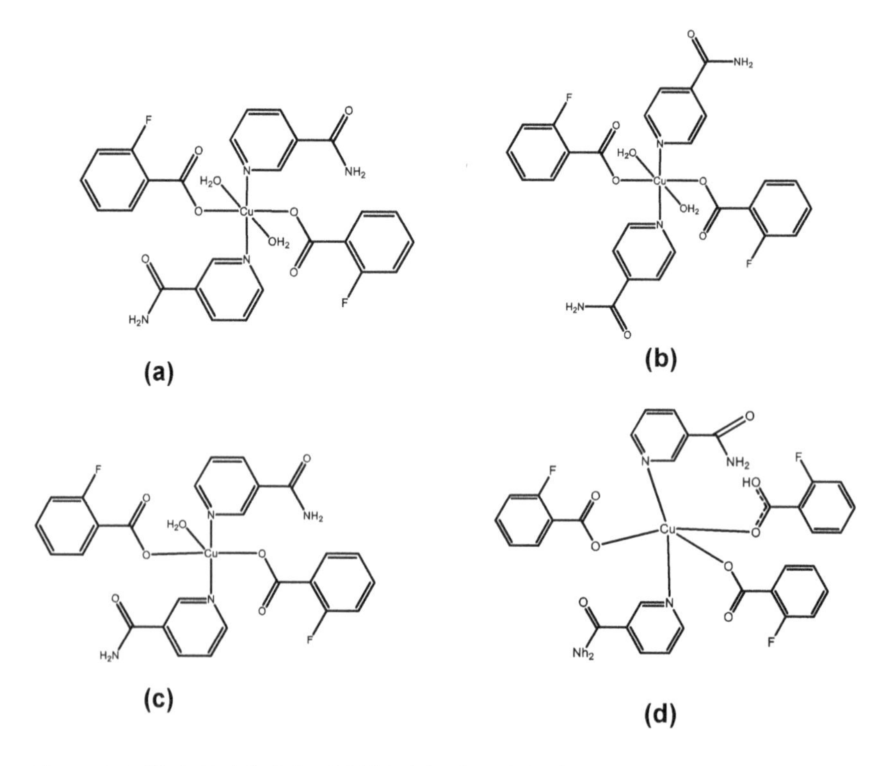

Figure 10.4: [Cu (2-FBz)$_2$(NA)$_2$(H_2O)$_2$] (a) [11], [Cu (2-FBz)$_2$(INA)$_2$(H_2O)$_2$] (b) [12], [Cu (2-FBz)$_2$(NA)$_2$(H_2O)] (c) [11] and [Cu(2-FBz)$_2$(2-FBA)(NA)$_2$] (d) [12] complexes' molecular structures.

Zheng et al. reported the crystal structures of [Cu(2-FBz)$_2$(2,2′-BPy)(H_2O)] and [Cu(2-FBz)$_2$(Phen)] complexes [13]. In both structures, 2-Fbz anions are monodentate coordinated. It has been determined that 2,2′-BPy and Phen ligands are bidentate chelate coordinated. The coordination number of copper in [Cu(2-FBz)$_2$(2,2′-BPy)(H_2O)] and [Cu(2-FBz)$_2$(Phen)] complexes is 5 and 4, respectively (Figure 10.5). The lengths of the Cu–O and Cu–N bonds in both complexes are 1.95 Å and 2.01 Å, respectively. The structures contain intermolecular O–H···O and C–H···O hydrogen bonds.

When the structures of monomeric Pb^{II} complexes were investigated, it was found that two different crystal structures had been reported in three independent studies (Figure 10.6). The crystal structure of the [Pb(2-FBz)$_2$(Phen)] complex was reported by Zhang et al. [14]. In this structure, 2-FBz anions and Phen molecules are coordinated in the form of bidentate chelate. The coordination number of the central metal cation is 6. Pb–N and Pb–O bond lengths were recorded in the range of 2.57–2.66 Å and 2.31–2.51 Å, respectively. In another study [15], the structure of the [Pb(2-FBz)$_2$(Phen)$_2$].2H_2O complex was reported. When the structure of the complex is examined, there are two bidentate coordinated 2-FBz anions, two bidentate chelate coordinated Phen molecules, and two uncoordinated water molecules around the eight-coordinated Pb^{II} cation (Figure 10.6b). There are O–H···O, O–H···F, and C–H···F hydrogen bonds in the structure. *d*(Pb–O) = 2.78 Å and *d*(Pb–N) = 2.63 Å. In the structure of the [Pb(2-FBz)$_2$(Phen)$_2$(H_2O)$_{0.5}$].2H_2O complex (Figure 10.6c), unlike the complex described in the previous structure, there is half occupied coordinated water molecule [16]. Bond lengths in this structure are reported to be *d*(Pb–O) = 2.68 Å, *d*(Pb–N) = 2.80 Å, and *d*(Pb–Ow) = 2.96 Å. When these three structures are evaluated, the Pb–N bond length increases, and the Pb–O bond length varies depending on whether the structure contains coordinated or uncoordinated water molecules.

The structure of the monomeric zinc 2-fluorobenzoate complex [Zn(2-FBz)$_2$(2-APy)] was reported by two different research groups in 2004 [17] and 2009 [18] (Figure 10.7). When the unit cell parameters of the complex with the same crystal structure were examined, it was determined that the β angle was reported as 93.707 (3)° in the study reported in 2004 and as 108.048 (2)° in the study reported in 2009. In both investigations,

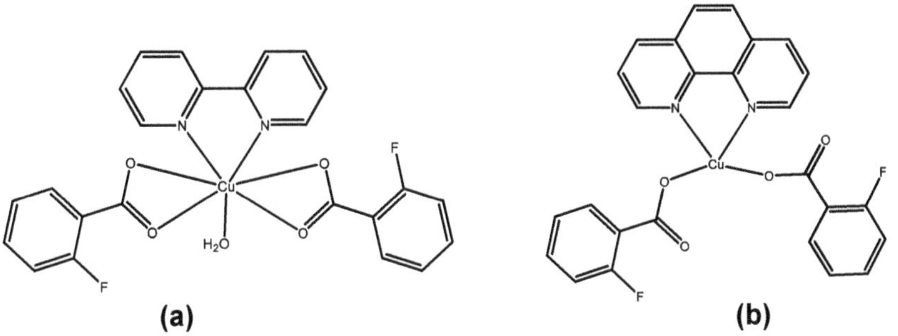

Figure 10.5: Molecular structures of [Cu(2-FBz)$_2$(2,2′-BPy)(H_2O)] (a) and [Cu(2-FBz)$_2$(Phen)] (b) [13].

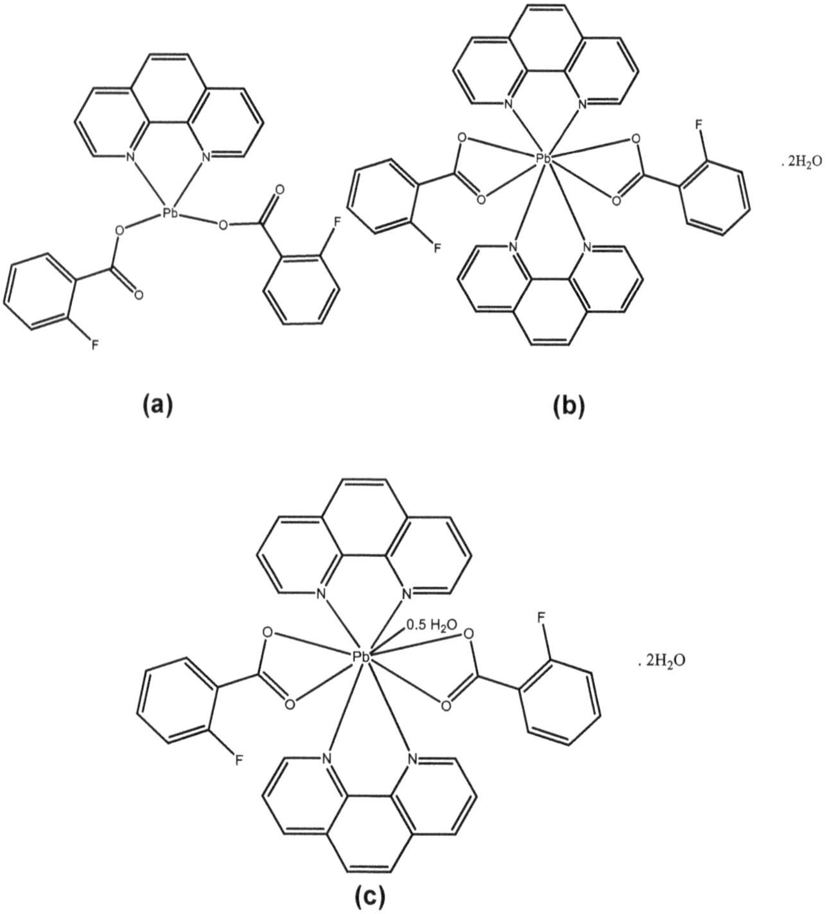

(a)

(b)

(c)

Figure 10.6: The structures of Pb(2-FBz)$_2$(Phen) (a) [14], [Pb(2-FBz)$_2$(Phen)$_2$].2H$_2$O (b) [15] and [Pb(2-FBz)$_2$(Phen)$_2$(H$_2$O)$_{0.5}$].2H$_2$O (c) [16].

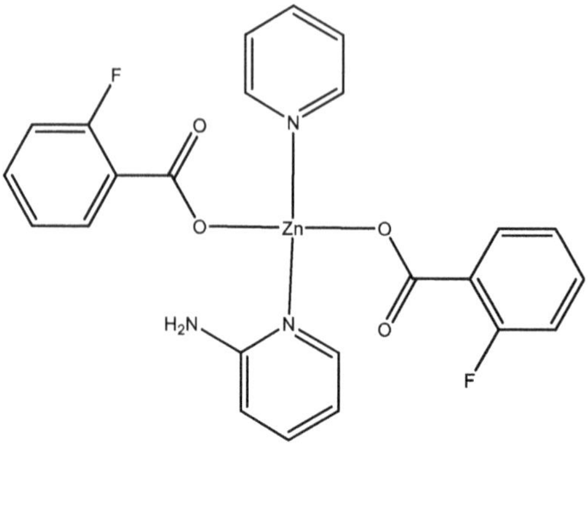

Figure 10.7: The molecular structure of [Zn(2-FBz)$_2$(2-APy)] [17, 18].

the mean bond lengths of Zn–O and Zn–N were 1.97 Å and 2.06 Å, respectively. The presence of N–H···O and N–H···F hydrogen bonds in three-dimensional networks was noted in the first study, while the presence of N–H···O, N–H···F, and C–H···F hydrogen bonds in 3D networks was mentioned in the second study.

10.2.2 Dimeric complexes

Five dimeric complexes of transition metal(II) 2-fluorobenzoate have been reported. In three of these complexes, the coordination center is cadmium, lead, and zinc, and in the other two complexes, copper. When the structure of the dimeric cadmium complex with the general formula [Cd(2-FBz)$_2$(Phen)] is examined, it is seen that the 2-FBz anions form bidentate and 1,10-phenanthroline ligands form bidentate chelate and the CdN_2O_5 polyhedron is formed (Figure 10.8). Average bond lengths are *d*(Cd–N) = 2.368 Å – 2.373 Å and *d*(Cd–O) = 2.325 Å – 2.441(2) Å. The oxygen atom of the 2-FBz anions forms a bridge between the Cd atoms and there is no data on this bond length [19].

One of the dimeric copper complexes does not have an N-donor ligand and instead coordinates as a 2-FBA N-donor ligand. *Paddle-wheel* units have been reported in [Cu(2-FBz)$_4$(2-FBA)$_2$ [20] and [Cu(2-FBz)$_4$(4-PM)$_2$] [21] complexes (Figure 10.9a and b). The Cu–O average bond length in these complexes is 1.96 Å. In the [Cu(2-FBz)$_4$(2-FBA)$_2$] complex, the bond length between the oxygen atom of the 2-FBA ligand and Cu(II) is 2.19 Å. The Cu–N average bond length in the [Cu(2-FBz)$_4$(4-PM)$_2$] complex is 2.15 Å.

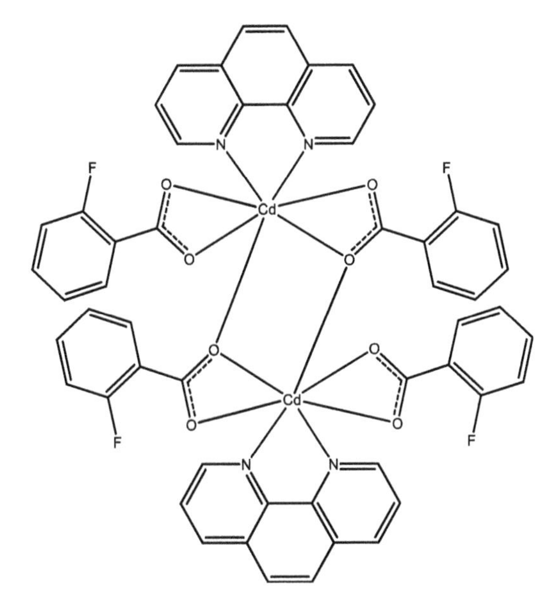

Figure 10.8: The molecular structure of Cd(2-FBz)$_2$(Phen) [19].

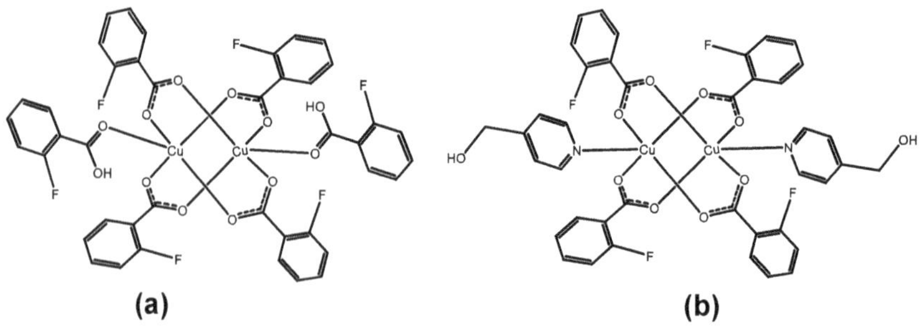

(a) (b)

Figure 10.9: The molecular structures of Cu(2-FBz)$_4$(2-FBA)$_2$ [20] and Cu(2-FBz)$_4$(4-PM)$_2$ [21].

Cu···Cu distance is ~2.65 Å in both complexes, which is consistent with the literature on *paddle-wheel* structures. The presence of O–H···O hydrogen bonds in the structure of both complexes has been reported.

The dimeric lead(II) 2-fluorobenzoate complex contains 2,2′-bipyridine ligand as co-ligand. When the structure of the [Pb(2-FBz)$_2$(BPy)] complex is examined (Figure 10.10) [22], there is a bidentate chelate coordinated 2,2′-bipyridine molecule around the Pb^{2+} cation and two oxygen atoms from two 2-fluorobenzoate anions (Figure 10.10). In the structure, the molecules form a 3D network through weak Pb–O, Pb···Pb, and hydrogen bond interactions. Average bond lengths were determined as d(Pb–N) = 2.521 Å – 2.637(6) Å and d(Pb–O) = 2.313 Å – 2.515 Å. These bond lengths are compatible with the Pb–O and Pb–N bond lengths of the monomeric anhydrous [Pb(2-FBz)$_2$(Phen)] complex.

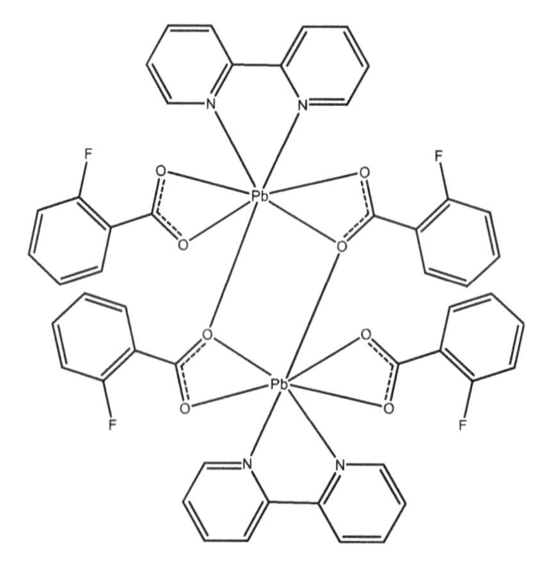

Figure 10.10: The molecular structure of Pb(2-FBz)$_2$(BPy) [22].

$[Zn_2(\mu\text{-}2\text{-}FBz)_2(2\text{-}FBz)_2(NA)_2].(2\text{-}FBA)$ complex, whose structure was determined by Hökelek et al., is also different from conventional dimeric complexes. In the structure of the complex, there are two bidentate bridge 2-FBz anions, one bidentate chelate and one monodentate 2-FBz anions, two NA molecules and a slightly distorted tetrahedral geometry around the zinc atom (Figure 10.11). There is also an uncoordinated 2-FBA molecule in the structure. The mean bond lengths of Zn–O and Zn–N are 2.02 Å and 2.03 Å, respectively [23].

10.2.3 Ionic complexes

It has been reported that two isostructure ionic complexes of transition metal(II) 2-fluorobenzoates. The metal atom in one of these complexes is Co(II), whereas in the other it is Ni(II). There are two nicotinamide ligands coordinated via nitrogen atoms and four water molecules coordinated via oxygen atoms in both structures around the metal atom, as well as two uncoordinated 2-FBz anions (Figure 10.12) [24, 25]. The Co–N and the Ni–N bond lengths have been reported as 2.14 Å and 2.08 Å, respectively. The hydrogen bonds (O–H···O, N–H···O, N–H···F, and C–H···F) connect molecules in 3D networks.

10.2.4 Polymeric complex

A polymeric complex of lead(II) 2-fluorobenzoate without an N-donor ligand has been reported. According to the literature review, this complex is the only example in which the 2-fluorobenzoate anion coordinates to the metal atom through the fluorine atom (Figure 10.13). There are PbO_7F and PbO_8 polyhedrons in a dimeric structure. In the structure where the fluorobenzoate anions are coordinated with the bidentate bridge and bidentate chelate, the Pb–O bond lengths range from 2.480 (6) to 2.766 (6) Å, and

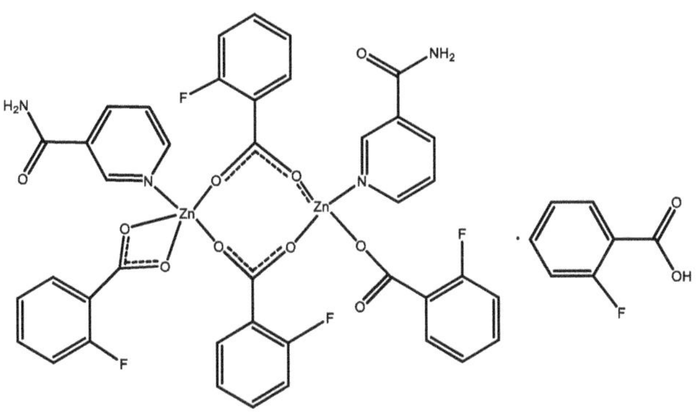

Figure 10.11: The molecular structure of $[Zn_2(\mu\text{-}2\text{-}FBz)_2(2\text{-}FBz)_2(NA)_2].(2\text{-}FBA)$ [23].

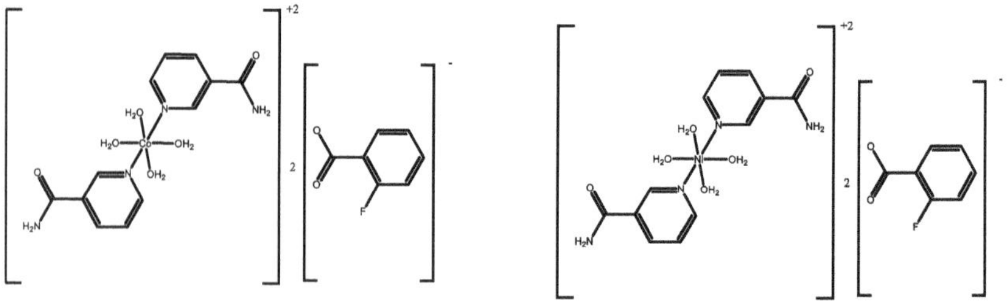

Figure 10.12: The molecular structures of $[Co(NA)_2(H_2O)_4].2(FBz)$ (a) [25] and $[Ni(NA)_2(H_2O)_4].2(FBz)$ (b) [24].

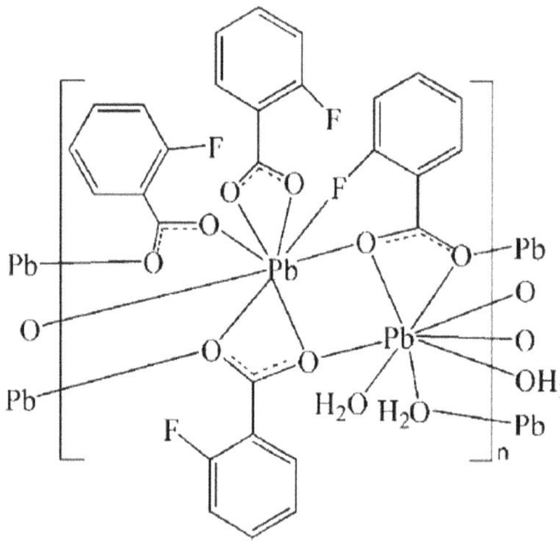

Figure 10.13: The polymeric structure of lead(II) 2-fluorobenzoate [26].

the Pb–F bond length is 2.856 (8) Å. The structure contains O–H···O and C—H···O hydrogen bonds and the polymeric chains are linked via C–H···O hydrogen bonds and π–π stacking interactions [26].

10.3 Conclusions

In this study, the crystal structures of the complexes of Cu(II), Zn(II), Cd(II), Pb(II), Co(II), and Ni(II) 2-fluorobenzoates with N- donor ligands such as 2,2′-bipyridine, 1,10′-phenanthroline, nicotinamide, isonicotinamide, 4-pyridylmethanol, and 2-aminopyridine were evaluated. It has been determined that these complexes have monomeric, dimeric, ionic, and polymeric structures. The number of monomeric complexes is higher than other complexes. Metal–O bond lengths are 1.96 Å – 2.76 Å in monomeric complexes and 2.02 Å – 2.60 Å in dimeric complexes. In monomeric, dimeric, and ionic complexes, the metal-N bond distance is longer than 2.00 Å. The variuos hydrogen bonds such as O–H···O, O–H···F, N–H···O, N–H···F, and C–H···F were found in reported crystal structures. As a result of this study, it is thought that many studies can be done with both N-donor ligand and different transition metals and new studies can be made on application areas. It is thought that new studies will make a great contribution to the coordination chemistry literature. Serial synthesis is also recommended in order to compare the similarities and differences of the structures and to synthesize the targeted molecules more easily.

References

1. Braga D. Crystal engineering, where from? where to? Chem Commun 2003;22:2751–4.
2. Mirzaei M, Eshtiagh-Hosseini H, Karrabi Z, Molčanov K, Eydizadeh E, Mague JT, et al. Crystal engineering with coordination compounds of NiII, CoII, and CrIII bearing dipicolinic acid driven by the nature of the noncovalent interactions. CrystEngComm 2014;16:5352–63.
3. Stachová P, Melník M, Korabik M, Mrozinski J, Koman M, Glowiak T, et al. Synthesis, spectral and magnetical characterization of monomeric [Cu(2-NO2bz)2(nia)2(H2O)2] and structural analysis of similar [Cu(RCOO)2(L–N)2(H2O)2] complexes. Inorg Chim Acta 2007;360:1517–22.
4. Beatty AM. Hydrogen bonded networks of coordination complexes. CrystEngComm 2001;3: 243–55.
5. Fujimura T, Seino H, Hidai M, Mizobe Y. One- or two-dimensional organometallic arrays containing PdIr2(μ3-S)2 mixed-metal sulfido cluster units connected via the nicotinamide or isonicotinamide ligands on their Pd sites through hydrogen-bonding interactions. J Organomet Chem 2004;689: 738–43.
6. Perec M, Baggio RF, Peña O, Sartoris RP, Calvo R. Synthesis and structures of four new compounds of the copper(II)–carboxylate–pyridinecarboxamide system. Inorg Chim Acta 2011;373:117–23.
7. Al-Asbahy WM, Usman M, Arjmand F, Shamsi M, Tabassum S. A dinuclear copper(II) complex with piperazine bridge ligand as a potential anticancer agent: DFT computation and biological evaluation. Inorg Chim Acta 2016;445:167–78.
8. Iqbal M, Ali S, Tahir MN. Octahedral copper(II) carboxylate complex: synthesis, structural description, DNA-binding and anti-bacterial studies. J Coord Chem 2018;71:991–1002.
9. Kubik S. Supramolecular chemistry: from concepts to applications. Berlin: De Gruyter; 2020.
10. Zhang BS, Zeng XR, Fang XN, Huang CF. Crystal structure of aqua(2, 2'-bipyridine-N, N') bis(2-fluoro-benzoato)cadmium(II), Cd(H2O)(C10H8N2)(C7H4FO2)2. Z Kristallogr N Cryst Struct 2005;220:141–2.
11. Jozef H, Danica Č, Lawson MK, Růžičková Z, Jorík V, Koman M, et al. Self-assembly hydrogen-bonded supramolecular arrays from copper(II) halogenobenzoates with nicotinamide: structure and EPR spectra. Chem Pap 2016;70:101–13.
12. Aakeröy CB, Beatty AM, Desper J, O'Shea M, Valdés-Martínez J. Directed assembly of dinuclear and mononuclear copper(II)-carboxylates into infinite 1-D motifs using isonicotinamide as a high-yielding supramolecular reagent. Dalton Trans 2003;20:3956–62.

13. Zheng M, Zheng YQ, Zhang BS. Synthesis, crystal structures, and characterization of copper(II) carboxylate complexes incorporating 1, 10-phenanthroline and bipyridine. J Coord Chem 2011;64: 3419–31.
14. Zhang BS, Zengn XR, Yu YY, Fangn XN, Huang CF. Crystal structure of (1, 10-phenanthroline-N, N') bis(2-fluorobenzoato)lead(II), Pb(FC6H4COO)2(C12H8N2). Z Kristallogr N Cryst Struct 2005;220: 75–6.
15. Zhang BS. Bis(2-fluorobenzoato-κ^2*O*, *O'*)bis(1, 10-phenanthroline-κ^2*N*, *N'*)lead(II) dihydrate. Acta Crystallogr E Struct Rep Online 2009;65:m1167–8.
16. Ye SF, Zhang BS. Hemiaquabis(2-fluorobenzoato-κ^2*O*, *O'*)bis(1, 10-phenanthroline-κ^2*N*, *N'*)lead(II) dihydrate. Acta Crystallogr E Struct Rep Online 2009;65:m936–7.
17. Yang HL, You ZL, Zhu HL. Bis(2-aminopyridine-κ*N*^{1})bis(2-fluorobenzoato-κ*O*)zinc(II). Acta Crystallogr E Struct Rep Online 2004;60:m1213–4.
18. Wang JQ, Zhang YW, Cheng L. Bis(2-fluorobenzoato-κ*O*)bis(pyridin-2-amine-κ*N*^{1})zinc(II). Acta Crystallogr E Struct Rep Online 2009;65:m950.
19. Lou QZ. Crystal structure of (1, 10-phenanthroline-N, N')-bis(2-fluorobenzoato) – cadmium(II), Cd(C7H4O2F)2(C12H8N2). Z Kristallogr N Cryst Struct 2007;222:105–6.
20. Valach F, Tokarcik M, Maris T, Watkin DJ, Prout CK. Bond-valence approach to the copper-copper and copper-oxygen bonding in binuclear copper(II) complexes: structure of tetrakis(2-fluorobenzoato-O, O')-bis(2-fluorobenzoate-O) dicopper(II). Z für Kristallogr – Cryst Mater 2000;215: 56–60.
21. Puchoňová M, Maroszová J, Mazúr M, Valigura D, Moncol J. Structures with different supramolecular interactions and spectral properties of monomeric, dimeric and polymeric benzoatocopper(II) complexes. Polyhedron 2021;197:115050.
22. Zhang BS. Crystal structure of (2, 2'-bipyridine-N, N)bis(2-fluorobenzoato)lead(II), Pb(C7H4O2F) 2(C10H8N2). Z Kristallogr N Cryst Struct 2006;221:355–6.
23. Hökelek T, Yılmaz F, Tercan B, Özbek FE, Necefoğlu H. Bis(μ-2-fluorobenzoato-1:2κ^2*O*: *O'*)(2-fluorobenzoato-1κ^2*O*, *O'*)(2-fluorobenzoato-2κ*O*)dinicotinamide-1κ*N*^{1}, 2κ*N*^{1} -dizinc(II)–2-fluorobenzoic acid (1/1). Acta Crystallogr E Struct Rep Online 2009;65:m1608–9.
24. Hökelek T, Dal H, Tercan B, Özbek FE, Necefoğlu H. Tetraaquabis(nicotinamide-κ*N*^{1})nickel(II) bis(2-fluorobenzoate). Acta Crystallogr E Struct Rep Online 2009;65:m1330–1.
25. Özbek FE, Tercan B, Şahin E, Necefoğlu H, Hökelek T. Tetraaquabis(nicotinamide-κ*N*^{1})cobalt(II) bis(2-fluorobenzoate). Acta Crystallogr E Struct Rep Online 2009;65:m341–2.
26. Zhang BS. *catena* -Poly[μ-aqua-2:1'κ^2*O*: *O* -aqua-2κ*O* -(2-fluorobenzoato-1κ^2*O*, *O'*)(μ$_2$ -2-fluorobenzoato-2':1κ^2*O*: *O'*)bis(μ$_3$ -2-fluorobenzoato)-2':1:2κ^4*O*: *O*, *O'*: *O'*;1:2:1'κ^5*F*, *O*: *O*, *O'*: *O'*-dilead(II)]. Acta Crystallogr E Struct Rep Online 2008;64:m1055–6.

Resha Kasim Vellattu Chola*, Farsana Ozhukka Parambil, Thasleena Panakkal, Basheer Meethale Chelaveettil, Prajitha Kumari and Sajna Valiya Peedikakkal

11 Clean technology for sustainable development by geopolymer materials

Abstract: Geopolymer materials have captivated as a promising material for building restoration due to their environmentally sustainable nature as well as their potential to use a variety of waste products as precursors. Numerous industrial, municipal and agricultural wastes can be used to create environmentally acceptable, sustainable, structurally sound geopolymer matrices. These new generation materials, fabricated by following the geopolymerisation reactions of alumino-silicate oxides with alkali activators, have the advantages of high mechanical strength, corrosion resistance, durability, fire resistance etc., and can serve as a substitute for construction materials like Ordinary Portland Cement since they carry enormous impact on the environment. This review presents the importance of geopolymeric materials and their role in sustainable development giving special emphasis to kaolin, metakaolin, zeolite, fly ash, dolomite, red mud and clay based geopolymer materials.

Keywords: clean technology; fly ash; geopolymer; kaolin; sustainable development; zeolite.

11.1 Introduction

Clean technology is a procedural approach for the prevention and minimization of negative environmental impact by reducing pollution and waste products through materials or processes. It aims to reduce risks to human health and the environment. Low and non-waste technology, recycle technology and waste utilization technology are the three broad categories of clean technology [1]. It can be applied in numerous systems such as environmental systems, water systems, industrial systems etc. In the environmental system, clean technology tries to minimize and utilize waste in proper ways. Advanced and affordable technologies

***Corresponding author: Resha Kasim Vellattu Chola,** Department of Chemistry, Pocker Sahib Memorial Orphanage College, Tirurangadi, Malappuram, Kerala, India, E-mail: chempsmotgi@gmail.com
Farsana Ozhukka Parambil, Thasleena Panakkal, Basheer Meethale Chelaveettil, Prajitha Kumari and Sajna Valiya Peedikakkal, Department of Chemistry, Pocker Sahib Memorial Orphanage College, Tirurangadi, Malappuram, Kerala, India

As per De Gruyter's policy this article has previously been published in the journal Physical Sciences Reviews. Please cite as: R. K. V. Chola, F. O. Parambil, T. Panakkal, B. M. Chelaveettil, P. Kumari and S. V. Peedikakkal "Clean technology for sustainable development by geopolymer materials" *Physical Sciences Reviews* [Online] 2022. DOI: 10.1515/psr-2022-0194 | https://doi.org/10.1515/9783110913361-011

are used for groundwater treatment. In industry minimum emission and significant energy savings are done by applying mathematical models, numerical simulations and utilization of industrial by-products as feedstock for other processes [2]. Fossil fuel consumption has increased exponentially due to population growth and technological advancement, which have a significant impact on the environment. Despite the rapid expansion of environmentally friendly, sustainable, and low-impact renewable energy sources, it is highly urged that the current process be made more efficient. In industries like cement industry, reusing and recycling industrial waste is seen to be an effective strategy for reducing power consumption [3]. Cement industry is a main global CO_2 emitter; therefore it is imperative to apply clean technology for a sustainable alternative to ordinary cement.

Since the early 1980s, Ordinary Portland Cement (OPC) is regarded as a significant building material in the construction field due to its excellent durability, cost-effectiveness and several advantages over other building materials. The OPC industry has significant negative influence on the environment and the production of OPC is very expensive and energy intensive. The production of OPC results in significant emissions of carbon dioxide (CO_2) into the atmosphere. Approximately 1.5 tonnes of raw materials are necessary for making one tonne of OPC, and 0.81 tonnes of CO_2 are released into the atmosphere during production. About 7% of the CO_2 produced globally is estimated to be attributable to the cement industry alone. OPC also has a negative environmental impact due to its high energy and resource consumption. It may not offer the properties needed for many types of constructions, such as high resistance to chemical attack and rapid mechanical strength development [4, 5]. Hence Engineers and scientists are searching to replace OPC with any other greener alternative.

In 1976, Joseph Davidovits introduced the concept of geopolymer [6]. Geopolymer materials are environmentally friendly cementitious materials with high strength, excellent acid and fire resistance, superior durability, lower shrinkage and permeability. The prime function of the geopolymer in the research field is the use of natural materials and industrially hazardous waste products as primary raw materials as well as the low discharge of carbon dioxide. Table 11.1 summarizes the basic differences between OPC and geopolymer concrete [7, 8]. Geopolymers have the advantages of high mechanical strength, artificial ageing resistance, excellent durability, high thermal resistance and chemical corrosion resistance [9].

The geopolymerisation reaction is an exothermic one having three major steps that are dissolution of aluminosilicate materials in alkaline solution, diffusion of silicon and aluminium and hardening of the gel phase to geopolymer. Sodium silicate and sodium hydroxide are the prominent alkaline solution for this purpose, where sodium hydroxide used for dissolution of the aluminosilicate source. Whereas sodium silicate is implemented as a binder, dispersant, plasticizer moreover alkali activator [10] i.e. geopolymerisation involves the activation of

Table 11.1: Differences between OPC and geopolymer concrete.

Ordinary Portland concrete	Geopolymer concrete
High CO_2 emission	Low CO_2 emission
Restricted to set times	Slow or fast set time based on alkali concentration and raw material reactivity
Moderate durability	More durable than ordinary Portland concrete
Liable to freezing and thawing cycles	Significant resistance to freeze and thaw cycling
Prone to acid attack	More resistant to organic and inorganic acids
Low thermal resistance	High thermal resistance without degradation
High energy consumption	Low energy consumption
Difficult to change modulus characteristics	Flexible and adjustable modulus characteristics
Restricted methods to increase strength	Strength can be increased by various techniques

different aluminosilicate-rich materials with a strong alkali activator solution forming a three-dimensional aluminosilicate network structure at ambient or slightly raised temperature [11, 12] (Figure 11.1) [13].

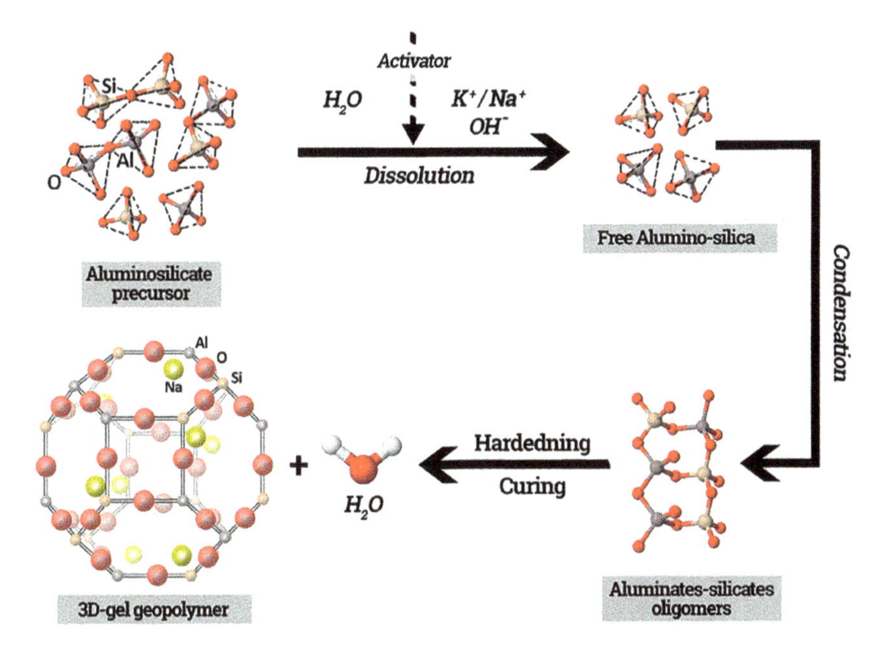

Figure 11.1: General geopolymerisation mechanism. Reproduced with permission from Safaa M et al. [13]. Creative Commons Attribution Non Commercial No Derivatives 2020.

During the manufacturing of geopolymer, the CO_2 emissions and energy consumption are very low, which offers a considerable reduction in global warming [14, 15]. This article provides a condensed overview on some of the precursors, with a focus on the abundant industrial wastes and low-cost raw materials that are commonly utilised to create geopolymeric materials and their other potential uses. They are kaolin, metakaolin, zeolite, fly ash, dolomite and red mud (Figure 11.2).

11.2 Geopolymer materials

There are several raw materials that may be involved in creating geopolymers with exciting uses. The use of waste products as raw materials for geopolymerisation can achieve the purpose of sustainable development. Fly ash, red mud and slags are industrially hazardous wastes that require special treatment due to their huge production and reduced reuse rate.

Kaolin is a mineral with a layered structure that forms by refined weathering processes. It is made up of tetrahedral sheets and octahedral sheets of silica and alumina by sharing an oxygen atom. By the exposed nature of the alumina-hydroxyl and silica-oxygen sheets, it interacts with various components. Moreover, the well-packed structure makes the layers inseparable and cannot be broken down. Kaolin may create a non-biodegradable barrier. Natural deposits and sediments with a high concentration of kaolin associated with other minerals are extremely effective at regulating the movement of dissolved organisms [16]. Metakaolin is a pozzolanic supplemental cementitious material and it is a calcined product of kaolin. Zeolites

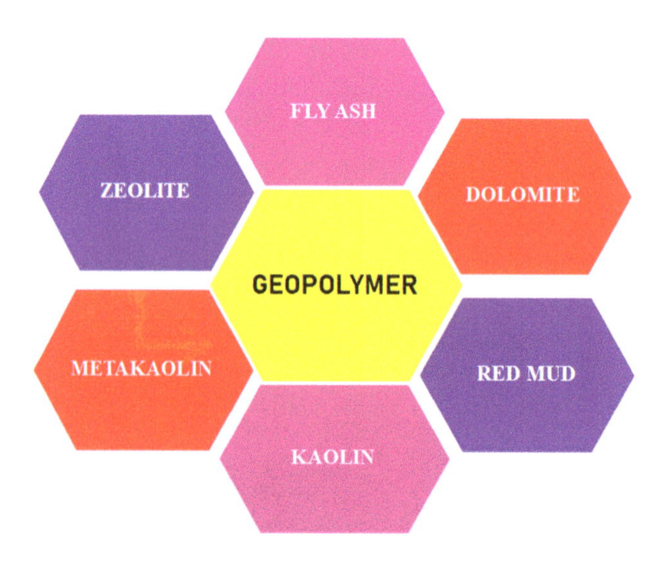

Figure 11.2: Major precursors used for the fabrication of geopolymeric materials.

are fascinating non-toxic materials, containing a large number of micropores with honeycomb framework into which molecules can be adsorbed to cleanse the environment and catalyse chemical processes. They are essential to green chemistry because they reduce the need for organic solvents [17]. A finely divided mineral waste called fly ash is created when coal is burned in power plants. Fly ash is used to substitute naturally occurring aggregates and minerals, which can drastically reduce aggregate demand. The use of fly ash has a favourable impact on the environment [18]. Due to its high alkalinity, red mud, a byproduct of the aluminium industry (Bayer method of alumina synthesis), has historically had trouble in disposal and utilization. The nature of red mud is most likely that of clay soil that is much stronger than typical clay soil [19].

Dolomite ($CaMg(CO_3)_2$) appears to develop in a variety of environments and can have a variety of structural, textural, and chemical properties. Dolomite is economically significant to man because of its diverse physical, mineralogical and chemical qualities, which make it an excellent construction material. Dolomite is a key raw material in many sectors, including paint, chemicals, pharmaceuticals and cosmetics [20]. The subsequent sections describe the use of each selected raw material for geopolymer synthesis, treatment parameters, variables affecting the physical qualities of the geopolymer, and applications in various domains. The fundamental as well as versatile raw materials, their unique properties and their application in various eras are given in Table 11.2.

Table 11.2: Properties and applications of geopolymeric materials.

Sl.No	Material	Properties	Applications	Reference
1.	Kaolin	– Chemical inertness in a wide range of pH – Non-abrasive materials – Insulators of electricity and heat. – Adequate compressive strength – Curing ability at room temperature is great, – Low shrinkage and permeability, – Good resistance to acid and heat – Excellent durability – Less energy consumption – Low carbon dioxide emission	– Formation of sustainable building and construction material – Coating material – Drug delivery system to tampering – Removal of heavy metals	[21] [22] [23] [24] [25]
2.	Meta kaolin	– Product of thermal treatment of kaolin – Resistant to acid	– Drug delivery for high potency drugs	[26] [27] [28]

Table 11.2: (continued)

Sl.No	Material	Properties	Applications	Reference
		– Thermal conductivity is poor – Rapidly develop the compressive strength	– Anti-corrosive cement fabrication – Thermal resistant and fire proof materials – Crack-free ceramic plate and bullet specimens – Manufacture tribological materials. – Oil-well cementing – Fabrication of biocompatible material	[29]
3.	Zeolite	– Hydrated alumino-silicates – Elevated pore volume – Remarkable surface area to volume ratio – Superior capacity in adsorption	– Effective in the adsorption and eleimination of dyes – Excellent in heavy metal ion removal	[30] [31]
4.	Bentonite	– Alkaline activation enhanced the ion adsorption capability – Excellent absorbing and rheological properties – Lower swelling capacity	– Decontamination of ammonium ions – Designing of high-performance materials for thermal insulation	[32] [33] [34]
5.	Fly ash	– Major component of coal combustion waste – Utilized directly as raw resource for geopolymer – Low-cost adsorbent material – Excellent durablity – High tensile and flexural strength – Exceptional compressive strength, – Good efflorescence with better resistance to Cl^-, SO_4^{2-} and acids	– Decontamination of radioactive and trace heavy metal ions – Fire proof material – Removal of methylene blue – Adsorptive removal of formaldehyde – Reduce the effects of global warming potentials – Development of high performance hybrid inorganic–organic composite	[35] [36] [37] [38] [39]
6.	Red mud	– By-product of alumina production – High alkalinity and high aluminium content – Sulphate resistance and fire resistance	– Geopolymer mortar – Used for load-bearing members in structures – Coating of concrete bricks	[40] [41] [42]
7.	Dolomite	– Dolomite is a double carbonate – Insoluble in dilute hydrochloric acid	– Fabrication of geopolymer mortar and concrete	[43] [44]

11.2.1 Kaolin

Kaolin is a soft aluminosilicate mineral with several applications. These are electrical and thermal insulators, non-abrasive materials and chemically inert over a wide pH range. In alkaline circumstances, kaoin minerals polycondensed, resulting in geopolymers with high compressive strength [21]. The amount of kaolin in relation to the alkaline activator solution affects the geopolymers' mechanical properties, hardness and flexural strength. Geopolymer displayed promising hardness and flexural strength while coating on low-grade wood substrate with a kaolin/alkaline activator ratio of 0.7 [22]. By occupying ions in tetrahedral cavities, the increased concentration of KOH and NaOH reduces compressive strength and eventually results in the formation of cracks [23].

Porosity and density have an impact on the geopolymer's strength; materials that are less porous and denser and have finer grains have more strength [45]. Furthermore, the pore size distribution affects the durability of geopolymer by entering the corrosive media into the pores and thereby causing damage [46]. However, the usage of aluminium powder as a foaming agent leads to kaolin geopolymer with excellent pore characteristics. This property generated an additional advantage for the geopolymer for heavy metal removal [25]. The production of high dense geopolymer with compact structure can be acquired by the addition of nanomaterials into the kaolin system. i.e., the addition of it fortifies the microstructure, and produces a homogeneous geopolymer with few points of defects [47].

11.2.2 Metakaolin

Thermally treated kaolin, i.e. metakaolin is also an important source for geopolymer synthesis. Three dimensional network structure of these geopolymers makes them anticorrosive materials [27]. The temperature treatment on the geopolymer alters the microscopic properties via thermal stress produced by changing the amorphous gel structure into a crystalline phase [48]. The introduction of calcium ions into the geopolymer imparts biocompatible properties to it. Moreover, it augments the durability and strength of geopolymers [28, 29]. The geopolymer with a higher density is acquired by the amorphous nature nano silica. Additionally, the incorporation of nano silica into metakaolin augments the thermal stability and compressive strength by improving the pore characteristics [49].

11.2.3 Zeolite

Another possible source for geopolymer synthesis has been reported as zeolite, crystalline aluminosilicate materials containing porous structures with high surface area. These materials are deemed as versatile adsorbents for toxic heavy metal ions due to their low price, unique adsorption properties and wide availability of raw materials [30]. It contains interchangeable cationic species such as alkali metals and alkaline earth elements with a large amount of water. The geopolymerisation reaction can be comparable to the synthesis of zeolite, since, under appropriate circumstances, zeolite phases are obtained along with geopolymerisation [50]. The natural zeolites can be converted into geopolymers with high compressive strength by the preliminary action of silicate activators and alkali activators on them. The variation in the concentration of sodium hydroxide and alkaline activator severely affected the compressive strength, whether it is natural zeolite or its any combination [51]. In addition to this high-temperature treatment also favours its reactivity and strength [52, 53]. Indeed, Sodium carbonate activation of natural zeolites for the geopolymer leads to the geopolymer formation to a lesser yield and contains some unreacted soda ash. These geopolymers show higher adhesive properties and are suitable for coating purposes like as kaolin-based ones [54, 55].

By utilising compressive strength measurement with XRD and SEM, S. Ozen et al. provided a thorough knowledge of the mechanical characteristics of natural zeolite-based geopolymers [56]. Three dimensionally interconnected structures with excellent pore characteristics were acquired by the effective incorporation of microporosity of zeolite and meso-macro porosity of geopolymer. The composite of geopolymer and Na13X zeolite offers high CO_2 adsorption properties [57]. The addition of zeolite into geopolymer accelerates the hydration of cement, whereas above a certain limit it reduces the compressive strength of geopolymer [58]. The photocatalytic and adsorption capabilities of geopolymers containing titanium dioxide and zeolite were studied by Kedsarin and co-workers. The composite shows high methylene blue removal efficiency since it is easy to handle and reusable. Here geopolymer acts as a binder to fix the zeolite and titanium oxide for this application [31].When transforming zeolite into the geopolymer, it contributes mechanical properties. Besides this, the zeolitic phase present in the geopolymer contributes some additional applications to these geopolymers such as heavy metal removal and carbon dioxide adsorption [59].

11.2.4 Other miscellaneous clays

Clays are commonly recognized as the foundation of geopolymer formation and reformation. Researchers have evaluated the properties of geopolymer derived from bentonite clay. The lower crystallinity, presence of heteroporosity, and improved capacity for sorption make this geopolymer effective for the removal of ammonia from the soil. The agglomerates of geopolymer permit the availability of spaces or voids, which leads to the better sorption of ammonia [32]. Besides this, the incorporation of bentonite into the geopolymer derived from fly ash and natural aggregates has excellent durability and rheological properties [60]. J. Jiang et al. attempted to synthesize porous geopolymer for enhancing the thermal insulation property. They converted the bentonite slurry to porous zeolite via geopolymer crystallization [33]. Porous geopolymers with lower thermal conductivity properties can also be prepared by combining swelled-bentonite slurry into it. Capillary pores in the bentonite slurry boost the number of pores and pore size distribution on low strength geopolymer matrix [34].

Illite is a typical clay mineral derived from the feldspar and felsic silicates as a weathering product. The substitution of one fourth of silicon by aluminium creates a charge deficiency in it. It is neutralized by potassium ions present between the layers. Montmorillonite has a similar structure except for the substitution of silicon with aluminium occurs in octahedral sheets. Additionally, every sixth aluminium is replaced by magnesium ions [61]. Uncalcined montmorillonite and illite are other potential precursor solutions for geopolymer synthesis. Since the alkali activation behaviour of these two precursors is a variant, their applications are also different. The geopolymers with montmorillonite clay are a more appropriate material for earth block construction than the geopolymers with illite clay. This is due to the structural breakdown that happened by thermal dehydroxylation or acid activation [62]. When transforming montmorillonite into the geopolymer, the microstructure changed drastically from the plate structure of its precursor solution [63].

Red clay or iron oxide-rich clays are natural and low-cost raw materials for geopolymers. P. As per Choeycharoen et al., the Si/Al ratio has an adverse effect on the strength of geopolymers. Higher Si/Al ratio and iron oxide-rich source materials appear in lower strength geopolymers [64].

11.2.5 Fly ash based geopolymer

Fly ash (also known as coal ash) is a byproduct from coal burning thermal power stations [65]. Type of coal used, conditions of coal combustion and emission control technology all have a significant impact on the physical and chemical characteristics of fly ash. It is an alkaline substance having a pH between 8.5 and 11.5. The average particle size is 1–100 µm. It has a fine structure with low to medium bulk density [35]. The major constituents of fly ash are oxides of silicon (SiO_2), calcium (CaO), aluminium (Al_2O_3) and iron (Fe_2O_3). In addition to these, it also contains toxic trace elements such as Hg, Pb, As, Sr, Mo, Ti, V, Cr, Mn and Co. Since it is a hazardous material proper disposal of fly ash is imperative and that requires a huge landfill [36].

Fly ash is effectively used in building and constructing fields for the production of cement, concrete, back fillings, blocks, bricks etc. [35]. It is a suitable substitute for ordinary Portland cement due to its abundant availability [66].

Fly ash is a geopolymer concrete binder along with alkali activator and aggregates. Thus, geopolymer concrete not only reduces the environmental problems due to the production of cement but also make the advantage of utilizing the waste (fly ash) from coal combustion. Common fly ash based geopolymers are geopolymer paste, concretes, and mortars. Quartz, sandstone, limestone, basalt and granite are the naturally occurring aggregates used in geopolymers. They are economically favorable, reduce pore density and crack formation and improve durability [67]. Fly ash based geopolymer technology introduces a greener approach to the environment and ecology since the emission of CO_2 during its production is very low when compared to the production of OPC [68] (Figure 11.3) [69].

The mechanical strength and durability of fly ash-based geopolymers are almost on par with those of hydrated Portland cement. Recent research put forward many advancements in the production of this geopolymer with high quality performance. Decomposition of aluminosilicate by alkali and subsequent polycondensation is the basic formation principle of fly ash based geopolymer. This reaction is energy efficient since it occurs at a mild temperature. Improper dumping of fly ash causes the release of toxic trace elements into the environment. If fly ash is used to produce geopolymer it can adsorb and immobilize the trace elements. Geopolymer cement

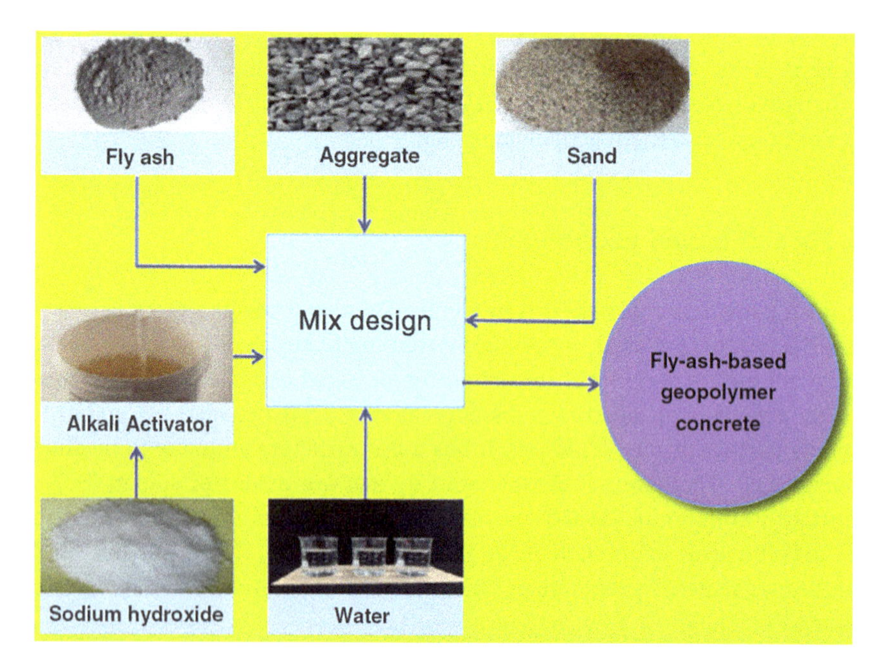

Figure 11.3: Schematic representation on preparation of geopolymer concrete derived from fly ash. Reproduced with permission from Han Q et al. [69]. Creative Commons Attribution 4.0 International License (2022).

made from fly ash can be used as a sealant of underground CO_2 storage, but the performance is not much satisfied. Composites prepared from geopolymer and biomass are lightweight and fireproof in nature [65]. Fly ash based geopolymers do not much deplete natural resources like dolomite, kaolin and metakaolin-based geopolymer materials [38].

Researchers have investigated the enhancement of geopolymers by incorporating bio admixtures into construction materials [70, 71]. Li et al. examined the possibility of using the chitosan derivative N-carboxymethyl chitosan to enhance the mechanical behaviour of fly ash-based geopolymers and obtained a denser network structure with improved mechanical property [39]. Kappa carrageenan (KC), a low-cost naturally occurring biopolymer, can positively affect the mechanical behaviour of fly ash-based geopolymers [72].

11.2.6 Dolomite

Anhydrous carbonate mineral composed of calcium, magnesium, and carbonate is known as Dolomite. Additionally, it has a high proportion of calicium oxide (CaO) and magnesium oxide (MgO) and a small amount of SiO_2 and Al_2O_3. It is an important geopolymer raw material [43]. Dolomite is cheap and abundant in the earth's crust.

As the Si and Al content increases, the strength, fire and heat resistance and adhesion to the steel substrate of geopolymer normally will increase. Application of dolomite in geopolymer is still at the beginning stage [73]. It is easy to grind into fine powder since it is a soft material. Dolomite is also known as algime (agricultural lime) because it is used by farmers to reduce soil acidity. It is also used for solving the magnesium deficiency of soil [38] and for the adsorption of heavy metals [74]. Dolomite is a double carbonate, with an alternate structural arrangement of both magnesium and calcium ions. Unlike calcite, rapid dissolution or effervescence formation (fizz) in dilute hydrochloric acid is absent for dolomite [75].

11.2.7 Red mud

Red mud is a hazardous industrial byproduct of alumina manufacturing. It is a nonferrous solid waste and dangerous material due to its strong alkalinity [40]. Therefore, red mud recycling is very urgent. Due to its high aluminium content and basicity, red mud is a suitable precursor for geopolymers. Three tonnes of bauxite ore produce 1.5–2.2 tonnes of red mud as byproduct for every 1 tonnes of aluminium. The setting time of geopolymer concrete is directly proportional to content of red mud whereas the compressive strength has an inverse relation to the content. Red mud based geopolymer concrete has higher fire and sulphate resistance than conventional concrete [41]. Red mud geopolymer paste is a good coating material for concrete wall. It has more stability than fly ash. The aesthetics of concrete blocks also increased by red mud [42].

The use of geopolymer technology is expanding in everyday tasks like creating ecologically acceptable construction materials and solving waste management problems. This technique has recently undergone commercial development and has been used for certain practical engineering tasks. Nevertheless, the widespread use of geopolymers faces some challenges. The high energy requirement of alkali activation during polymerization, the possibility of efflorescence formation, difficulty in finding suitable parameters for mix design, cost and environmental impact of activators, limited knowledge about kinetics of geopolymerisation, and limited studies on the durability are the main obstacles of real-world application of geopolymers [76].

11.3 Conclusion and future perspectives

Increased greenhouse gas emissions and environmental catastrophes force researchers to implement green technologies for sustainable development. Naturally occurring clays, aluminosilicate materials and hazardous solid wastes from industries like red mud and fly ash are potential precursor materials for the effective synthesis of geopolymer and the augmentation of its properties. Since they succeed in the purpose of converting waste into new products, they greatly assist the sustainable development goals outlined by the United Nations. Moreover they promote a circular economy. These materials not only transformed into geopolymer but also enhance stability, compressive strength, mechanical properties and thermal insulation properties. Due to this, geopolymers have been effectively utilized in different eras for emerging applications of global concern.

The behaviour of geopolymers depends on various factors. For understanding the mechanical properties of this material a thorough effort is required. Bulk of prior studies has mostly concentrated on the mixing of precursors, the relationship of activators with aluminosilicates, and the impact of the extra fillers. However, the impact of different types of aluminosilicates with desirable characterestics, which are crucial components, has not frequently been studied. Therefore, more study is required to examine the impact of different aluminosilicates on the various properties of geopolymers. Investigations should be done to create more affordable, environmentally friendly precursors, ideally made from waste materials.

References

1. Muralikrishna IV, Manickam V. Environmental management: science and engineering for industry. Oxford, UK: Butterwoth-Heinemann; 2017.

2. Mikulcic H, Baleta J, Klemes JJ. Cleaner technologies for sustainable development. Clean Eng Technol 2022;7:100445.
3. Shehata N, Mohamed OA, Sayed ET, Abdelkareem MA, Olabi AG. Geopolymer concrete as green building materials: recent applications, sustainable development and circular economy potentials. Sci Total Environ 2022;836:155577.
4. Kumar MP. Reducing the environmental impact of Concrete. Concr Int 2005;10:61–6.
5. Huseien GF, Mirza J, Ismail M, Ghoshal SK, Hussein AA. Geopolymer mortars as sustainable repair material: a comprehensive review. Renew Sustain Energy Rev 2017;80:54–74.
6. Davidovits J. Geopolymers - inorganic polymeric new materials. J Therm Anal Calorim 1991;37: 1633–56.
7. Amran YHM, Alyousef R, Alabduljabbar H, El-Zeadani M. Clean production and properties of geopolymer concrete; A review. J Clean Prod 2020;251:119679.
8. Kumar S, Kumar R. Geopolymer: cement for low carbon economy. Indian Concr J 2014;88:29–37.
9. Amer H, Mohammed A, Shariq M. A review of properties and behaviour of reinforced geopolymer concrete structural elements- a clean technology option for sustainable development. J Clean Prod 2020;245:118762.
10. Ramasamy S, Hussin K, Abdullah MMAB, Ghazali CMR, Sandu AV, Mohammed B, et al. Recent dissertations on kaolin based geopolymer materials. Rev Adv Mater Sci 2015;42:83–91.
11. Cong P, Cheng Y. Advances in geopolymer materials: a comprehensive review. J Traffic Transport Eng 2021;8:283–314.
12. Lingyu T, Dongpo H, Jianing Z, Hongguang W. Durability of geopolymers and geopolymer concretes: a review. Rev Adv Mater Sci 2021;60:1–14.
13. Safaa M, Samira M, El Aiman M, Taha Y, Mostafa B, Rachid H. Mine wastes based geopolymers: a critical review. Cleaner Eng Techn 2020;2020:100014.
14. Olivia M, Jingga H, Toni N, Wibisono G. Biopolymers to improve physical properties and leaching characteristics of mortar and concrete: a review. IOP Conf Ser Mater Sci Eng 2018;345:012028.
15. Burduhos Nergis DD, Abdullah MMAB, Vizureanu P, Mohd Tahir MF. Geopolymers and their uses: review. IOP Conf Ser Mater Sci Eng 2018;374:10.
16. Miranda-Trevino JC, Coles CA. Kaolinite properties, structure and influence of metal retention on pH. Appl Clay Sci 2003;23:133–9.
17. Rhodes CJ. Properties and applications of zeolites. Sci Prog 2010;93:223–84.
18. Ismail KN, Hussin K, Idris MS. Physical, chemical & mineralogical properties of fly-ash. J Nucl Relat Technol 2007;4:47–51.
19. Reddy PS, Reddy NG, Serjun VZ, Mohanty B, Das SK, Reddy KR, et al. Properties and assessment of applications of red mud (bauxite residue): current status and research needs. Waste Biomass Valorization 2021;12:1185–217.
20. Akande JM, Agbalajobi SA. Analysis on some physical and chemical properties of Oreke dolomite deposit. J Miner Mater Char Eng 2013;01:33–8.
21. Heah CY, Kamarudin H, Mustafa Al Bakri AM, Bnhussain M, Luqman M, Khairul Nizar I, et al. Kaolin-based geopolymers with various NaOH concentrations. Int J Miner Metall Mater 2013;20:313–22.
22. Ramasamy S, Hussin K, Abdullah MMAB, Ghazali CMR, Binhussain M, Sandu AV. Interrelationship of kaolin, alkaline liquid ratio and strength of kaolin geopolymer. IOP Conf Ser Mater Sci Eng 2016; 133:012004.
23. Nmiri A, Hamdi N, Yazoghli-Marzouk O, Duc M, Srasra E. Synthesis and characterization of kaolinite-based geopolymer: alkaline activation effect on calcined kaolinitic clay at different temperatures. J Mater Environ Sci 2017;8:276–90.
24. Cai B, Engqvist H, Bredenberg S. Evaluation of the resistance of a geopolymer-based drug delivery system to tampering. Int J Pharm 2014;465:169–74.

25. Ariffin N, Abdullah MMAB, Postawa P, Zamree Abd Rahim S, Mohd Arif Zainol MRR, Jaya RP, et al. Effect of aluminium powder on kaolin-based geopolymer characteristic and removal of Cu2+. Materials (Basel) 2021;14:1–19.
26. Jämstorp E, Forsgren J, Bredenberg S, Engqvist H, Strømme M. Mechanically strong geopolymers offer new possibilities in treatment of chronic pain. J Controlled Release 2010;146:370–7.
27. Zhang ZH, Zhu HJ, Zhou CH, Wang H. Geopolymer from kaolin in China: an overview. Appl Clay Sci 2016;119:31–41.
28. Pangdaeng S, Sata V, Aguiar JB, Pacheco-Torgal F, Chindaprasirt J, Chindaprasirt P. Bioactivity enhancement of calcined kaolin geopolymer with $CaCl_2$ treatment. ScienceAsia 2016;42:407–14.
29. Mackenzie KJD, Rahner N, Smith ME, Wong A. Calcium-containing inorganic polymers as potential bioactive materials. J Mater Sci 2010;45:999–1007.
30. Ren Z, Wang L, Li Y, Zha J, Tian G, Wang F, et al. Synthesis of zeolites by in-situ conversion of geopolymers and their performance of heavy metal ion removal in wastewater: a review. J Clean Prod 2022;349:131441.
31. Pimraksa K, Setthaya N, Thala M, Chindaprasirt P. Geopolymer/Zeolite composite materials with adsorptive and photocatalytic properties for dye removal. PLoS One 2020;15:1–20.
32. Jaimes JE, Montao AM, González CP. Geopolymer derived from bentonite: structural characterization and evaluation as a potential sorbent of ammonium in waters. J Phys Conf Ser 2020;1587:1–5.
33. Jiang J, Yang Y, Hou L, Lu Z, Li J, Niu Y. Facile preparation and hardened properties of porous geopolymer-supported zeolite based on swelled bentonite. Constr Build Mater 2019;228:117040.
34. Yang Y, Jiang J, Hou L, Lu Z, Li J, Wang J. Pore structure and properties of porous geopolymer based on pre-swelled bentonite. Constr Build Mater 2020;254:119226.
35. Vilakazi AQ, Ndlovu S, Chipise L, Shemi A. The recycling of coal fly ash: a review on sustainable developments and economic considerations. Sustain 2022;14:1–32.
36. Zhuang XY, Chen L, Komarneni S, Zhou CH, Tong DS, Yang HM, et al. Fly ash-based geopolymer: clean production, properties and applications. J Clean Prod 2016;125:253–67.
37. Adewuyi YG. Recent advances in fly-ash-based geopolymers: potential on the utilization for sustainable environmental remediation. ACS Omega 2021;6:15532–42.
38. Zain H, Abdullah MMAB, Hussin K, Ariffin N, Bayuaji R. Review on various types of geopolymer materials with the environmental impact assessment. MATEC Web Conf 2017;97:01021.
39. Li Z, Chen R, Zhang L. Utilization of chitosan biopolymer to enhance fly ash-based geopolymer. J Mater Sci 2013;48:7986–93.
40. Liang X, Ji Y. Mechanical properties and permeability of red mud-blast furnace slag-based geopolymer concrete. SN Appl Sci 2021;3:1–10.
41. Singh S, Biswas RD, Aswath MU. Experimental study on red mud based geopolymer concrete with fly ash & GGBS in ambient temperature curing. Int J Adv Mech Civ Eng 2016;2394–827.
42. Singh S, Basavanagowda SN, Aswath MU, Ranganath RV. Durability of bricks coated with red mud based geopolymer paste. IOP Conf Ser Mater Sci Eng 2016;149:012070.
43. Azimi EA, Abdullah MMAB, Ming LY, Yong HC, Hussin K, Aziz IH. Review of dolomite as precursor of geopolymer materials. MATEC Web Conf 2016;78:01090.
44. Prabha VC, Revathi V, Sivamurthy RS. Ambient cured high calcium fly ash geopolymer concrete with dolomite powder. Eur J Environ Civil Eng 2021;2012262.
45. Liew YM, Heah CY, Mohd Mustafa AB, Kamarudin H. Structure and properties of clay-based geopolymer cements: a review. Prog Mater Sci 2016;83:595–629.
46. Wen J, Zhou Y, Ye X. Study on structure and properties of kaolin composites-based geopolymers. Chem Eng Trans 2018;66:463–8.
47. Hassaan MM, Khater HM, El-Mahllawy MS, El Nagar AM. Production of geopolymer composites enhanced by nano-kaolin material. J Adv Ceram 2015;4:245–52.

48. Adjei S, Elkatatny S, Ayranci K. Effect of elevated temperature on the microstructure of metakaolin-based geopolymer. ACS Omega 2022;7:10268–76.
49. Zidi Z, Ltifi M, Zafar I. Characterization of nano-silica local metakaolin based-geopolymer: microstructure and mechanical properties. Open J Civ Eng 2020;10:143–61.
50. Ma G, Bai C, Wang M, He P. Effects of Si/Al ratios on the bulk-type zeolite formation using synthetic metakaolin-based geopolymer with designated composition. Crystals 2021;11:1310.
51. Ibrahim R. Mechanical strength variation of zeolite-fly ash geopolymer mortars with different activator concentrations. Chall J Concr Res Lett 2021;12:96–103.
52. Ozen S, Uzal B. Effect of characteristics of natural zeolites on their geopolymerization. Case Stud Constr Mater 2021;15:e00715.
53. Villa C, Pecina ET, Torres R, Gómez L. Geopolymer synthesis using alkaline activation of natural zeolite. Constr Build Mater 2010;24:2084–90.
54. Nikolov A, Rostovsky I, Nugteren H. Geopolymer materials based on natural zeolite. Case Stud Constr Mater 2017;6:198–205.
55. Ruzaidi CM, Al Bakri AMM, Binhusain M, Salwa MS, Alida A, Muhammad Faheem MT, et al. Study on properties and morphology of kaolin based geopolymer coating on clay substrates. Key Eng Mater 2014;594–595:540–5.
56. Özen S, Alam B. Compressive strength and microstructural characteristics of natural zeolite-based geopolymer. Period Polytech Civ Eng 2018;62:64–71.
57. Papa E, Medri V, Amari S, Manaud J, Benito P, Vaccari A, et al. Zeolite-geopolymer composite materials: production and characterization. J Clean Prod 2017;171:76–84.
58. Saraya ME, Sayed M. Characterization and evaluation of natural zeolite as a pozzolanic material. Al Azhar Bull Sci 2018;29:17–34.
59. Rozek P, Krol M, Mozgawa W. Geopolymer-zeolite composites: a review. J Clean Prod 2019;230: 557–79.
60. Waqas RM, Faheem B, Danish MA, Alqurashi M, Mosaberpanah MA, Masood B, et al. Influence of bentonite on mechanical and durability properties of high-calcium fly ash geopolymer concrete with natural and recycled aggregates. Materials 2021;7790:1–23.
61. Okashah AM, Zainal FF, Hayazi NF, Nordin MN, Abdullah A. Pozzolanic properties of calcined clay in geopolymer concrete: a review. AIP Conf Proc 2021;2339:020234.
62. Marsh A, Heath A, Patureau P, Evernden M, Walker P. Alkali activation behaviour of un-calcined montmorillonite and illite clay minerals. Appl Clay Sci 2018;166:250–61.
63. Marsh A, Heath A, Patureau P, Evernden M, Walker P. Stabilisation of clay mixtures and soils by alkali activation Alastair. In: International symposium on earthen dwellings and structures; 2019: 15–26 p.
64. Choeycharoen P, Sornlar W, Shongkittikul W, Wannagon A. Superior properties and structural analysis of geopolymer synthesized from red clay. Chiang Mai J Sci 2019;46:1234–48.
65. Nath P, Sarker PK. Fly ash based geopolymer concrete: a review. Silicon 2013;14:2453–72.
66. Saravanan G, Jeyasehar CA, Kandasamy S. Flyash based geopolymer concrete-a state of the art review. J Eng Sci Technol Rev 2013;6:25–32.
67. Temuujin J, Van RA, MacKenzie KJD. Preparation and characterisation of fly ash based geopolymer mortars. Constr Build Mater 2010;24:1906–10.
68. Davidovits J. False values on CO_2 emission for geopolymer cement/concrete. Geopolymer Inst Libr Tech Pap 2015;24:1–9.
69. Han Q, Zhang P, Wu J, Jing Y, Zhang D, Zhang T. Comprehensive review of the properties of fly ash-based geopolymer with additive of nano-SiO_2. Nanotechnol Rev 2022;11:1478–98.
70. Gupta NS. Plant biopolymer–geopolymer: organic diagenesis and kerogen formation. Front Mater 2015;2:2007–10.

71. Abdollahnejad Z, Kheradmand M. Mechanical properties of hybrid cement based mortars containing two biopolymers. Int Sch Sci Res Innov 2017;11:1223–6.
72. Li Z, Zhang L. Fly ash-based geopolymer with kappa-carrageenan biopolymer. Biopolym Biotech Admixtures Eco-Efficient Constr Mater 2016;173–92.
73. Aizat EA, Bakri AL, Liew YM, Heah CY. Chemical composition and strength of dolomite geopolymer composites. AIP Conf Proc 2017;1885:1–5.
74. Ariffin N, Abdullah MMAB, Zainol RRMA, Murshed MF. Geopolymer as an adsorbent of heavy metal: a review. AIP Conf Proc 2017;1885:020030.
75. Kranjc A, Hacquet B. The pioneer of karst geomorphologists. Acta Carsol 2006;35:163–8.
76. Musab A, Ashraf A, Gurkan Y, Alper A, Mustafa S. Properties of geopolymers sourced from construction and demolition waste: a review. J Build Eng 2022;50:104104.

Subbulakshmi Ganesan*, Gopalakrisnan Padmapriya, Sanduni Anupama De Zoys and Izegaegbe Daniel Omoikhoje

12 Biofuel as an alternative energy source for environmental sustainability

Abstract: Organic carbon fixation is the primary source of biofuels energy. Plant biomass and municipal and industrial waste are used to make gasoline that is renewable and biodegradable. Biofuel could be beneficial due to the reduced dependency on fuel, to lower the reliance on overseas oil, to lower the emissions of greenhouse gases and it provides job opportunities for the rural people. Moreover, due to the very high demand for fuel, because most of the machines, vehicles use fuel to conduct day today activities. Due to this high demand, the price is also very high. Not only that, the burning of fossil fuel normally release carbon dioxide, carbon monoxide and nitrogen dioxide which are the greenhouse gases. But although biogas too releases those gases, they are released in very lower amounts. Therefore, fuel can act as a good solution for the problems that occur due to global warming and many other environmental problems.

Keywords: carbon fixation; global warming; greenhouse gases; renewable.

12.1 Introduction

Rudolf Diesel was the first to use vegetable oil to make biodiesel in the late 1800s. A petroleum-free alternative to gasoline was made possible in 1998 by an EPA decision to allow the commercial production of biofuel. Biofuel production topped 105 billion liters worldwide in 2010 [1]. In 2011, Europe accounted for 53% of the world's production of biodiesel. The International Energy Agency has set a target of reducing petroleum and coal use and switching to biofuels by 2050 [2]. Corn, sugar cane, sugar beet, camellia, algae, palm oil, animal fat and agricultural wastes can all be used to produce biofuel.

***Corresponding author: Subbulakshmi Ganesan**, Department of Chemistry, Jain University, Bangalore, Karnataka, India, E-mail: g.subbulakshmi@jainuniversity.ac.in
Gopalakrisnan Padmapriya, Department of Chemistry, Jain University, Bangalore, Karnataka, India
Sanduni Anupama De Zoys and Izegaegbe Daniel Omoikhoje, Department of Life Science, Jain University, Bangalore, Karnataka, India

As per De Gruyter's policy this article has previously been published in the journal Physical Sciences Reviews. Please cite as: S. Ganesan, G. Padmapriya, S. A. De Zoys and I. D. Omoikhoje "Biofuel as an alternative energy source for environmental sustainability" *Physical Sciences Reviews* [Online] 2022. DOI: 10.1515/psr-2022-0209 | https://doi.org/10.1515/9783110913361-012

Biofuel can be grouped as,

(1) First era biofuel
   These are known as ordinary biofuels. It in cooperates sugar, starch and vegetables [3].
(2) Second era biofuel
   These are known as the high-level biofuels, and they are comprised of various kinds of biomass. The biomass comprised of lignocellulosoic material like wood, straw and waste plastic.
(3) Third era biofuel
   These are separated principally from marine algae. The utilization of biofuels assists with keeping the environment cleans by:
   (1) Regulating the carbon dioxide rates in the atmosphere.
   (2) By maintaining up with the carbon cycle.
   (3) By making a method for diminishing global warming.

E.g. A harvest is used to produce a large amount of biofuel that will absorb the same amount of carbon dioxide emitted consuming the barrel created (Figure 12.1).

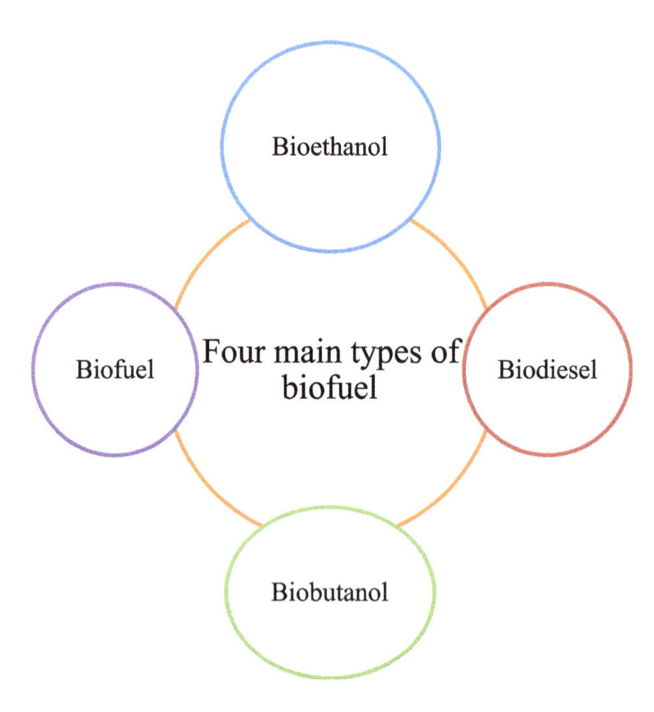

Figure 12.1: Types of Biofuel.

12.1.1 Bioethanol

This is produced with the help carbohydrate sources such as corn, sugarcane, etc. [4].

Bioethanol production procedure

(1) At first, the sugar cane stem, juice and bagasse were separated and this step is called as milling.
(2) Next the sugar was converted into alcohol/ethanol and it was called as fermentation.
(3) Then distillation was carried out, which is the separation of ethanol.
(4) The last step is dehydration.

$$C_{12}H_{22}O_{11}+H_2O \longrightarrow {}^{Invertase}C_6H_{12}O_6+C_6H_{12}O_6$$

$$C_6H_{12}O_6 \longrightarrow {}^{Zymase}C_2H_5OH + CO_2$$

Fermentation is carried out between 250 °C and 300 °C from about 2–3 days.

12.1.2 Biodiesel

It is made up of vegetable oils and animal fats which are renewable sources. Transesterification is carried out in order to produce biodiesel.

Transesterification

R1— COO — CH_2
R2 — COO — CH $+3CH_3OH$ ⟹ R1 — $COOCH_3$
R3 — COO — CH_2 R2 — $COOCH_3$
R3 — $COOCH_3$

Methyl esters

+

OH — CH_2
OH — CH
OH — CH

Glycerol

12.1.3 Algae are a source of biofuel

The algae that are used as biofuel production is usually aquatic green algae. Most of the algae are single celled organisms which are growing in marine or freshwater environments [5]. Similar to the other plants, the algae too will convert the water, CO_2 and sunlight in order to produce starch and oxygen. The vegetable oil obtained from algae can be used directly or those fuels can be refined to form various biofuels [6, 7]. The carbohydrates that are obtained from the algae can be fermented into biofuels that include ethanol as well as butanol. The biomass that is obtained from algae can be used for the pyrolysis oil or the combined heat and for the generation of power.

12.2 Methodology

Biofuel can be produced with the help of animal or plant sources.

This is a cost effective methodology where this fuel can be used as a vehicle fuel. Biofuels are prepared under two scales:
(1) Lab scale- For the research purposes
(2) Large scale

Raw materials used;
(1) Animal source: Animal body fat
(2) Plant source: Seed oil/edible oil. Mustard oil/olive oil
(3) Methanol: It will interact with plant as well as with animal oil
(4) KOH: As a catalyst (Figure 12.2)

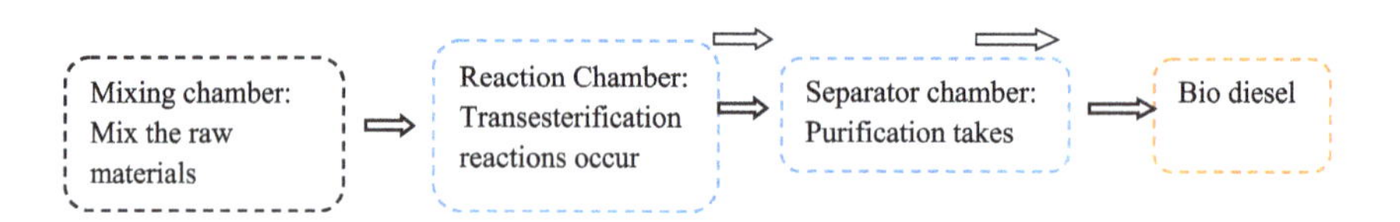

Figure 12.2: Methodology.

All the raw materials are added into the mixing chamber. Methanol too was added into it. H2SO4 too was added in to it. Then these will help methanol and the other raw materials to mix well. All the substances are transferred into the reactor methanol and KOH are added into the reaction chamber. In the separation chamber 2 sensors are present where one sensor will attract glycerol [8–10]. Glycerol remains in the bottom and diesel will come to the top. The counter chamber will drag all the glycerol which is produced and some other compounds too are added which will convert glycerol to glycerine. Glycerol is also having methanol/methanol related by products. Therefore, it was purified.

12.2.1 Methodology for biodiesel production

Step 1: Cultivation

For growing algae, different methods are used as,

(1) Raceway pond:
Long, shallow pool of water which contains paddle wheels with lot of surface area for sun to shine.

(2) Bioreactor:
This method is more expensive and therefore the productivity is high.

Step 2: Harvesting

This is called as dewatering too. This will concentrate the biofuel from 1–50%. There are 2 methods in this step:

(1) Floatation.
(2) Reverse osmosis (This consumes lot of energy, which is highly expensive).

Step 3: Oil extraction

There are two types of steps used here,

(1) Mechanical steps- The micro-algae were subjected to mechanical pressure and then break open the cells and oil was taken.
(2) Chemical steps- Chemicals are used here in order to break open the cells.

During the separation process, it will help to separate water and oil.

Step 4: Conversion

Transesterification reactions take place here.

12.3 Results and discussion

There are advantages as well as disadvantages of biofuel.

12.3.1 Advantages

(1) Algae are renewable sources in making biofuels.
(2) Low cost- Algae can be grown on both brackish and salt water.
(3) It does not need to displace the farmland which is used for the growth of food sources.
(4) Efficient recycling of atmospheric carbon.

(5) Algae contribute to the production of oxygen more than 80%.
(6) Low amount of greenhouse gas emission.

12.3.2 Disadvantages

(1) Not efficient comparatively
(2) Use land that could be used for the growth of food.
(3) High labour power.
(4) Cannot be used to run the cars without any modification.

12.3.3 The effect of temperature with biofuel

The reaction temperature is one of the major factors that effects the production of biofuel. Here when the temperature increases, the production too will be increased (Figure 12.3).

12.3.3.1 Reaction time

When the reaction time increases, the yield will also increase.

12.3.3.2 Catalyst amount

The rate of the reaction increases when the activation energy decreases and can be carried out both in acidic and alkali conditions.

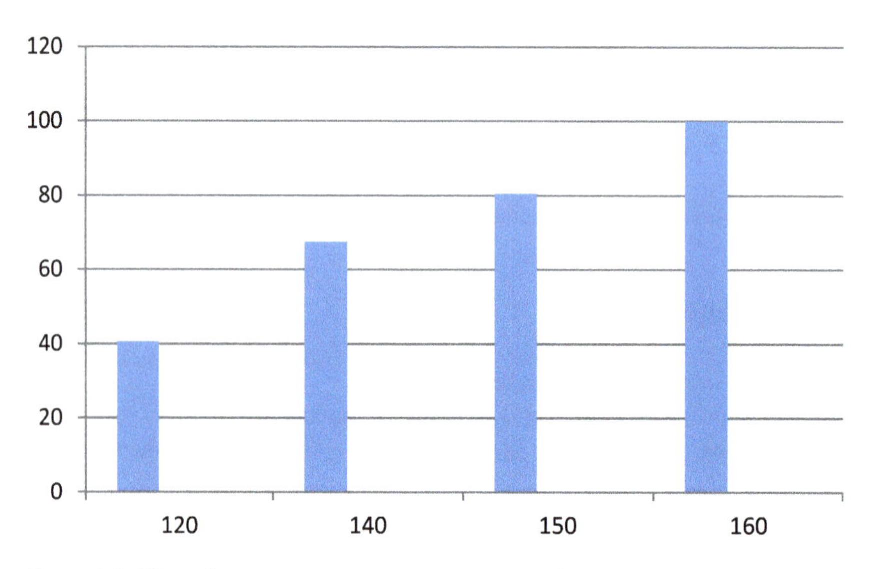

Figure 12.3: Effect of temperature: X axis-Temperature in Celsius; Y axis-% Conversion of biofuel.

12.4 Conclusions

Biofuel is a renewable source of energy which can act as a good solution for the energy crisis. Micro-algae types such as *chlorella, Dunaliella, Salia, Spirulina platensis* can be used mostly in order to make different types of biofuels. The factors that enhance the yield of oil are solvent ratio, biomass size and contact time. In order to obtain the maximum yield of oil, e biomass size should be 0.5 mm with 24 h which is the contact time, temperature, catalysts, oil to methanol ratio, are the other factors that affect the yield. It was observed that at 100 degrees Celsius, catalyst amount 0.5% weight of oil, reaction titre of 25 min can yield 100% of biofuel.

References

1. Marker J, Raven R, Gruffer B. Sustainability transitions: an emerging field of research and its prospects. Res Policy 2012;416:955–67.
2. Ozesmi U, Ozesmi SL. Ecological models based on people's knowledge: a multi-step fuzzy cognitive mapping approach. Ecol Model 2004;176:43–64.
3. Mattes J, Huber A, Koehrsen J. Energy transitions in small-scale regions – what we can learn from a regional innovation systems perspective. Energy Pol 2015;78:255–64.
4. Anu A, Kumar A, Jain KK, Singh B. Process optimization for chemical pretreatment of rice straw for bioethanol production. Renew Energy; 2020:156:1233–43.
5. Scharlemann JPW, Laurance WF. How green are biofuels? Science 2008;319:43–4.
6. Buran. Environmental benefits of implementing alternate energy technologies in developing countries. Appl Energy 2003;76:89–100.
7. Slade R, Bauen A. Micro-algae cultivation for biofuels: cost, energy balance, environmental impacts and future prospects. Biomass Bioenergy 2013;53:29–38.
8. Zaimes GG, Khanna V. Assessing the critical role of ecological goods and services in microalgal biofuel life cycles. RSC Adv 2014;4:44 980–90.
9. Esteves VPP. Land use change (LUC) analysis and life cycle assessment (LCA) of Brazilian soybean biodiesel. Clean Technol Environ Policy 2016;18:1655–73.
10. Kalaivani K, Ravikumar G, Balasubramanian N. Environmental impact studies of biodiesel production from Jatropha curcas in India by life cycle assessment. Environ Prog Sustain Energy 2014;33:1340–9.

Index

https://doi.org/10.1515/9783110913361-011

www.ingramcontent.com/pod-product-compliance
Ingram Content Group UK Ltd.
Pitfield, Milton Keynes, MK11 3LW, UK
UKHW051426310725
7179UKWH00004B/11